jacaranda *plus*

Next generation teaching and learning

Access all formats of your online Jacaranda resources in three easy steps!

To access your resources:

1 ▶ go to **www.jacplus.com.au**

2 ▶ log in to your existing account, or create a new account

3 ▶ enter your unique registration code(s).

Note

- Only one JacPLUS account is required to register all your Jacaranda digital products.
- By registering the code(s) within your JacPLUS bookshelf, you are agreeing to purchase the resource(s). Please view the terms and conditions when registering.

REGISTRATION CODE

Electronic versions of this title are available online; these include eBookPLUS and PDFs. Your unique registration codes for this title are:

GKXG67VGXQF

H2M63UL39PF

GH9G7QY2JKQ

Each code above provides access for one user to the eBookPLUS and PDFs.

NEED HELP?

If you would like to discuss specific digital licensing options or request digital trials, or if you require any other assistance, email support@jacplus.com.au or telephone 1800 JAC PLUS (1800 522 7587).

T0362714

A Wiley Brand

CONTENTS

ABOUT THIS RESOURCE

Jacaranda Maths Quest 12 Essential Mathematics Units 3 & 4 for Queensland is expertly tailored to address comprehensively the intent and structure of the new syllabus. The *Jacaranda Maths Quest for Queensland* series provides easy-to-follow text and is supported by a bank of resources for both teachers and students. At Jacaranda we believe that every student should experience success and build confidence, while those who want to be challenged are supported as they progress to more difficult concepts and questions.

Preparing students for exam success

Chapter openers provide students with their learning sequence and syllabus outcomes.

A variety of online resources are available in the eBookPLUS; these include video eLessons, interactivities, SkillSHEETS and weblinks.

Each subtopic concludes with a carefully graded exercise which provides opportunities for success and challenge.

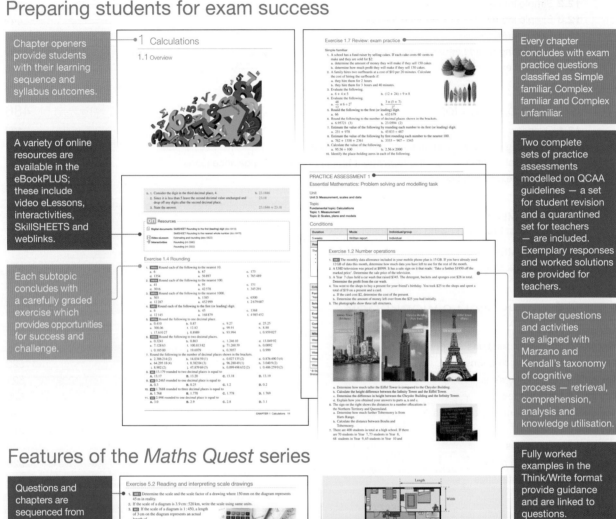

Every chapter concludes with exam practice questions classified as Simple familiar, Complex familiar and Complex unfamiliar.

Two complete sets of practice assessments modelled on QCAA guidelines — a set for student revision and a quarantined set for teachers — are included. Exemplary responses and worked solutions are provided for teachers.

Chapter questions and activities are aligned with Marzano and Kendall's taxonomy of cognitive process — retrieval, comprehension, analysis and knowledge utilisation.

Features of the *Maths Quest* series

Questions and chapters are sequenced from lower to higher levels of complexity; ideas and concepts are logically developed and questions are carefully graded, allowing every student to achieve success.

An extensive glossary of mathematical terms is provided in print and as a hover-over feature in the eBookPLUS.

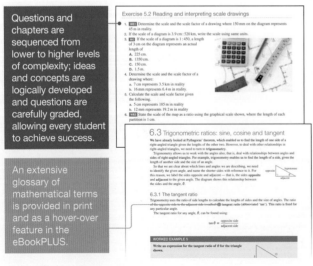

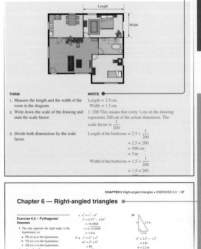

Fully worked examples in the Think/Write format provide guidance and are linked to questions.

Free fully worked solutions are provided, enabling students to get help where they need it, whether at home or in the classroom — help at the point of learning is critical. Answers are provided at the end of each chapter in the print and offline PDF.

eBookPLUS features

Fully worked solutions for every question

Digital documents: downloadable SkillSHEETS to support skill development and SpreadSHEETS to explore mathematical relationships and concepts

A downloadable PDF of the entire chapter of the print text

Interactivities and video eLessons placed at the point of learning to enhance understanding and correct common misconceptions

In the Prelims section of your eBookPLUS

A downloadable PDF of the entire solutions manual, containing worked solutions for every question in the text

A set of four practice assessments: two problem solving and modelling tasks and two examination-style assessments

A downloadable PDF of the entire print text

Additional resources for teachers available in the eGuidePLUS

In the Resources tab of every chapter there are two chapter tests in downloadable, customisable Word format with worked solutions.

In the Prelims section of the eGuidePLUS

Work programs are provided to assist with classroom planning.

Practice assessments: in addition to the four provided in the eBookPLUS, teachers have access to a further four quarantined assessments. Modelled on QCAA guidelines, the problem solving and modelling tasks are provided with exemplary responses while the examination-style assessments include annotated worked solutions. They are downloadable in Word format to allow teachers to customise as they need.

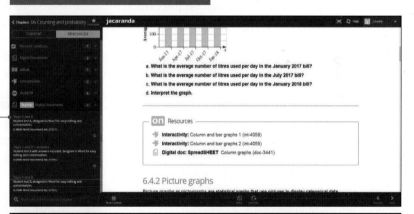

About eBookPLUS

jacaranda *plus*

This book features eBookPLUS: an electronic version of the entire textbook and supporting digital resources. It is available for you online at the JacarandaPLUS website (**www.jacplus.com.au**).

Join **thousands** of other students and teachers in discovering the **next generation** in **teaching** and **learning solutions** ...

Using JacarandaPLUS

To access your eBookPLUS resources, simply log on to **www.jacplus.com.au** using your existing JacarandaPLUS login and enter the registration code. If you are new to JacarandaPLUS, follow the three easy steps below.

Step 1. Create a user account

The first time you use the JacarandaPLUS system, you will need to create a user account. Go to the JacarandaPLUS home page (**www.jacplus.com.au**), click on the button to create a new account and follow the instructions on screen. You can then use your nominated email address and password to log in to the JacarandaPLUS system.

Step 2. Enter your registration code

Once you have logged in, enter your unique registration code for this book, which is printed on the inside front cover of your textbook. The title of your textbook will appear in your bookshelf. Click on the link to open your eBookPLUS.

Step 3. Access your eBookPLUS resources

Your eBookPLUS and supporting resources are provided in a chapter-by-chapter format. Simply select the desired chapter from the table of contents. Digital resources are accessed within each chapter via the resources tab.

> Once you have created your account, you can use the same email address and password in the future to register any JacarandaPLUS titles you own.

Using eBookPLUS references

eBookPLUS logos are used throughout the printed books to inform you that a digital resource is available to complement the content you are studying.

Searchlight IDs (e.g. **INT-0001**) give you instant access to digital resources. Once you are logged in, simply enter the Searchlight ID for that resource and it will open immediately.

Minimum requirements

JacarandaPLUS requires you to use a supported internet browser and version, otherwise you will not be able to access your resources or view all features and upgrades. The complete list of JacPLUS minimum system requirements can be found at **http://jacplus.desk.com**.

Troubleshooting

- Go to **www.jacplus.com.au** and click on the Help link.
- Visit the JacarandaPLUS Support Centre at **http://jacplus.desk.com** to access a range of step-by-step user guides, ask questions or search for information.
- Contact John Wiley & Sons Australia, Ltd. Email: support@jacplus.com.au Phone: 1800 JAC PLUS (1800 522 7587)

ACKNOWLEDGEMENTS

The authors and publisher would like to thank the following copyright holders, organisations and individuals for their assistance and for permission to reproduce copyright material in this book.

Images

• Alamy Australia Pty Ltd: **13**/© Tom Griffiths/ZUMA Press, Inc.; **38** (centre)/imageBROKER; **91** (top)/domonabike; **111**/Neil Tingle; • Corbis Royalty Free: **221** (bottom)/© Corbis Images; **368**/Corbis Royalty Free; • Desmos: **342** (top left), (top right), (centre), (bottom), **343** • Digital Stock: **207** (bottom)
• fncalculator.com: **470**, **471** (top), (bottom), **472** (top), (bottom), **473**, **483**, **488** (top), **489** (top), (bottom), **490**, **491** (top), (bottom), **492** (top), (bottom), **493** (top), (bottom), **497** (top), (centre), (bottom), **498** (top), **500** (top), (bottom)/© Copyright 2012–2019 All Rights Reserved www.fncalculator.com • Getty Images: **14** (centre), **519**/Blend Images • Getty Images Australia: **155** (top)/Steve0/iStockphoto/Getty Images Australia; **155** (centre)/Rony Zmiri; **155** (bottom)/Jeremy Samuelson • iStockphoto: **407**/adventtr • John Wiley & Sons Australia: **8** (centre), **105**, **413** (bottom), **418** (bottom), **521** (left), (right)/© John Wiley & Sons Australia/Taken by Kari-Ann Tapp; **14** (bottom), **21**, **23** (centre)/John Wiley & Sons Australia; **420** (top)/© John Wiley & Sons Australia/Jennifer Wright; • Microsoft Excel: **468**/© Microsoft Excel • Newspix: **259** (upper centre)/Kym Smith/Newspix • Photodisc: **3** (centre right), **211** (bottom), **222** (bottom), **409**/© Photodisc; **4** (bottom), **34**, **39**, **285** (bottom), **385**/© PhotoDisc, Inc; **259** (bottom)/ © 1999 PhotoDisc, Inc; **375**/Photodisc/Getty Images
• Science Photo Library: **530**/Jim Varney • Shutterstock: **1**/zphoto; **3** (centre left)/ChameleonsEye; **3** (centre middle)/Daniel Vine Photography; **3** (bottom)/fritz16; **4** (top)/OSTILL is Franck Camhi; **4** (centre)/Alexey Boldin; **8** (bottom)/Kovaleva_Ka; **12**/Jimmie48 Photography; **14** (top)/Rawpixel.comShutterstock; **22**/Ilin Sergey; **23** (top)/Dusan Pavlic; **25** (top)/Ruth Black; **25** (centre)/Chanwoot_BoonsuyaShutterstock; **29**/Yurii Andreichyn; **38** (bottom)/Venot; **40** (left)/itechno; **40** (right)/Valeriy Lebedev; **41** (top)/Christoprudov Dmitriy; **41** (upper centre)/Wilm Ihlenfeld; **41** (lower centre)/Bryan Solomon; **41** (bottom)/StudioSmart; **42** (top)/Casper1774 Studio; **42** (centre)/FocusDzignShutterstock; **42** (bottom)/Neale Cousland; **43**/FreshStudio; **49** (top)/Skocko; **49** (centre)/FreshPaint; **49** (centre)/Margarita Borodina; **49** (bottom)/Neale Cousland; **51**/beeboys; **57**/Mikhail Bakunovich; **73**/Anita Ponne; **74** (top)/nikiteev_konstantin; **74** (centre left)/Chris Hellyar; **74** (centre middle left)/WIN12_ET; **74** (centre middle right)/Nicholas Piccillo; **74** (centre right)/Dan Breckwoldt; **76** (top left)/pisaphotography; **76** (top right)/dotshock.com; **76** (centre left)/David Bostock; **76** (centre right)/Thomas Bethge; **76** (bottom)/Adam Fahey Designs; **77** (top)/Yuganov Konstantin; **77** (bottom)/Visual Collective; **86** (top)/Orgus88; **86** (bottom)/DJIdroneMarvin; **89**/Chris D573; **90**/IZO; **91** (bottom)/Jahanzaib Naiyyer; **92** (top)/Piotr Wawrzyniuk; **92** (bottom)/kitzcorner; **99** (bottom)/Neil Balderson; **99** (top)/Denis Tabler; **108**/tohzoli001; **110**/Yurii Vydyborets; **125** (top)/Pavel L Photo and Video; **125** (bottom)/Vanessa Belfiore; **128**/Guenter Albers; **129** (top)/Ramona Kaulitzki; **129** (bottom)/Alexey Boldin; **130**/RAFDC; **133**/MaraZe; **134** (top)/Olaf Speier; **134** (bottom)/s-ts; **140** (top)/Olaf Speier; **140** (upper centre)/photoiconix; **140** (lower centre)/Sararoom Design; **140** (upper bottom)/Gena73; **140** (lower bottom)/klikkipetra; **141**/Abdoabdalla; **145** (top)/Kues; **145** (bottom)/Mayer George; **146** (top)/Martin Gardeazabal; **146** (centre)/Axel Sandesh; **146** (bottom)/iurii; **147** (top left)/Mitch Gunn; **147** (top right)/Quang Ho; **147** (bottom left)/Eric Broder Van Dyke; **147** (bottom right)/FabrikaSimf; **148** (top left)/photogal; **148** (top right)/Sea Wave; **148** (centre left)/Lenscap Photography; **148** (centre right)/Lipowski Milan; **148** (bottom)/D. Pimborough; **149** (top)/Chih Hsuan Peng; **149** (bottom)/SaGa Studio; **150**/Fer Gregory; **152**/Dawid Rojek; **153** (top)/Olga Selyutina; **153** (bottom)/Jason Patrick Ross; **154** (top)/vvoe; **154** (bottom)/dragon_fang; **158**/Franck Boston; **159**/Andy Dean Photography; **161**/shooarts; **162**/Shutter Baby photo; **163** (top)/Maxx-Studio; **163** (centre)/New Vibe; **163** (bottom)/Peshkova; **164**/charobnica; **165**/visivastudio; **166**/elfishes; **167**/Inna Ogando; **171** (top)/Irina Rogova; **171** (upper centre)/freesoulproduction; **171** (centre)/Aarcady; **171** (lower centre)/Canoneer; **171** (bottom)/elfishes; **172** (top)/Ana Vasileva; **172** (bottom)/Ksander; **173** (top)/Inna Ogando; **173** (bottom)/Evgeny Karandaev; **174**/Seregam; **179** (top)/HSNphotography; **179** (bottom)/Syda Productions; **180** (top)/Alex Farias; **180** (centre)/MvanCaspel; **180** (bottom)/HST6; **181** (top)/OtmarW; **181** (bottom)/Lev Kropotov; **182**/RossHelen; **183** (top)/Michael Smolkin; **183** (bottom)/Irina Rogova; **184** (top)/cybrain; **184** (upper centre)/Ljupco Smokovski;

184 (lower centre)/oksana2010; **184** (bottom)/burnel1; **185** (top)/handmadee3d; **185** (bottom)/Ana Vasileva; **186/**jazman; **189/**Adha Ghazali; **194** (top)/petcharaPJ; **194** (bottom)/Brett Rabideau; **196/**GOLFX; **203/**kirill guzhvinsky; **206/**Orla; **207** (top)/Aleksey Stemmer; **207** (centre)/GoTaR; **211** (top)/Diane Diederich; **212/**Kzenon; **217** (bottom)/Sergii Korshun; **217** (top)/Zhukova Valentyna; **221** (top)/hxdyl; **221** (upper centre)/sirtravelalot; **221** (lower centre)/ElRoi; **222** (top)/NadyaEugene; **225/**M-SUR; **226/**Inspiring; **227/**Olga Dmitrieva; **233/**Iakov Filimonov; **234/**Demkat; **236** (top)/Monkey Business ImagesUnknown; **236** (bottom)/VladKol; **237/**Arkadi Bulva; **238** (top)/simez78; **238** (centre)/Africa Studio; **242/**Ekkachai; **245/**xpixel; **246/**M. Unal Ozmen; **248/**Martin Novak; **249/**Pavel1964; **250** (top)/Bikeworldtravel; **250** (bottom)/Kucher Serhii; **251** (top)/Mehmet Cetin; **251** (centre)/alison1414; **251** (bottom)/James Watts; **253/**wavebreakmedia; **255** (top)/Mikbiz; **255** (centre)/Microgen; **255** (bottom)/Africa Studio; **256/**Andy Dean Photography; **258** (top)/Lemau Studio; **258** (bottom)/Hurst Photo; **259** (top)/loveaum; **259** (lower centre)/ArTDi101; **260** (top)/ChameleonsEye.com; **260** (bottom)/mangostock; **261/**Katherine Welles; **262/**Travnikov Studio; **263** (top)/Denys Prykhodov; **263** (upper centre)/Sararwut Jaimassiri; **263** (lower centre)/paulista; **263** bottom)/M. Unal Ozmen; **264/**Chris Parypa Photography; **270/**george studio; **275** (top)/PHILIPPE MONTIGNY; **275** (bottom)/SPF; **276** (top)/nikamo; **276** (bottom)/MaraZe; **277** (top)/nata-lunata; **277** (bottom)/Christine Pedretti; **279** (top)/Dmitrijs Dmitrijevs; **279** (upper centre)/Elnur; **279** (lower centre)/cobalt88; **279** (bottom)/siixth; **280** (top)/FamVeld; **280** (bottom)/Roman Samokhin; **281/**NadyaEugene; **282/**Rido; **283/**Nils Versemann; **284/**Katrina Elena; **285** (top)/bikeriderlondon; **286** (top)/spass; **286** (bottom)/Marlon Lopez MMG1 Design; **287** (top)/ESB Professional; **287** (bottom)/Pete Niesen; **288** (top)/Holli; **288** (bottom)/Monkey Business Images; **289/**Quang Ho; **291** (top)/NaMaKuKi; **291** (bottom)/VLADYSLAV DANILIN; **292/**Herbert Kratky; **295/**Iryna Inshyna; **297/**Chris Parypa Photography; **298** (top)/Fakhrul Najmi; **298** (bottom)/Toa55; **299** (top)/nd3000; **299** (bottom)/A.J. Pictures; **315/**Melodist; **324/**violetkaipa; **328** (top)/llucky78; **328** (bottom)/gualtiero boffi; **330/**elenabsl; **335/**Stock Creative; **349/**Panumas Yanuthai; **357/**marekuliaszShutterstock; **359/**Moremar; **361/**michaeljung; **363/**wavebreakmedia; **364** (top)/Bojan Milinkov; **364** (bottom)/Photographee.eu; **365** (top)/Crdjan; **365** (bottom)/Serhi Bobyk; **366/**zstock; **372/**TK Kurikawa; **376/**Iulian Valentin; **377** (top)/Stefan Schurr; **377** (bottom)/Aleksey Stemmer; **379/**Holmes Su; **381/**Max Topchii; **382/**TK Kurikawa; **383** (top)/Robyn Mackenzie; **383** (bottom)/Minerva Studio; **384** (top)/wavebreakmedia/Denis Belitsky; **384** (bottom)/ Photology1971; **385** (top)/badahos.com; **385** (bottom)/SpeedKingz; **386** (top)/Dmitry Sheremeta; **386** (bottom)/Tyler Olson; **391** (top)/Bobex-73; **391** (bottom)/Natalia K; **392** (top)/Gilmanshin; **392** (bottom)/ SOMKKU.com; **395/**Rawpixel.com; **397/**wavebreakmedia; **398** (top)/Eugen Thome; **398** (bottom)/Nata-Lia; **399/**Evikka; **410/**Francesco Abrignani; **411/**Olesia Bilkei; **412/**bekulnis; **413** (top)/Becris; **414** (top)/ NatalyaBond; **414** (bottom)/Eivaisla; **415/**Alliya2; **418** (top left)/Ailisa; **418** (top right)/M. Unal Ozmen; **419** (top)/© Dmitriy Shironosov.com; **419** (centre)/hddigital; **419** (bottom)/Eivaisla; **420** (bottom)/Malyugin; **422/**Patryk Kosmider; **424/**Vereshchagin Dmitry.com; **427** (top)/Pavel Nesvadba; **427** (bottom)/bbernard; **428** (top)/Africa Studio; **428** (centre)/Olga Listopad; **428** (bottom)/Joe Gough; **429/**SpeedKingz; **433** (top)/Akhenaton Images; **433** (upper centre)/Dreamsquare; **433** (centre)/koya979; **433** (lower centre)/Rhonda Roth; **433** (bottom)/successo images; **434** (top)/SIAATH; **434** (bottom)/Pixel Embargo; **441/**TZIDO SUN; **444/**WAYHOME studio; **448** (top)/Roman Samborskyi; **448** (bottom)/Max Topchii **462/**Atstock Productions; **464/**ESB Professional; **475** (top)/UfaBizPhoto; **475** (bottom)/TK Kurikawa; **486/**Watchara Ritjan; **488** (bottom)/I viewfinder; **494/**Maksym Povozniuk; **495/**Artazum; **496/**Kwangmoozaa; **502** (top)/S_Photo; **502** (bottom)/ pixelheadphoto digitalskillet; **504/**baranq; **512/**Kumar Jatinder; **527/**Simone Andress • Spatial Vision: **160**

Text

Creative Commons: **426/**© Commonwealth of Australia, Australian Bureau of Statistics; **430/**Department of Education and Training; *Maths Quest 12 Essential Mathematics for Queensland Units 3 & 4*, 2019 v1.1 syllabus content © State of Queensland (Queensland Curriculum & Assessment Authority) 2019.

Every effort has been made to trace the ownership of copyright material. Information that will enable the publisher to rectify any error or omission in subsequent reprints will be welcome. In such cases, please contact the Permissions Section of John Wiley & Sons Australia, Ltd.

1 Calculations

1.1 Overview

LEARNING SEQUENCE

CONTENT

In this chapter, students will learn to:
- solve practical problems requiring basic number operations
- apply arithmetic operations in their correct order
- ascertain the reasonableness of answers to arithmetic calculations
- use leading-digit approximation to obtain estimates of calculations
- use a calculator for multi-step calculations
- check results of calculations for accuracy
- recognise the significance of place value after the decimal point
- evaluate decimal fractions to the required number of decimal places
- round up or round down numbers to the required number of decimal places
- apply approximation strategies for calculations.

Fully worked solutions for this chapter are available in the Resources section of your eBookPLUS at www.jacplus.com.au.

1.2 Number operations

1.2.1 Practical numerical problems

Throughout the day we are always solving practical numerical problems in our head without really knowing that we are doing it. It could be calculating what time you need to leave home to get to school on time, or calculating how much money you need to purchase food from the school canteen. Life constantly requires us to solve these numerical problems, so it is important that we can do so quickly and accurately.

WORKED EXAMPLE 1

A USB stick can hold 512 MB of data. If 386 MB of the USB stick is already filled, determine how much space is left on the USB stick.

THINK	WRITE
1. What is the total storage space?	Space available $= 512$ MB
2. How much space has been used?	Space used $= 386$ MB
3. The space left is the difference between the total space and the space used.	Space left $= 512 - 386 = 126$ MB
4. State the answer.	The space left on the USB stick is 126 MB.

WORKED EXAMPLE 2

Nathan has a part-time job that pays $15.50 per hour. Nathan gets paid time and a half for hours worked on Saturdays and double time for hours worked on Sundays. Use your calculator to calculate how much Nathan gets paid in a week when he works 5 per hours on Friday, 4 per hours on Saturday and 5.5 per hours on Sunday.

THINK	WRITE
1. Calculate the amount Nathan earned on Friday.	Money earned Friday $= 5 \times \$15.50$ $= \$77.50$
2. Calculate the amount Nathan earned on Saturday.	Money earned Saturday $= 4 \times \$15.50 \times 1.5$ $= \$93.00$
3. Calculate the amount Nathan earned on Sunday.	Money earned Sunday $= 5.5 \times \$15.50 \times 2$ $= \$170.50$
4. Calculate Nathan's weekly pay by adding the amounts earned on Friday, Saturday and Sunday.	Total pay $= \$77.50 + \$93.00 + \$170.50$ $= \$341.00$
5. State Nathan's weekly pay.	Nathan earned $341.00 for work that week.

Exercise 1.2 Number operations

1. **WE1** The monthly data allowance included in your mobile phone plan is 15 GB. If you have already used 12 GB of data this month, determine how much data you have left to use for the rest of the month.

2. A UHD television was priced at $8999. It has a sale sign on it that reads: 'Take a further $1950 off the marked price'. Determine the sale price of the television.

3. A Year 7 class held a car wash that raised $345. The detergent, buckets and sponges cost $28 in total. Determine the profit from the car wash.

4. You went to the shops to buy a present for your friend's birthday. You took $25 to the shops and spent a total of $19 on a present and a card.
   a. If the card cost $2, determine the cost of the present.
   b. Determine the amount of money left over from the $25 you had initially.

5. The photographs show three tall structures.

Infinity Tower (Brisbane)
249 metres high

Chrysler Building (New York)
319 metres high

Eiffel Tower (Paris)
324 metres high

   a. Determine how much taller the Eiffel Tower is compared to the Chrysler Building.
   b. Calculate the height difference between the Infinity Tower and the Eiffel Tower.
   c. Determine the difference in height between the Chrysler Building and the Infinity Tower.
   d. Explain how you obtained your answers to parts a, b and c.

6. The sign on the right shows the distances to a number of locations in the Northern Territory and Queensland.
   a. Determine how much further Tobermorey is from Harts Range.
   b. Calculate the distance between Boulia and Tobermorey.

7. There are 400 students in total at a high school. If there are 70 students in Year 7, 73 students in Year 8, 68 students in Year 9, 65 students in Year 10 and 72 students in Year 11, determine the number of students there are in Year 12.

PLENTY HIGHWAY 12
Jct. Sandover Hwy 27
Harts Range 145
Tobermorey 496
Boulia 743
Central Australian Gemfields 77

8. Consider the digits 2, 3, 4 and 5.
   a. Construct the largest possible number using these digits.
   b. Construct the smallest possible number using these digits.
   c. Calculate the difference between the numbers from parts b and c.

9. Assume your seventeenth birthday is today.
   a. Determine your age in weeks. (There are 52 weeks in a year.)
   b. Determine your age in days. (Assume 365 days in a year.)

10. While training for a triathlon, your friend follows a specific training program where she runs 15 km, cycles 25 km and swims 10 km each week. If your friend trains for 8 weeks, calculate the total distance she will travel in her training.

11. The photograph on the right shows a mobile phone and its associated costs.
   a. Determine the cost for a 1-minute call on this phone.
   b. Determine the cost for a 35-minute call on this phone.
   c. If you make five 3-minute calls each day, calculate how much the calls will cost you for a year using this phone. (There are 365 days in a standard year.)
   d. Determine the cost to send 20 text messages from this phone.
   e. Calculate the difference in cost between sending 100 text messages and sending 50 picture messages using this phone.

Voice calls:
47c per 30 seconds
Text messages:
15c per message
Picture messages:
50c per message

12. **WE2** Sarah has a part-time job that pays $18.50 per hour. Sarah gets paid time and a half for hours worked on Saturdays and double time for hours worked on Sundays. Use your calculator to calculate how much Sarah gets paid in a week when she works 4 hours on Friday, 5 hours and 30 minutes on Saturday and 8 hours on Sunday.

13. Assuming that each egg in the photograph on the right is the same size, determine the mass of each egg.

Total mass of eggs = 650 grams

14. In a class of 24 students, each plays cricket, or soccer, or both. If 16 students play cricket and 10 students play soccer, determine the number of students who play both sports.

1.3 Order of operations

The **order of operations** is a set of rules that determines the order in which mathematical operations are to be performed.

Consider $8 + 12 \div 4$. If you perform the addition first, the answer is $20 \div 4 = 5$. If you perform the division first, the answer is $8 + 3 = 11$. The correct answer to $8 + 12 \div 4$ is 11, because the order of operations specifies that division should be performed before addition.

1.3.1 Rules for the order of operations

1. Evaluate any calculations inside brackets.
2. Evaluate any power or root calculations (indices).
3. Evaluate any multiplication *and* division calculations, working in the order they appear *from left to right*.
4. Evaluate any addition *and* subtraction calculations, working in the order they appear *from left to right*.

Order of operations

The order of operations rules can be remembered as **BIDMAS**:

Brackets

Indices (powers and roots)

Division *and* **M**ultiplication (working from left to right)

Addition *and* **S**ubtraction (working from left to right)

WORKED EXAMPLE 3

Use the order of operations rules to calculate each of the following.
a. $6 \div 3 \times 9 + 3$
b. $12 \div 2 + 4 \times (4 + 6)$
c. $\{4 + [3 \times 2 + (15 - 8)]\} + 5$

THINK	WRITE
a. 1. Write the question.	a. $6 \div 3 \times 9 + 3$
2. No brackets or powers are included in this question. Evaluate the division and multiplication from left to right.	$= 2 \times 9 + 3$ $= 18 + 3$
3. Evaluate the addition to calculate the answer.	$= 21$
b. 1. Write the question.	b. $12 \div 2 + 4 \times (4 + 6)$
2. Evaluate the addition inside the brackets first.	$= 12 \div 2 + 4 \times 10$
3. Evaluate the division and multiplication, working from left to right.	$= 6 + 4 \times 10$ $= 6 + 40$
4. Evaluate the addition to calculate the answer.	$= 46$
c. 1. Write the question.	c. $\{4 + [3 \times 2 + (15 - 8)]\} + 5$
2. Remove the innermost brackets by calculating the answer to $15 - 8$.	$= \{4 + [3 \times 2 + 7]\} + 5$
3. Remove the next pair of brackets by first evaluating the multiplication and then the addition.	$= \{4 + [6 + 7]\} + 5$ $= \{4 + 13\} + 5$
4. Remove the last pair of brackets by calculating the answer to $4 + 13$.	$= 17 + 5$
5. Evaluate the addition to calculate the answer.	$= 22$

Use the order of operations rules to calculate each of the following.

a. $[(4+2)^2 + 6] \div 7$

b. $\dfrac{3 \times (\sqrt{40-4} - 3)^2 + (10 \div 2)}{12 \div 3}$

THINK	WRITE
a. 1. Write the question.	a. $[(4+2)^2 + 6] \div 7$
2. Evaluate the inner brackets first by adding 4 and 2.	$= [6^2 + 6] \div 7$
3. Evaluate the indices by calculating 6^2.	$= [36 + 6] \div 7$
4. Evaluate the addition in the remaining brackets.	$= 42 \div 7$
5. Evaluate the division to calculate the answer.	$= 6$
b. 1. Write the question.	b. $\dfrac{3 \times (\sqrt{40-4} - 3)^2 + (10 \div 2)}{12 \div 3}$
2. Evaluate any expressions in brackets, In the first bracket evaluate the expression under the square root first, then complete the subtraction.	$= \dfrac{3 \times (\sqrt{36} - 3)^2 + (10 \div 2)}{12 \div 3}$
	$= \dfrac{3 \times (6 - 3)^2 + (10 \div 2)}{12 \div 3}$
	$= \dfrac{3 \times 3^2 + 5}{12 \div 3}$
3. Evaluate the power term by squaring 3.	$= \dfrac{3 \times 9 + 5}{12 \div 3}$
4. Evaluate the multiplication and then the addition in the numerator. Evaluate the division in the denominator.	$= \dfrac{32}{4}$
5. Calculate the answer.	$= 8$

Using a calculator to evaluate expressions involving order of operations

Many scientific and graphic calculators can input and evaluate expressions in one line.

In Worked example 4a we are asked to evaluate $[(4+2)^2 + 6] \div 7$. To determine its value on a calculator, follow the steps below.

- Input the expression $((4+2)^2 + 6) \div 7$ in one continuous line into your calculator using only round brackets.
- Press = or ENTER to evaluate.

$\therefore [(4+2)^2 + 6] \div 7 = 6$

In Worked example 4b we are asked to evaluate $\dfrac{3 \times \left(\sqrt{40-4} - 3\right)^2 + (10 \div 2)}{12 \div 3}$. To determine its value on a calculator, follow the steps below.

- If the calculator has a fraction template access this function. This will set up two entry boxes; one for the numerator and one for the denominator.
- Input $3 \times \left(\sqrt{40-4} - 3\right)^2 + (10 \div 2)$ into the entry box for the numerator.
- Input $12 \div 3$ into the entry box for the denominator.

- Press = or ENTER to evaluate.

$$\therefore \frac{3 \times \left(\sqrt{40 - 4} - 3 \right)^2 + (10 \div 2)}{12 \div 3} = 8$$

WORKED EXAMPLE 5

Sean buys 4 oranges at \$0.70 each and 10 pears at \$0.55 each. Calculate the average price Sean paid for a piece of fruit.

THINK	WRITE
1. Calculate the total amount of money spent on fruit.	$(4 \times \$0.70) + (10 \times \$0.55) = \$2.80 + \$5.50$ $= \$8.30$
2. Count the total number of pieces of fruit that Sean purchased.	$4 \text{ oranges} + 10 \text{ pears} = 14 \text{ pieces of fruit}$
3. The average price paid for a piece of fruit is equal to the total cost divided by the total number of pieces.	$\text{Average price} = \dfrac{\text{total cost}}{\text{total number of pieces}}$ $= \dfrac{\$8.30}{14}$ $= \$0.59$
4. State the answer.	The average price paid for a piece of fruit is \$0.59.

 Resources

📄 **Digital document** SkillSHEET Order of operations and directed numbers (doc-6392)

🎞 **Video eLesson** BIDMAS (eles-1883)

🧩 **Interactivity** Order of operations (int-3707)

Exercise 1.3 Order of operations

1. **WE3** Use the order of operations rules to calculate each of the following.
   a. $3 \times 4 \div 2$
   b. $6 \div 3 \times 3 \div 2$
   c. $24 \div (12 - 4)$
   d. $12 \div 4 + 36 \div 6 + 2$
   e. $12 \times (20 - 12)$
   f. $(18 - 15) \div 3 \times 27$
   g. $52 \div 13 + 75 \div 25$
   h. $(12 - 3) \times 8 \div 6$
   i. $\{[(16 + 4) \div 4] - 2\} \times 6$
   j. $5^2 - 15 + 10 \div 5$

2. **WE4** Use the order of operations rules to calculate each of the following.
   a. $13 + 2^3 \div \left(\sqrt{64} - 1 \right) + 2$
   b. $50 - \left(\sqrt{100 \div 4} + 2 \right)^2$
   c. $[(5^3 - 25) \div 5^2]^2$
   d. $\dfrac{36 + \sqrt{36} \div 6}{7}$
   e. $\{[55 - (16 \div 4)^2] \div 13\} + 10$

3. Insert one set of brackets in the appropriate place to make each of these statements true.
   a. $12 - 8 \div 4 = 1$
   b. $4 + 8 \times 5 - 4 \times 5 = 40$
   c. $3 + 4 \times 9 - 3 = 27$
   d. $3 \times 10 - 2 \div 4 + 4 = 10$
   e. $10 \div 5 + 5 \times 9 \times 9 = 81$
   f. $18 - 3 \times 3 \div 5 = 9$

4. Your friend is having trouble with the order of operations. A sample of
   her work is shown at the right.

$120 \div 6 \times 2$	$18 - 6 + 5$
$= 120 \div 12$	$= 18 - 11$
$= 10 \times$	$= 7 \times$

   a. Explain what your friend is doing incorrectly.
   b. Show the correct solutions to the questions.
5. Use BIDMAS to evaluate each of the following.

   a. $(6 + 3) - (2 + 5)$

   b. $72 + \left(\dfrac{1}{2} \times 40\right) - 3 \times 4$

   c. $56 - 6 \times 4 \div (13 - 7)$

   d. $23 \times 3 - 12 \times 3$

   e. $\dfrac{32}{8} \times 3 + 6^2$

   f. $(4 + 5)^2 - (3 \times 2)$

   g. $\dfrac{[7 + 2 + (3 \times 4)]}{3}$

   h. $\sqrt{(12 - 7) \times 5}$

   i. $[(27 \div 3)^2 + 4] \div 7$

6. Evaluate each of the following, leaving your answer as a fraction where appropriate.

   a. $\dfrac{13 - 4}{7 + 1}$

   b. $\dfrac{(2 + 3) \times 5}{(12 - 3) \div 3}$

   c. $\dfrac{(6 - 3)^3 - 20}{6^2}$

   d. $\dfrac{4 \times [(5 - 3)^3 - (4 \div 2)]}{15 \div 3 \times 2}$

7. For a birthday party, you buy two packets of
   paper plates at $2 each, three bags of chips at
   $4 each, and three boxes of party pies at
   $5 each.

   a. How does the order of operations help you
      find the correct total cost of these items?
   b. Write an equation to show the operations required
      to calculate the total cost.
   c. Calculate the total cost.

Box of party pies — $5

Bag of chips — $4

Packet of paper plates — $2

8. Use the digits $1, 2, 3$ and 4 and the operators $+, -, \times,$
   and $\div$ to construct equations that result in the numbers
   1 to 5 (the numbers 2 and 4 are already done for you). You must use each digit in each expression, and
   you may not use any digit more than once. You may combine the digits to form larger numbers (like 21).
   You may also use brackets to make sure the operations are done in the correct order.
   $1 =$
   $2 = 4 - 3 + 2 - 1$
   $3 =$
   $4 = 4 - 2 + 3 - 1$
   $5 =$

9. **WE5** Pat buys 5 bananas at $0.30 each and 8 apples at $0.25 each.
   Calculate the average price that was paid for a piece of fruit.

1.4 Rounding

1.4.1 Rounding integers

Numbers can be rounded to different degrees of accuracy.

Rounding to the nearest 10

To round to the nearest 10, think about which multiple of 10 the number is closest to. For example, if 34 is rounded to the nearest 10, the result is 30 because 34 is closer to 30 than it is to 40.

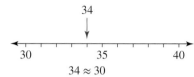

$$34 \approx 30$$

(*Note:* The symbol ≈ represents 'is approximately equal to'.)

Rounding to the nearest 100

To round to the nearest 100, think about which multiple of 100 the number is closest to.

Leading-digit approximation (rounding to the first digit)

To round to the first (or leading) digit, use the following guidelines:

- Consider the digit after the leading one (i.e. the second digit).
- If the second digit is 0, 1, 2, 3 or 4, the first digit stays the same and all the following digits are replaced with zeros.
- If the second digit is 5, 6, 7, 8 or 9, the first digit is raised by 1 (rounded up) and all the following digits are replaced with zeros.

For example, if 2345 is rounded to the first digit, the result is 2000 because 2345 is closer to 2000 than it is to 3000.

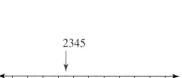

$$2345 \approx 2000$$

WORKED EXAMPLE 6

Round the number 23 743 to the following.
a. **Nearest 10**
b. **Nearest 100**
c. **Nearest 1000**

THINK	WRITE
a. 1. Consider the number starting in the tens position. The number is 43.	a. 23 743
2. Decide what multiple of 10 the number 43 is closest to.	It is closest to 40.
3. Write the rounded number by replacing 43 with 40.	23 743 ≈ 23 740
b. 1. Consider the number starting in the hundreds position. The number is 743.	b. 23 743

▶

2. Decide what multiple of 100 the number 743 is closest to.	It is closest to 700.
3. Write the rounded number by replacing 743 with 700.	$23\,743 \approx 23\,700$
c. 1. Consider the number starting in the thousands position. The number is 3743.	c. 23 743
2. Decide what multiple of 1000 the number 3743 is closest to.	It is closest to 4000.
3. Write the rounded number by replacing 3743 with 4000.	$23\,743 \approx 24\,000$

WORKED EXAMPLE 7

Round each of the following numbers to the first (or leading) digit.
a. 2371   **b. 872**

THINK

a. Since the second digit (3) is less than 5, leave the leading digit unchanged and replace all other digits with zeros.

b. The second digit (7) is greater than 5. Add 1 to the leading digit and replace all other digits with zeros.

WRITE

a. $2371 \approx 2000$

b. $872 \approx 900$

1.4.2 Rounding decimals

When rounding a decimal, a similar approach is used. If a number is to be rounded to two decimal places, use the following guidelines:
- Look at the third decimal place.
- If the digit in the third decimal place is less than 5 {0, 1, 2, 3 or 4}, leave the second decimal value unchanged and drop off all digits after the second decimal place.
- If the digit in the third decimal place is greater than or equal to five {5, 6, 7, 8 or 9}, add 1 to the second decimal value, then drop off all digits after the second decimal place.

WORKED EXAMPLE 8

Round 23.1846 to the following.
a. One decimal place   **b. Two decimal places**

THINK

a. 1. Consider the digit in the second decimal place, 8.

2. Since it is greater than or equal to 5 add 1 to the first decimal value and drop off any digits after the first decimal place.

3. State the answer.

WRITE

a. 23.1846

23.2

$23.1846 \approx 23.2$

b. 1. Consider the digit in the third decimal place, 4.	**b.** 23.1846
2. Since it is less than 5 leave the second decimal value unchanged and drop off any digits after the second decimal place.	23.18
3. State the answer.	$23.1846 \approx 23.18$

on Resources

Digital documents SkillSHEET Rounding to the first (leading) digit (doc-6418)

SkillSHEET Rounding to the nearest whole number (doc-6476)

Video eLesson Estimating and rounding (eles-0822)

Interactivities Rounding (int-3980)

Rounding (int-3932)

Exercise 1.4 Rounding

1. **WE6a** Round each of the following to the nearest 10.
   a. 6   b. 67   c. 173
   d. 1354   e. 56 897   f. 765 489
2. **WE6b** Round each of the following to the nearest 100.
   a. 41   b. 91   c. 151
   d. 3016   e. 42 578   f. 345 291
3. **WE6c** Round each of the following to the nearest 1000.
   a. 503   b. 1385   c. 6500
   d. 12 287   e. 452 999   f. 2 679 687
4. **WE7** Round each of the following to the first (or leading) digit.
   a. 6   b. 45   c. 1368
   d. 12 145   e. 168 879   f. 4 985 452
5. **WE8a** Round the following to one decimal place.
   a. 0.410   b. 0.87   c. 9.27   d. 25.25
   e. 300.06   f. 12.82   g. 99.91   h. 8.88
   i. 17.610 27   j. 0.8989   k. 93.994   l. 0.959 027
6. **WE8b** Round the following to two decimal places.
   a. 0.3241   b. 0.863   c. 1.246 10   d. 13.049 92
   e. 7.128 63   f. 100.813 82   g. 71.260 39   h. 0.0092
   i. 0.185 00   j. 19.6979   k. 0.3957   l. 0.999
7. Round the following to the number of decimal places shown in the brackets.
   a. 2.386 214 (2)   b. 14.034 59 (1)   c. 0.027 135 (2)   d. 0.876 490 3 (4)
   e. 64.295 18 (4)   f. 0.382 04 (3)   g. 96.280 49 (1)   h. 3.040 9 (2)
   i. 8.902 (2)   j. 47.879 69 (3)   k. 0.099 498 632 (2)   l. 0.486 259 0 (2)
8. **MC** 13.179 rounded to two decimal places is equal to
   A. 13.17   **B.** 13.20   **C.** 13.18   **D.** 13.19
9. **MC** 0.2465 rounded to one decimal place is equal to
   A. 0.3   **B.** 0.25   **C.** 1.2   **D.** 0.2
10. **MC** 1.7688 rounded to three decimal places is equal to
    A. 1.768   **B.** 1.770   **C.** 1.778   **D.** 1.769
11. **MC** 2.998 rounded to one decimal place is equal to
    A. 3.0   **B.** 2.9   **C.** 2.8   **D.** 3.1

12. Round the following to the nearest unit.
  a. 10.7   b. 8.2   c. 3.6   d. 92.7
  e. 112.1   f. 21.76   g. 42.0379   h. 2137.50
  i. 0.12   j. 0.513   k. 0.99   l. 40.987

1.5 Estimation and approximation strategies

Estimating is useful when an accurate answer is not necessary. When you do not need to know an exact amount, an estimate or **approximation** is enough.

An estimate is based on information, so it is not the same as a guess. You can estimate the number of people in attendance at a sportsground based on an estimate of the fraction of seats filled.

Estimation using rounding can help you check if your calculations are correct.

WORKED EXAMPLE 9

Estimate the number of people in attendance at Rod Laver Arena using the information in the photograph.

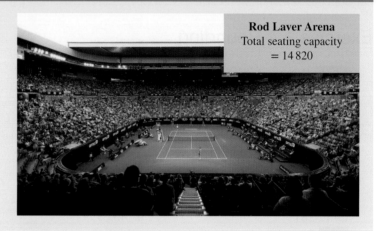

Rod Laver Arena
Total seating capacity
= 14 820

THINK	WRITE
1. Estimate the fraction of seats that are filled in the photograph.	Approximately 4 tenths of the seats are filled.
2. Calculate 4 tenths of the total seating capacity of Rod Laver Arena.	4 tenths of $14\,820 = \dfrac{4}{10} \times 14\,820$ $= 0.4 \times 14\,820$ $= 5928$
3. State the answer.	The estimated number of people in attendance is 5928.

WORKED EXAMPLE 10

Estimate the value of 85 697 × 248 by rounding each number to its first (or leading) digit.

THINK	WRITE
1. Write the question.	$85\,697 \times 248$
2. Round the first number to its first digit. The second digit in 85 697 is 5, so round 85 697 up to 90 000.	$\approx 90\,000 \times 248$

3. Round the second number to its first digit. The second digit in 248 is 4, so round 248 down to 200.

$$\approx 90\,000 \times 200$$

4. Complete the multiplication.

$$\approx 180\,000$$

 **Resources**

 Interactivity Estimation (int-4318)

Exercise 1.5 Estimation and approximation strategies

1. **WE9** Estimate the number of people in attendance at Hisense Arena using the information in the photograph.

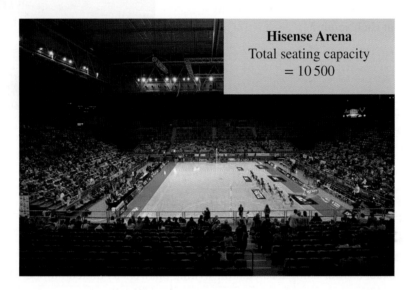

Hisense Arena
Total seating capacity
= 10 500

2. **WE10** Estimate the value of the following by rounding each number to the first (or leading) digit.
   a. $487 + 962$
   b. $184\,029 + 723\,419$
   c. $942\,637 - 389\,517$
   d. 87×432
   e. $623 \times 12\,671$
   f. $69\,241 \div 1297$
3. Give three examples of situations in which it is suitable to use an estimate or a rounded value instead of an exact value.
4. A website tells you that it will take 30 minutes to drive from your house to your cousin's house.
   a. Will it take you exactly 30 minutes?
   b. How might the website designers have estimated this amount of time?
5. In your own words, describe the difference between an estimate and a guess.

6. Describe one way to estimate how many students there are at your school.

7. Sports commentators often estimate the crowd size at sports events. They estimate the percentage of the seats that are occupied, and then use the venue's seating capacity to estimate the crowd size.

   a. The photograph on the right shows the crowd at a soccer match at the Brisbane Cricket Ground (the Gabba). Use the information given with the photograph to calculate the estimated number of people in attendance.

   b. Choose two well-known sports venues and research the maximum seating capacity for each.

   c. If you estimate that three out of every four seats are occupied at each of the venues chosen in part b for particular sports events, how many people are in attendance at each venue?

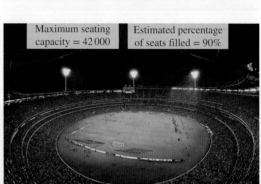

| Maximum seating capacity = 42 000 | Estimated percentage of seats filled = 90% |

8. i. Estimate the answer to each of the following expressions by rounding to the first (or leading) digit.
   ii. Calculate the exact answer using a calculator.
   - a. $46 + 85$
   - b. $478 + 58 + 2185$
   - c. $37 - 25$
   - d. $54 - 28$
   - e. $2458 - 1895$
   - f. 25×58
   - g. 197×158
   - h. $10\,001 - 572$
   - i. $23\,547 \times 149$

9. i. Estimate the answer to each of the following by rounding to the nearest 10.
   ii. Calculate the exact answer using a calculator, writing answers as mixed numbers where appropriate.
   - a. $68 \div 8$
   - b. $158 \div 8$
   - c. $425 \div 11$
   - d. $3694 \div 6$
   - e. $\dfrac{245}{5}$
   - f. $\dfrac{168}{7}$

10. Use the photo on the right to estimate the following.

    a. The number of apples
    b. The number of grapes
    c. The number of peaches and nectarines

11. Three 97-cm lengths of wood are needed to make a bookcase. When buying the wood at a hardware store, should the lengths be rounded in any way? If so, how should they be rounded?

1.6 Decimal fractions

1.6.1 Decimal place value

In a decimal number, the whole number part and the fractional part are separated by a **decimal point**.

Whole number part · 73.064 · Fractional part · Decimal point

A place-value table can be extended to include decimal place values. It can be continued to an infinite number of **decimal places**.

Thousands	Hundreds	Tens	Ones	Tenths	Hundredths	Thousandths	Ten-thousandths
1000	100	10	1	$\dfrac{1}{10}$	$\dfrac{1}{100}$	$\dfrac{1}{1000}$	$\dfrac{1}{10\,000}$

The decimal number 73.064 represents 7 tens, 3 ones, 0 tenths, 6 hundredths and 4 thousandths.
In **expanded form**, 73.064 is written as:

$$(7 \times 10) + (3 \times 1) + \left(0 \times \frac{1}{10}\right) + \left(6 \times \frac{1}{100}\right) + \left(4 \times \frac{1}{1000}\right)$$

When reading decimals, the whole number part is read normally and each digit to the right of the decimal point is read separately. For example, 73.064 is read as 'seventy-three point zero six four'.

The number of decimal places in a decimal is the number of digits after the decimal point. The number 73.064 has three decimal places.

The zero (0) in 73.064 means that there are no tenths. The zero must be written to hold the place value; otherwise the number would be written as 73.64, which does not have the same value. This zero is called a 'place-holding zero'.

Any whole number can be written as a decimal number by showing the empty decimal places after the decimal point. For example, the number 2 can be written as 2.0, 2.00 or with any number of zeros after the decimal point. These are not place-holding zeros; they are trailing zeros.

WORKED EXAMPLE 11

Write the value of the digit 5 in each of the following decimal numbers.
a. **2.005** b. **498.57** c. **0.05**

THINK	WRITE
a. The digit 5 is the third decimal place to the right of the decimal point, so the value is thousandths.	a. $\dfrac{5}{1000}$
b. The digit 5 is the first decimal place to the right of the decimal point, so the value is tenths.	b. $\dfrac{5}{10}$
c. The digit 5 is the second decimal place to the right of the decimal point, so the value is hundredths.	c. $\dfrac{5}{100}$

1.6.2 Comparing decimals

Decimals are compared using digits with the same place value.

The decimal with the largest number in the highest place-value column is the largest decimal, *regardless of the number of decimal places.*

For example, 15.71 is larger than 15.702 because the first place value with different digits (moving from left to right) is hundredths, and 1 is greater than 0; that is, 15.71 > 15.702.

Tens	Ones	• Tenths	Hundredths	Thousandths
1	5	• 7	1	
1	5	• 7	0	2

1.6.3 Adding and subtracting decimals

A decimal number has an infinite number of decimal place values in its fractional part.

Trailing zeros are zeros that appear to the right of the decimal number and have no digits other than zero following them.

When decimals with different numbers of decimal places are added or subtracted, trailing zeros can be written so that both decimals have the same number of decimal places.

Decimal numbers can be added and subtracted in a similar manner to whole numbers. Set out the numbers to be added or subtracted in vertical columns so that the decimal points are lined up.

Answers to decimal addition and subtraction may be checked mentally by rounding each decimal number to the nearest whole number and then adding or subtracting them.

12.300 000
↖
Trailing zeros

WORKED EXAMPLE 12

Calculate the value of 197.76 + 52.9. Check your answer using a calculator.

THINK	WRITE
1. Set out the problem in vertical columns so that the decimal points are lined up. Fill in the empty place values with trailing zeros, as shown in pink.	197.76 +52.90
2. Add the digits as you would for whole numbers. Write the decimal point in the answer in line with the decimal points in the question.	^{1}1^{1}9^{1}7.76 + 5 2.90 2 5 0.66
3. State the answer and check using a calculator.	197.76 + 52.9 = 250.66

WORKED EXAMPLE 13

Calculate the value of 125.271 − 85.08. Check your answer using a calculator.

THINK	WRITE
1. Set out the subtraction problem in vertical columns. Line up the decimal points so that the digits of the same place value are underneath each other. Fill in the empty place values with trailing zeros, as shown in pink.	125.271 − 85.080

2. Subtract the digits as you would for whole numbers. Write the decimal point in the answer in line with the decimal point in the question.

$$^0 \cancel{1}^1 25.^1 \cancel{2}^1 71$$
$$- \quad 85.\ 0\ \ 80$$
$$\overline{\quad 40.\ 1\ \ 91}$$

3. State the answer and check using a calculator.

$$125.271 - 85.08 = 40.191$$

1.6.4 Multiplying using powers of 10

Powers of 10 are $10, 100, 1000, 10\,000$ and so on. When you multiply a decimal by a power of 10, you move the position of the decimal point to the right by the number of zeros in the power of 10.

Multiples of 10 are $10, 20, 30, 40, \dots, 120, \dots, 400, \dots, 16\,000$ and so on. Multiplying by 400 is the same as multiplying by 4 and then multiplying by 100.

Powers

$10^1 = 10$

$10^2 = 100$

$10^3 = 1000$

$10^4 = 10\,000$

Multiples

$1 \times 100 = 100$
$2 \times 100 = 200$
$3 \times 100 = 300$
$4 \times 100 = 400$

WORKED EXAMPLE 14

Calculate the value of the following.

a. **1.34 × 100**   b. **64.7 × 1000**

THINK	WRITE
a. 1. Write the number 1.34 and identify where the decimal point is.	a. 1.34
2. When multiplying by 100, there are 2 zeros, so move the decimal point 2 places to the right.	1.34 $\rightarrow 134$
3. State the answer.	$1.34 \times 100 = 134$
b. 1. Write the number 64.7 and identify where the decimal point is.	b. 64.7
2. When multiplying by 1000, there are 3 zeros, so move the decimal point 3 places to the right. Add zeros where necessary.	64.700 $\rightarrow 64\,700$
3. State the answer.	$64.7 \times 1000 = 64\,700$

WORKED EXAMPLE 15

Calculate the value of 2.56 × 7000. Check your answer using a calculator.

THINK	WRITE
1. Multiplying by 7000 is the same as multiplying by 7 and then multiplying by 1000.	2.56×7000 $= 2.56 \times 7 \times 1000$
2. Complete the first multiplication 2.56×7.	$\begin{array}{r} 2.56 \\ \times 7 \\ \hline 17.92 \end{array}$
3. Rewrite the question showing the result of the first multiplication.	$2.56 \times 7 \times 1000$ $= 17.92 \times 1000$
4. Complete the second multiplication 17.92×1000 by moving the decimal point in 17.92 three places to the right. Add zeros where necessary.	$= 17\,920$
5. Write the answer and check using a calculator.	$2.56 \times 7000 = 17\,920$

1.6.5 Dividing by powers of 10

When you divide a decimal by a power of 10, you move the position of the decimal point to the left by the number of zeros in the power of 10.

WORKED EXAMPLE 16

Calculate the value of the following.
a. **234.25 ÷ 100**
b. **4.28 ÷ 1000**

THINK	WRITE
a. 1. Write the number 234.25 and identify where the decimal point is.	a. 234.25
2. When dividing by 100, there are 2 zeros, so move the decimal point 2 places to the left.	$\overset{\curvearrowleft\curvearrowleft}{234.25}$ $\to 2.3425$
3. State the answer.	$234.25 \div 100 = 2.3425$
b. 1. Write the number 4.28 and identify where the decimal point is.	b. 4.28
2. When dividing by 1000, there are 3 zeros, so move the decimal point 3 places to the left. Add zeros before the 4 where necessary.	$\overset{\curvearrowleft\curvearrowleft\curvearrowleft}{0004.28}$ $\to 0.00428$
3. State the answer.	$4.28 \div 1000 = 0.00428$

1.6.6 Converting decimals to fractions

Decimals can be written as fractions by using place values.

WORKED EXAMPLE 17

Convert each of the following decimals to fractions in simplest form.
a. **0.7**   b. **3.25**

THINK   **WRITE**

a.   The digit 7 is in the tenths column, so it represents   a. $0.7 = \dfrac{7}{10}$
     7 tenths, which is written as $\dfrac{7}{10}$.

Ones	·	Tenths	Hundredths
0	·	7	0

b. 1. The number 3.25 consists of 3 ones, 2 tenths $\left(\dfrac{2}{10}\right)$ b. $3.25 = 3 + \dfrac{2}{10} + \dfrac{5}{100}$
      and 5 hundredths $\left(\dfrac{5}{100}\right)$.

Ones	·	Tenths	Hundredths
3	·	2	5

2. Add 2 tenths and 5 hundredths by writing $\dfrac{2}{10}$ as a $3.25 = 3 + \dfrac{20}{100} + \dfrac{5}{100}$
   fraction with a denominator of 100.

$$= 3 + \dfrac{25}{100}$$

3. Simplify $\dfrac{25}{100}$ by cancelling by a factor of 25. $= 3 + \dfrac{{}^1\cancel{25}}{\cancel{100}^4}$

$$= 3 + \dfrac{1}{4}$$

$$= 3\dfrac{1}{4}$$

1.6.7 Converting fractions to decimals

A fraction can be expressed as a decimal by dividing the numerator by the denominator.

$$\frac{5}{8} = 5 \div 8$$

When a division results in a remainder, add a trailing zero and continue the division until there is no remainder or until a pattern can be seen.

If a division reaches a stage where there is no remainder, the decimal is called a **finite decimal**.

If a division continues endlessly with a **repeating pattern**, it is called an **infinite recurring decimal**.

Infinite recurring decimals can be written in an abbreviated form by placing a dot above the repeating pattern. When more than one digit is repeated, a dot is placed above the first digit and last digit in the repeating pattern or a bar is placed above the entire repeating pattern.

Common recurring decimals

$$\frac{1}{3} = 0.333\,333\,... = 0.\dot{3}$$

$$\frac{1}{6} = 0.166\,666\,... = 0.1\dot{6}$$

$$\frac{3}{7} = 0.428\,574\,285\,7\,... = 0.\dot{4}285\,7\dot{7} \text{ or } 0.\overline{428\,57}$$

WORKED EXAMPLE 18

Write $\dfrac{5}{8}$ as a decimal. Check your answer using a calculator.

THINK	WRITE
1. The fraction $\dfrac{5}{8}$ can be expressed as $5 \div 8$. Write the division.	$8\overline{)5}$
2. Divide 8 into 5; the result is 0 remainder 5. Write the 0 above the 5 and write the remainder as shown in pink. Add a decimal point and trailing zero next to the 5.	$\dfrac{0}{8\overline{)5.^50}}$
3. Divide 8 into 50; the result is 6 remainder 2. Write the decimal point in the answer above the decimal point in the question. Write the 6 above the 0 and write the remainder as shown in red. Add another trailing zero.	$\dfrac{0\ 6}{8\overline{)5.^50^20}}$
4. Divide 8 into 20; the result is 2 remainder 4. Write the 2 above the 0 and write the remainder, as shown in green. Add another trailing zero.	$\dfrac{0.6\ 2}{8\overline{)5.^50^20^40}}$
5. Divide 8 into 40; the result is 5. Write the 5 above the 0, as shown in pink. There is no remainder, so the division is finished.	$\dfrac{0.6\ 2\ 5}{8\overline{)5.^50^20^40}}$
6. Write the answer and check using a calculator.	$\dfrac{5}{8} = 0.625$

WORKED EXAMPLE 19

Write $6\dfrac{2}{3}$ as a decimal. Check your answer using a calculator.

THINK	WRITE
1. The mixed number $6\dfrac{2}{3}$ can be converted to a decimal by converting the fractional part.	$6\dfrac{2}{3} = 6 + \dfrac{2}{3}$
2. The fraction $6\dfrac{2}{3}$ can be expressed as $2 \div 3$. Write the division.	$3\overline{)2}$
3. Divide 3 into 2; the result is 0 remainder 2. Write 0 above the 2 and write the remainder beside the next smallest place value, as shown in pink. Write the decimal point in the answer above the decimal point in the question and add a trailing zero.	$\dfrac{0}{3\overline{)2.^20}}$

4. Divide 3 into 20; the result is 6 remainder 2. Write 6 above the 0 and write the remainder, as shown in red. Add a trailing zero.

$$3\overline{)2.^20^20}$$
$$\quad\ 0.\,6$$

5. Divide 3 into 20; the result is 6 remainder 2. This remainder is the same as the previous remainder. The remainder will continue to be the same. The decimal is a recurring decimal. Write the fraction and its equivalent decimal.

$$3\overline{)2.^20^20^20}$$
$$\quad\ 0.\,6\ 6$$

$$\frac{2}{3} = 0.\dot{6}$$

6. Write the fractional part of the mixed number as a decimal.

$$6\frac{2}{3} = 6 + \frac{2}{3}$$
$$= 6 + 0.\dot{6}$$

7. Add the whole number and the decimal part. Write the answer and check using a calculator.

$$= 6.\dot{6}$$

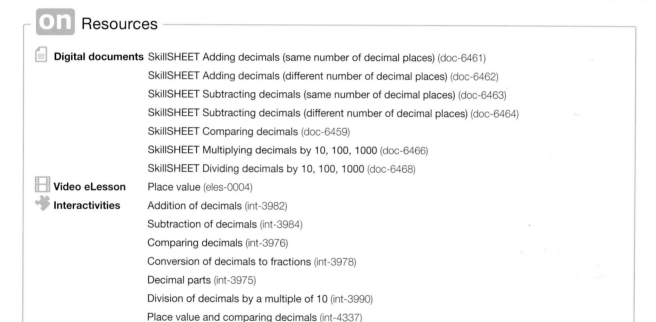

on Resources

Digital documents   SkillSHEET Adding decimals (same number of decimal places) (doc-6461)

SkillSHEET Adding decimals (different number of decimal places) (doc-6462)

SkillSHEET Subtracting decimals (same number of decimal places) (doc-6463)

SkillSHEET Subtracting decimals (different number of decimal places) (doc-6464)

SkillSHEET Comparing decimals (doc-6459)

SkillSHEET Multiplying decimals by 10, 100, 1000 (doc-6466)

SkillSHEET Dividing decimals by 10, 100, 1000 (doc-6468)

Video eLesson   Place value (eles-0004)

Interactivities   Addition of decimals (int-3982)

Subtraction of decimals (int-3984)

Comparing decimals (int-3976)

Conversion of decimals to fractions (int-3978)

Decimal parts (int-3975)

Division of decimals by a multiple of 10 (int-3990)

Place value and comparing decimals (int-4337)

Exercise 1.6 Decimal fractions

1. State the number of decimal places in each of the following decimals.
   - a. 548.5845
   - b. 0.007
   - c. 1.1223
   - d. 15.001
   - e. 4.1
   - f. 13.42
   - g. 105.021
   - h. 3.880 99

2. a. How many decimal places do we use in our currency?
   - b. Write the amount of money shown in the photograph:
     - i. in words
     - ii. in numbers.

3. Identify the place-holding zeros in each of the following.
   - a. 10.23
   - b. 105.021
   - c. 11.010
   - d. 0.001
   - e. 282.0001
   - f. 15.00

4. Write each of the following in expanded form.
   - a. 16.02
   - b. 11.046
   - c. 222.03
   - d. 15.11
   - e. 4.701
   - f. 68.68

5. Write the decimal number represented by each of the following.
   - a. $(4 \times 10) + (2 \times 1) + \left(5 \times \dfrac{1}{10}\right) + \left(0 \times \dfrac{1}{100}\right) + \left(4 \times \dfrac{1}{1000}\right)$
   - b. $\left(2 \times \dfrac{1}{10}\right) + \left(7 \times \dfrac{1}{100}\right) + \left(2 \times \dfrac{1}{1000}\right)$
   - c. $1 + \left(9 + \dfrac{1}{1000}\right)$
   - d. $(3 \times 10) + \dfrac{9}{10} + \dfrac{3}{100} + \dfrac{4}{1000}$

6. 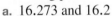 **WE11** Write the value of the digit 7 in each of the following decimal numbers.
   - a. 2.075
   - b. 15.701
   - c. 12.087
   - d. 93.1487
   - e. 17.16
   - f. 73.064
   - g. 24.025 75
   - h. 1.0077

7. Use an appropriate method to determine which decimal number is larger in each of the following pairs.
   - a. 16.273 and 16.2
   - b. 137.02 and 137.202
   - c. 95.89 and 95.98
   - d. 0.001 and 0.0001
   - e. 0.123 and 0.2
   - f. 0.0101 and 0.012

8. **WE12** Calculate the value of each of the following. Check your answers using a calculator.
   - a. 14.23 + 254.52
   - b. 79.58 + 18.584
   - c. 99.999 + 0.01
   - d. 58.369 + 86.12 + 78
   - e. 485.5846 + 5 + 584.58 + 0.57
   - f. 34.2 + 7076 + 20.5604 + 1.53

9. **WE13** Calculate the value of each of the following. Check your answers using a calculator.
   - a. 25.3458 − 25.2784
   - b. 848.25 − 68.29
   - c. 58.8 − 24.584
   - d. 470 − 28.57
   - e. 15.001 − 0.007
   - f. 35.1 − 9.007 51

10. Calculate each of the following to two decimal places by:
    - i. rounding all numbers first, then adding or subtracting
    - ii. adding or subtracting the numbers first, then rounding.
    - a. 4.457 + 5.386
    - b. 47.589 − 35.410
    - c. 126.917 − 35.492
    - d. 168.268 + 21.253

11. **WE14** Calculate the value of each of the following
    - a. 6.284 × 100
    - b. 5.3 × 1000

12. **WE15** Calculate the value of each of the following. Check your answers using a calculator.
    - a. 1.2345 × 50
    - b. 1.2345 × 600
    - c. 1.2345 × 7000
    - d. 1.2345 × 20 000

13. **WE16** Calculate the value of each of the following.
    a. $3.45 \div 10$
    b. $123.98 \div 100$
    c. $1245.37 \div 1000$
    d. $3.569 \div 10$
    e. $0.246 \div 1000$
    f. $2.48 \div 100$

14. **WE17** Convert each of the following decimals to fractions in simplest form.
    a. 0.1
    b. 0.5
    c. 0.8
    d. 0.12
    e. 0.21
    f. 0.84
    g. 0.05
    h. 0.625
    i. 3.8
    j. 2.13
    k. 12.42
    l. 10.0035

15. **WE18 & 19** Write each of the following fractions as a decimal. Check your answers using a calculator.
    a. $\dfrac{3}{8}$
    b. $\dfrac{4}{5}$
    c. $\dfrac{5}{4}$
    d. $\dfrac{3}{4}$
    e. $2\dfrac{1}{4}$
    f. $3\dfrac{1}{20}$
    g. $\dfrac{13}{2}$
    h. $\dfrac{1}{5}$

16. Having purchased the items shown, what change would you expect from $50?

1.7 Review: exam practice

1.7.1 Calculations: summary

Number operations
- Solving practical numerical problems requires us to use addition, subtraction multiplication and division skills quickly and accurately.
- When calculating answers to mathematical problems it is always important to understand the question so you have an idea of what a reasonable answer should be.

Order of operations
- When evaluating arithmetic calculations it is important to follow the correct order of operations, as described by BIDMAS.

    Brackets
    Indices (powers and roots)
    Division *and* **M**ultiplication (working from left to right)
    Addition *and* **S**ubtraction (working from left to right)

Rounding
- To round to the nearest 10, think about which multiple of 10 the number is closest to.
  For example, if 34 is rounded to the nearest 10, the result is 30 because 34 is closer to 30 than it is to 40.

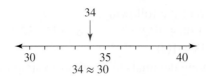

$$34 \approx 30$$

- To round to the nearest 100, think about which multiple of 100 the number is closest to.
- To round to the first (or leading) digit, use the following guidelines.
  - If the second digit is $0, 1, 2, 3$ or 4, the first digit stays the same and all the following digits are replaced with zeros.
  - If the second digit is $5, 6, 7, 8$ or 9, the first digit is raised by 1 (rounded up) and all the following digits are replaced with zeros.
- When rounding a decimal, look at the first digit after the number of decimal places required.
  - If this digit is less than 5, leave the digit in the last required decimal place unchanged and drop off all digits that follow it.
  - If this digit is greater than or equal to 5, increase the digit in the last required decimal place by 1 and drop off all digits that follow it.

Estimation and approximation strategies

- An estimation is based on information, so it is not merely a guess.
- The number of people in attendance at a sportsground can be estimated based on an estimate of the fraction of seats filled.
- Estimation using rounding can be useful to check if calculations are correct.

Decimal fractions

- A place-value table can be extended to include decimal place values. It can be continued to an infinite number of decimal places.

Thousands	Hundreds	Tens	Ones	•	Tenths	Hundredths	Thousandths	Ten-thousandths
1000	100	10	1	•	$\dfrac{1}{10}$	$\dfrac{1}{100}$	$\dfrac{1}{1000}$	$\dfrac{1}{10\,000}$

- Decimals can be compared by looking for the decimal with the largest number in the highest place-value column, regardless of the number of decimal places.
- Trailing zeros are zeros that appear to the right of the decimal point that have no digits other than zeros following them.
- Decimals can be added or subtracted by setting out the numbers in vertical columns with the decimal points lined up.
- To multiply or divide a decimal number by a power of 10, move the position of the decimal point to the right (for multiplication) or left (for division) by the number of zeros in the power of 10.
- To multiply a decimal number by a multiple of 10, we can use the fact that, for example, multiplying by 400 is the same as multiplying by 4 and then multiplying by 100.
- A decimal can be expressed as a fraction by writing the digits after the decimal point as a fraction using their place value.
- A fraction can be expressed as a decimal by dividing the numerator by the denominator.
- If a decimal does not terminate and has decimal numbers repeating in a pattern, it is called a recurring decimal.

Exercise 1.7 Review: exam practice

Simple familiar

1. A school has a fund raiser by selling cakes. If each cake costs 60 cents to make and they are sold for $2:
   a. determine the amount of money they will make if they sell 150 cakes
   b. determine how much profit they will make if they sell 150 cakes.

2. A family hires two surfboards at a cost of $10 per 20 minutes. Calculate the cost of hiring the surfboards if:
   a. they hire them for 2 hours
   b. they hire them for 3 hours and 40 minutes.

3. Evaluate the following.
   a. $6 + 4 \times 5$
   b. $(12 + 24) \div 9 \times 8$

4. Evaluate the following.
   a. $\dfrac{48}{12} \times 6 \div 2^2$
   b. $\dfrac{3 \times (5 + 7)}{2^2}$

5. Round the following to the first (or leading) digit.
   a. 66
   b. 432 679

6. Round the following to the number of decimal places shown in the brackets.
   a. 6.95721 (3)
   b. 23.0594 (2)

7. Estimate the value of the following by rounding each number to its first (or leading) digit.
   a. $251 + 978$
   b. $45\,833 \div 487$

8. Estimate the value of the following by first rounding each number to the nearest 100.
   a. $762 + 1358 + 2361$
   b. $3333 - 967 - 1545$

9. Calculate the value of the following.
   a. $95.56 \div 100$
   b. 2.56×2000

10. Identify the place-holding zeros in each of the following.
    a. 26.07
    b. 10.967

11. i. Estimate the answer to each of the following expressions by rounding to the first (or leading) digit.
    ii. Calculate the exact answer using a calculator.
    a. $58 + 89$
    b. $211 - 58$
    c. $169 + 239$
    d. $1234 - 456$

12. Round each of the following to the nearest unit.
    a. 121.60
    b. 0.512
    c. 79.4
    d. 9.6

Complex familiar

13. Andrew fills his car with petrol once a week. Over the last 4 weeks he paid the following prices per litre for his petrol: $1.24, $1.39, $1.46 and $1.40. What was the average price, to two decimal places Andrew paid for his petrol over the 4 weeks?

14. Cara, James and Lisa saved their coins and when they counted them they had $8.70, $12.55 and $14.45 respectively.
    a. Estimate their total savings by first rounding each amount to the nearest dollar.
    b. Calculate their actual total savings.

15. Calculate the value of 4.56×3000.

16. a. Determine the value of the following, writing your answer as a fraction.

$$\frac{2\left[\left(4^3 \div 2^3\right) + 3\left(24 - 18\right)\right]}{(3 + 6 \div 2)}$$

    b. Using a calculator, write your answer from part a as a decimal.

Complex unfamiliar

17. Convert each of the following decimals to fractions in simplest form.

    a. 0.6   **b.** 5.75

18. Write $3\dfrac{7}{8}$ as a decimal.

19. Sharyn goes to the supermarket to get some snacks to take to the netball. She purchases 2 packets of chips at $4.50 each, 3 packets of lollies at $3.25 each and 1 packet of biscuits at $5.80 each.

    a. How much did Sharyn spend?

    b. If Sharyn paid with $30, how much change did she receive?

20. Joel is going to paint sections of his house. He will use three different types of paint. The prices and quantities of each type required are:

- 3 litres of Paint A at $21.75 per litre
- 0.75 litres of Paint B at $72.50 per 2 litres
- 7.5 litres of Paint C at $12.75 per 0.5 litre.

Determine the total cost of the paint required to complete the project.

Answers

1 Calculations

Exercise 1.2 Number operations

1. 3 GB
2. $7049
3. $317
4. a. $17   b. $6
5. a. 5 m
   b. 75 m
   c. 70 m
   d. To find the difference, subtract the height of the smaller building from the taller building.
6. a. 351 km   b. 247 km
7. 52
8. a. 5432   b. 2345   c. 3087
9. a. 884   b. 6205
10. 400 km
11. a. $0.94   b. $32.90   c. $5146.50
    d. $3   e. $10
12. $522.63
13. $54\frac{1}{6}$ grams
14. 2 students

Exercise 1.3 Order of operations

1. a. 6   b. 3   c. 3   d. 11   e. 96
   f. 27   g. 7   h. 12   i. 18   j. 12
2. a. $16\frac{1}{7}$ b. 1   c. 16
   d. $\frac{37}{7}$ e. 13
3. a. $(12 - 8) \div 4 = 1$
   b. $(4 + 8) \times 5 - 4 \times 5 = 40$
   c. $3 + 4 \times (9 - 3) = 27$
   d. $3 \times (10 - 2) \div 4 + 4 = 10$
   e. $10 \div (5 + 5) \times 9 \times 9 = 81$
   f. $(18 - 3) \times 3 \div 5 = 9$
4. a. When the only operations are multiplication and division, they should be done from left to right. When the only operations are addition and subtraction, they should be done from left to right.
   b. 40 and 17
5. a. 2   b. 80   c. 52
   d. 33   e. 48   f. 75
   g. 7   h. 5   i. $12\frac{1}{7}$
6. a. $1\frac{1}{8}$ b. $8\frac{1}{3}$ c. $\frac{7}{36}$ d. $2\frac{2}{5}$
7. a. Use multiplication first to calculate the cost of each type of food, and then add these costs together to calculate the total cost of all the food.
   b. Cost = $(2 \times \$2) + (3 \times \$4) + (3 \times \$5)$
   c. $31
   $1 = (4 - 3) \div (2 - 1)$
   $2 = 4 - 3 + 2 - 1$
8. $3 = 21 \div (3 + 4)$
   $4 = 4 - 2 + 3 - 1$
   $5 = 3 + 4 - (2 \times 1)$
9. $0.27

Exercise 1.4 Rounding

1. a. 10   b. 70   c. 170
   d. 1350   e. 56 900   f. 765 490
2. a. 0   b. 100   c. 200
   d. 3000   e. 42 600   f. 345 300
3. a. 1000   b. 1000   c. 7000
   d. 12 000   e. 453 000   f. 2 680 000
4. a. 6   b. 50   c. 1000
   d. 10 000   e. 200 000   f. 5 000 000
5. a. 0.32   b. 0.86   c. 1.25
   d. 13.05   e. 7.13   f. 100.81
   g. 71.26   h. 0.01   i. 0.19
   j. 19.70   k. 0.40   l. 1.00
6. a. 0.4   b. 0.9   c. 9.3
   d. 25.3   e. 300.1   f. 12.8
   g. 99.9   h. 8.9   i. 17.6
   j. 0.9   k. 94.0   l. 1.0
7. a. 2.39   b. 14.0   c. 0.03
   d. 0.8765   e. 64.2952   f. 0.382
   g. 96.3   h. 3.04   i. 8.90
   j. 47.880   k. 0.10   l. 0.49
8. C
9. D
10. D
11. A
12. a. 11   b. 8   c. 4
    d. 93   e. 112   f. 22
    g. 42   h. 2138   i. 0
    j. 1   k. 1   l. 41

Exercise 1.5 Estimation and approximation strategies

1. 6300
2. a. 1500   b. 900 000   c. 500 000
   d. 36 000   e. 6 000 000   f. 70
3. Adding the cost of groceries when shopping; calculating the cost of petrol for a trip; determining the size of a crowd.
4. a. Possibly, but most likely not.
   b. They may have divided the distance by the average speed.
5. An estimate is based on information.
6. Multiply the number of students in an average class by the number of classes in the school.
7. a. 37 800
   b. Sample responses can be found in the Worked Solutions in the online resources.
   c. Sample responses can be found in the Worked Solutions in the online resources.
8. a. i. 140   ii. 131
   b. i. 2560   ii. 2721
   c. i. 10   ii. 12
   d. i. 20   ii. 26
   e. i. 0   ii. 563
   f. i. 1800   ii. 1450
   g. i. 40 000   ii. 31126
   h. i. 9400   ii. 9429
   i. i. 2 000 000   ii. 3 508 503
9. a. i. 7   ii. $8\frac{1}{2}$
   b. i. 16   ii. $19\frac{3}{4}$

c. i. 43 ii. $38\frac{7}{11}$

d. i. 369 ii. $615\frac{2}{3}$

e. i. 25 ii. 49

f. i. 17 ii. 24

10. a. Approximately 90 **b.** Approximately 2000

c. Approximately 80

11. If material lengths are to be rounded it should be to a longer length than required or the purchased wood will be too short.

Exercise 1.6 Decimal fractions

1. a. 4 **b.** 3 **c.** 4

d. 3 **e.** 1 **f.** 2

g. 3 **h.** 5

2. a. 2

b. i. Seventy-eight dollars and fifty-five cents.

ii. $78.55

3. a. Ones

b. Tens, tenths

c. Tenths

d. Ones, tenths, hundredths

e. Tenths, hundredths, thousandths

f. There are no place-holding zeros.

4. a. $(1 \times 10) + (6 \times 1) + \left(0 \times \frac{1}{10}\right) + \left(2 \times \frac{1}{100}\right)$

b. $(1 \times 10) + (1 \times 1) + \left(0 \times \frac{1}{10}\right) + \left(4 \times \frac{1}{100}\right) +$
$\left(6 \times \frac{1}{1000}\right)$

c. $(2 \times 100) + (2 \times 10) + (2 \times 1) + \left(0 \times \frac{1}{10}\right) + \left(3 \times \frac{1}{100}\right)$

d. $(1 \times 10) + (5 \times 1) + \left(1 \times \frac{1}{10}\right) + \left(1 \times \frac{1}{100}\right)$

e. $(4 \times 1) + \left(7 \times \frac{1}{10}\right) + \left(0 \times \frac{1}{100}\right) + \left(1 \times \frac{1}{1000}\right)$

f. $(6 \times 10) + (8 \times 1) + \left(6 \times \frac{1}{10}\right) + \left(8 \times \frac{1}{100}\right)$

5. a. 42.504 **b.** 0.272

c. 1.009 **d.** 30.934

6. a. $\frac{7}{100}$ **b.** $\frac{7}{10}$

c. $\frac{7}{1000}$ **d.** $\frac{7}{10\,000}$

e. 7 **f.** 70

g. $\frac{7}{10\,000}$ **h.** $\frac{7}{1000}, \frac{7}{10\,000}$

7. a. 16.273 **b.** 137.202 **c.** 95.98

d. 0.001 **e.** 0.2 **f.** 0.012

8. a. 268.75 **b.** 98.164 **c.** 100.009

d. 222.489 **e.** 1075.7346 **f.** 7132.2904

9. a. 0.0674 **b.** 779.96 **c.** 34.216

d. 441.43 **e.** 14.994 **f.** 26.092 49

10. a. i. 9.85 ii. 9.84

b. i. 12.18 ii. 12.18

c. i. 91.43 ii. 91.43

d. i. 189.52 ii. 189.52

11. a. 628.4 **b.** 5300

12. a. 61.725 **b.** 740.7

c. 8641.5 **d.** 24 690

13. a. 0.345 **b.** 1.2398 **c.** 1.245 37

d. 0.3569 **e.** 0.000 246 **f.** 0.0248

14. a. $\frac{1}{10}$ **b.** $\frac{1}{2}$ **c.** $\frac{4}{5}$

d. $\frac{3}{25}$ **e.** $\frac{21}{100}$ **f.** $\frac{21}{25}$

g. $\frac{1}{20}$ **h.** $\frac{5}{8}$ **i.** $3\frac{4}{5}$

j. $2\frac{13}{100}$ **k.** $12\frac{21}{50}$ **l.** $10\frac{7}{2000}$

15. a. 0.375 **b.** 0.8 **c.** 1.25

d. 0.75 **e.** 2.25 **f.** 3.05

g. 6.5 **h.** 0.2

16. Cost = $40.54

Change = $9.46

1.7 Review: exam practice

1. a. $300 **b.** $210

2. a. $120 **b.** $220

3. a. 26 **b.** 32

4. a. 6 **b.** 9

5. a. 70 **b.** 400 000

6. a. 6.957 **b.** 23.06

7. a. 1300 **b.** 100

8. a. 4600 **b.** 800

9. a. 0.9556 **b.** 5120

10. a. Tenths **b.** Units

11. i. a. 150 **b.** 140

c. 400 **d.** 500

ii. a. 147 **b.** 153

c. 408 **d.** 778

12. a. 122 **b.** 1 **c.** 79 **d.** 10

13. $1.37

14. a. $36 **b.** $35.70

15. 13 680

16. a. $\frac{26}{3}$ **b.** $8.\dot{6}$

17. a. $\frac{3}{5}$ **b.** $5\frac{3}{4}$

18. 3.875

19. a. $24.55 **b.** $5.45

20. $283.69

2 Geometry

2.1 Overview

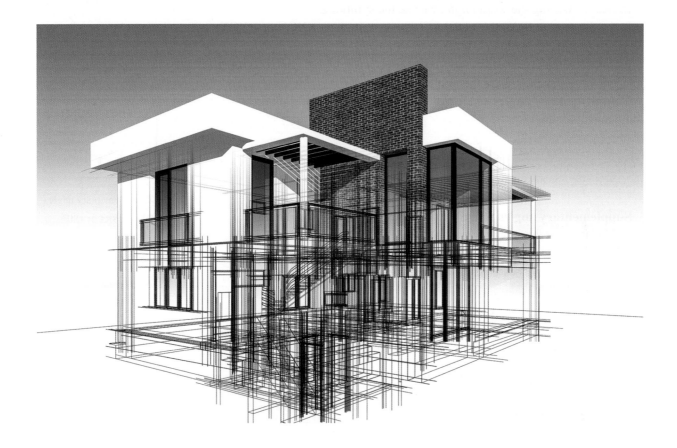

LEARNING SEQUENCE

2.1 Overview
2.2 Review of angles
2.3 Two-dimensional geometric shapes
2.4 Three-dimensional objects
2.5 Two-dimensional representations of three-dimensional objects
2.6 Review: exam practice

CONTENT

In this chapter, students will learn to:
- recognise the properties of common two-dimensional geometric shapes, including squares, rectangles and triangles, and three-dimensional solids, including cubes, rectangular-based prisms and triangular-based prisms
- interpret different forms of two-dimensional representations of three-dimensional objects, including nets of cubes, rectangular-based prisms and triangular-based prisms [complex].

Fully worked solutions for this chapter are available in the Resources section of your eBookPLUS at www.jacplus.com.au.

2.2 Review of angles

2.2.1 Angles

Angles are formed at the point of **intersection** between two straight lines

Angles can be named using capital letters of the English alphabet. A common way to name an angle is to use three letters: two letters to represent the arms of the angle, and a third letter to represent its **vertex**. The letter representing the vertex is always placed in the middle (between the two letters representing the arms).

Instead of writing the word *angle*, we use the symbol ∠.

In the diagram at right, the angle at O can be written as ∠XOY or ∠YOX.

Alternatively, the angle at O could be written ∠O.

Adjacent angles share a common side (arm). ∠MNO and ∠ONP below right are adjacent angles as they share the arm $\overline{NO}$.

Complementary angles add to 90°. ∠MNO and ∠ONP at right are complementary angles.

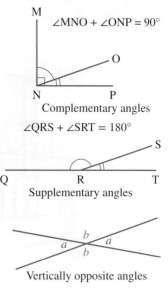

∠MNO + ∠ONP = 90°

Complementary angles

Supplementary angles add to 180°. ∠QRS and ∠SRT at right are supplementary angles.

∠QRS + ∠SRT = 180°

Supplementary angles

Vertically opposite angles are formed when two lines intersect. The intersection creates two sets of equal sized angles, as shown at right.

Vertically opposite angles

Angles can be classified according to their size.

An **acute angle** is greater than 0°, but less than 90°.

A **right angle** is an angle that equals exactly 90°.

An **obtuse angle** is greater than 90° but less than 180°.

A **straight angle** equals exactly 180°.

A **reflex angle** is greater than 180° but less than 360°.

A **revolution** or a **perigon** is an angle that equals exactly 360° (a full circle).

Parallel lines never intersect; the distance between them is constant along their lengths. Parallel lines are marked with identical arrowheads.

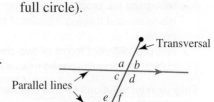

Parallel lines

Transversal

A straight line that crosses two parallel lines is called a **transversal**. A transversal creates three sets of angle pairs as follows.

Angle pair	Positions	Relationship
Corresponding angles (F angles)		The same size. For example, in the parallel lines preceding this table: $a = e$; $b = f$; $d = h$; $c = g$.
Alternate angles (Z angles)		The same size. For example, in the parallel lines preceding this table: $c = f$; $d = e$.
Co-interior angles (C angles)		Supplementary. For example, in the parallel lines preceding this table: $c + e = 180°$; $d + f = 180°$.

WORKED EXAMPLE 1

Find the value of the pronumerals in the diagram.

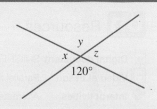

THINK

1. The angles y and $120°$ are vertically opposite, so they are equal.

2. The angle z and the angle marked $120°$ form a straight line, so they are supplementary angles (i.e. they add to $180°$). Subtract $120°$ from $180°$ to find z.

3. The angles x and z are vertically opposite, so they are equal.

WRITE

$y = 120°$

$z = 180° - 120°$
$\quad = 60°$

$x = 60°$

WORKED EXAMPLE 2

Calculate the value of the pronumerals in the diagram.

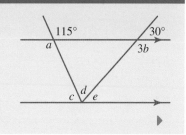

THINK	WRITE
1. Angle a is vertically opposite to 115°; so a is equal to 115°.	$a = 115°$
2. $3b$ and 30° are supplementary (add to 180°). Write an equation to represent this.	$3b + 30° = 180°$
3. Solve the equation for b by subtracting 30° from both sides, then dividing both sides by 3.	$3b + 30° - 30° = 180° - 30°$ $$3b = 150°$$ $$\frac{3b}{3} = \frac{150°}{3}$$ $$b = 50°$$
4. a and c are co-interior and so add to 180°. Write an equation to represent this.	$a + c = 180°$
5. Substitute $a = 115°$ into the equation and solve for c by subtracting 115° from both sides.	$$115° + c = 180°$$ $$115° + c - 115° = 180° - 115°$$ $$c = 65°$$
6. e and 30° are corresponding and hence are equal.	$e = 30°$
7. c, d and e form a straight line so they add to 180°. Write an equation to represent this.	$c + d + e = 180°$
8. Substitute $c = 65°$ and $e = 30°$. Solve for d by adding 65° and 30°, then subtracting 95° from both sides.	$$65° + d + 30° = 180°$$ $$d + 95° = 180°$$ $$d + 95° - 95° = 180° - 95°$$ $$d = 85°$$

on Resources

Digital document	SkillSHEET Classifying angles (doc-6452)
Video eLesson	Parallel lines (eles-2309)
Interactivities	Alternate angles (int-3971)
	Calculation of angles associated with parallel lines (int-3972)
	Co-interior angles (int-3970)
	Corresponding angles (int-3969)
	Vertically opposite and adjacent angles (int-3968)

2.2.2 Angles around a point

Angles with a common vertex that form a circle add to 360°. In the example at right, $x + y + z = 360°$.

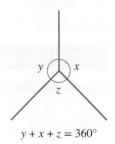

$y + x + z = 360°$

Find the value of the pronumerals in each of the following.

a.

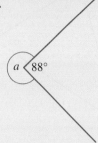

b.

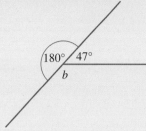

c.

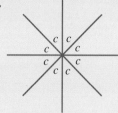

THINK	WRITE
a. 1. These angles form a circle, so they must add to 360°. Write an equation to represent this.	a. $a + 88° = 360°$
2. Subtract 88° from 360° to determine the value of a.	$a + 88° - 88° = 360° - 88°$ $a = 272°$
b. 1. These angles form a circle, so they must add to 360°. Write an equation to represent this.	b. $b + 180° + 47° = 360°$
2. Add 180° and 47° to simplify the left side of the equation.	$b + 227° = 360°$
3. Subtract 180° and 47° from both sides to determine the value of b.	$b = 360° - 227°$ $b = 133°$
c. 1. These angles form a circle, so they must add to 360°. Write an equation to represent this.	c. $c + c + c + c + c + c + c + c = 360°$
2. There are eight equal angles in this diagram. Simplify the equation by collecting like terms.	$8c = 360°$
3. Divide both sides by 8 to determine the value of c.	$\dfrac{8c}{8} = \dfrac{360°}{8}$ $c = 45°$

 Resources

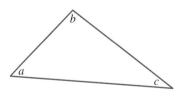 **Interactivity** Angles at a point (int-6157)

2.2.3 Angles and triangles

Interior angles

In the triangle at right, $a + b + c = 180°$, where a, b and c represent the three interior angles of the triangle.

Sum of interior angles of a triangle

The sum of the interior angles of a triangle is 180°.

This rule can be used to find unknown values of interior angles of triangles.

Calculate the value of the pronumeral in the triangle shown.

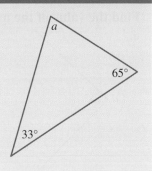

THINK	WRITE
1. The sum of the three interior angles is 180°. Write this as an equation.	$a + 33° + 65° = 180°$
2. Simplify the equation by adding 33° and 65°.	$a + 98° = 180°$
3. Solve the equation for a by subtracting 98° from both sides of the equation.	$a + 98° - 98° = 180° - 98°$ $a = 82°$

Exterior angles

If one edge of a triangle continues past the vertex, the angle formed is called an **exterior angle**.

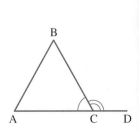

Exterior angle

The interior angle and the exterior angle next (or adjacent) to it are supplementary angles; they add to 180°. For the triangle formed by the house:

$$\angle ACB + \angle BCD = 180°$$
$$\angle ACB = 180° - \angle BCD$$
$$\angle BCD = 180° - \angle ACB$$

Determine the value of the exterior angle ∠BCD in the diagram shown.

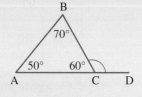

THINK	WRITE
1. The sum of the supplementary angles, ∠ACB and ∠BCD, is 180°.	$\angle ACB + \angle BCD = 180°$
2. ∠ACB is 60°. Subtract this value from 180° to determine the value of ∠BCD.	$\angle BCD = 180° - \angle ACB$ $= 180° - 60°$ $= 120°$

The sum of the interior angles of a triangle is 180°, and there are 180° in a straight angle. Therefore, we can establish the following rule.

Exterior angle of a triangle
An exterior angle of a triangle is equal to the sum of the two opposite interior angles.

In Worked example 5, the size of the exterior angle ∠BCD could have been calculated by adding the two opposite interior angles, ∠BAC and ∠ABC:

$$\angle BCD = \angle BAC + \angle ABC$$
$$= 50° + 70°$$
$$= 120°$$

 Resources

 Interactivities Angles in a triangle (int-3965)

                            Interior and exterior angles of a triangle (int-3966)

Exercise 2.2 Review of angles

1. Consider the angles shown.

   a. Name each angle using:
     i. three letters   ii. a single letter.
   b. Name the *type* of each angle.

2. Copy and complete the following table by placing the listed angles in the appropriate places.

23°	92°	122°	154°	179°	5°	90°	110°	45°	55°	270°	14°	160°
100°	78°	160°	69°	89°	190°	145°	300°	359°	80°	2°	92°	181°

Acute angles	Obtuse angles	Right angles	Reflex angles

3. Consider the diagram shown.
   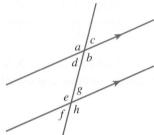
   a. Name the six interior angles in three-letter format.
   b. Name the type of angle for each of the following:
      i. a ii. b iii. c
      iv. d v. e vi. f
   c. List the angle pairs that are complementary.
   d. List the angle pairs that are alternate.
   e. If $e = 30°$, determine the values of the remaining pronumerals.
4. Copy and complete the following sentences using the diagram shown.

   a. Angle a is vertically opposite angle _____.
   b. Angle a is a corresponding angle with angle _____.
   c. Angle b is co-interior with angle _____.
   d. Angle g is alternate to angle _____.
   e. Angle h is vertically opposite angle _____.
   f. Angle b is a corresponding angle with angle _____.

5. a. Draw two angles, XOY and VOW, so that they are vertically opposite each other.
   b. From your angles in part a, list another pair of vertically opposite angles.
6. **WE1** Find the value of the pronumerals in each of the following.

a.
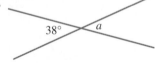

b.
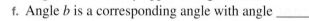

c.
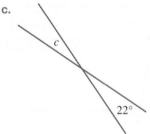

d.
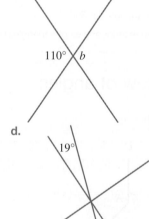

e.

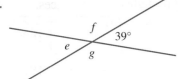

f.

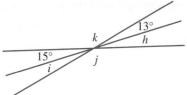

7. **WE2** Calculate the value of the pronumerals in the following diagrams.

a.

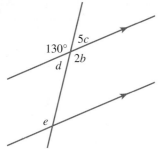

b.

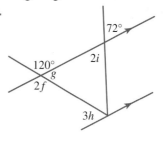

8. **WE3** Find the value of the pronumeral in each of the following.

a.

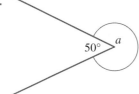

b.

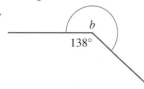

c.

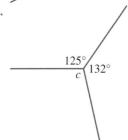

d.

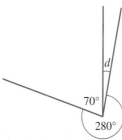

e.

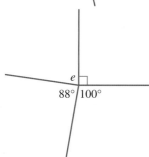

f.

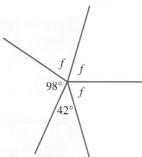

9. Find the value of the pronumeral in each of the following.

a.

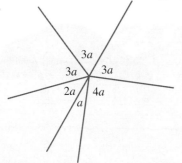

b.

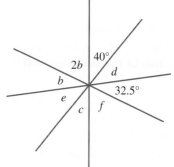

10. **WE4** Calculate the value of the pronumeral in each of the following triangles.

a.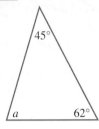

b.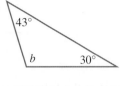

c.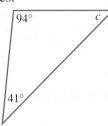

d.

e.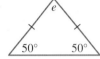

f.

11. **WE5** Calculate the value of the pronumerals in each of the following diagrams.

a.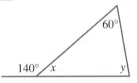

b.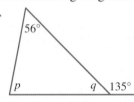

c.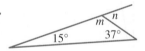

d.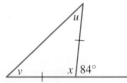

12. Calculate the value of the pronumeral in the photograph.

13. A long ladder is placed against the side of a building. The angle that the ladder makes with the ground is $55.5°$.
   a. Draw a diagram of the ladder against the building and label it.
   b. Name the type of the triangle created by the ladder, the building and the ground.
   c. Calculate the exterior angle that the ladder makes with the ground. Show this on your diagram.

14. Identify three different angles in the photograph.
    a. From the photograph, choose one group of angles that are situated around a common point. Draw this group of angles in your workbook.
    b. State the relationships between the angles you drew in part **a**.

15. Calculate the value of the pronumerals in the diagram.

2.3 Two-dimensional geometric shapes

Geometry or Euclidian Geometry is the branch of Mathematics concerned with the study of shapes and their properties and the relationships between shapes. The word geometry has its roots in the Greek word γεομετρια which means 'earth (geo-) measurement (-metron)'.

The rigorous mathematical rules of Geometry that we use today were first written in the book 'Elements' by the Greek mathematician Euclid, also known as the 'Father of Geometry'.

Two-dimensional shapes are recognised by their property of having length and width only and no thickness. A two-dimensional shape can be drawn on a piece of paper because the surface of a piece of paper has only two dimensions: length and width.

2.3.1 Polygons

A **polygon** is a closed two-dimensional shape formed by a set number of straight lines. The straight lines, called **sides**, are connected to each other to form a vertex. The minimum number of sides a polygon can have is three. The number of sides of a polygon is always equal to the number of vertices. The number of sides also determines the name of the polygon.

Polygons can be either **concave** or **convex**. A **convex polygon** has all interior angles less than 180°. All its vertices point outside the polygon as shown in the diagram.

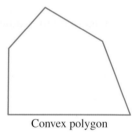

Convex polygon

A **concave polygon** has at least one reflex angle. This vertex points inside the polygon. If the arms of the reflex angle are extended, they will pass through the inside of the polygon as shown in the diagram.

The vertices of a polygon are labelled using upper case (capital) letters. The polygon in the diagram at right is the hexagon ABCDEF.

The angles formed at the ends of a line segment inside a polygon such as $\angle A$ and $\angle B$ are said to be **consecutive** angles or consecutive vertices. Line segments that meet at one vertex, such as AB and BC, are also consecutive.

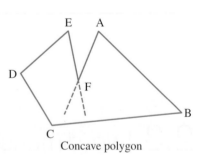

Concave polygon

Polygons can be either **regular polygons** or **irregular polygons**. Regular polygons have equal sides and equal angles.

The sum of the internal angles of a polygon $= (n - 2) \times 180°$
where $n =$ the number of sides.

Polygon	Number of sides and vertices	Regular polygon	Sum of angles in the polygon	Size of one angle in the regular polygon
Triangle	3	Equilateral triangle	180°	60°
Quadrilateral	4	Square	360°	90°
Pentagon	5	Regular pentagon	540°	108°
Hexagon	6	Regular hexagon	720°	120°

(Continued)

Polygon	Number of sides and vertices	Regular polygon	Sum of angles in the polygon	Size of one angle in the regular polygon
Heptagon	7	Regular heptagon	900°	$128\frac{4}{7}°$
Octagon	8	Regular octagon	1080°	135°

WORKED EXAMPLE 6

Name the polygons shown in the diagram, with mathematical reasons, and calculate the size of the unknown angles.

a. $\angle A = 120°$, $\angle B = 90°$, $\angle C = 132°$ and $\angle D = 90°$

---

b. ∠A = ∠C = 90°, ∠B = 63°, ∠D = 270° and ∠E = 115°.

THINK

a. 1. State the name of the polygon.

2. Determine the sum of the interior angles of a polygon with 5 sides.

3. Write an equation connecting all the angles of the polygon.

4. Substitute any known values.

5. Solve the equation to determine the value of the unknown angle.

b. 1. State the name of the polygon.

2. Determine the sum of the interior angles of a polygon with 6 sides.

3. Write an equation connecting all the angles of the polygon.

4. Substitute any known values.

5. Solve the equation to determine the value of the unknown angle.

WRITE

a. The polygon is an irregular pentagon because:
- it has five sides and five vertices (pentagon)
- the angles are not equal in size (irregular).

$$\text{angle sum} = (n-2) \times 180°$$
$$= (5-2) \times 180°$$
$$= 3 \times 180°$$
$$= 540°$$

$$∠A + ∠B + ∠C + ∠D + ∠E = 540°$$

$$120° + 90 + 132 + 90° + x° = 540°$$

$$432° + x° = 540°$$
$$x° = 540° - 432°$$
$$x° = 108°$$
$$∴ ∠E = 108°$$

b. The polygon is an irregular concave hexagon because:
- it has six sides and six vertices (hexagon)
- the angles are not equal in size (irregular)
- one of the vertices points inside the polygon

$$\text{angle sum} = (n-2) \times 180°$$
$$= (6-2) \times 180°$$
$$= 4 \times 180°$$
$$= 720°$$

$$∠A + ∠B + ∠C + ∠D + ∠E + ∠F = 720°$$

$$90° + 63° + 90° + 270° + 115° + ∠F = 720°$$

$$628° + ∠F = 720°$$
$$∠F = 720° - 628°$$
$$∴ ∠F = 92°$$

2.3.2 Triangles

A **triangle** is a three-sided polygon. The name of the triangle in the diagram shown is ΔABC. Alternative notations for this triangle are ΔBCA and ΔCAB.

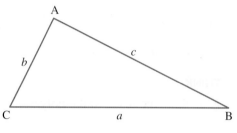

Notation:

- The sides of a triangle are labelled using lower case letters.
- The vertices of a triangle are labelled using upper case letters.
- The same letter is used to label a vertex and its opposite side, such as vertex A and opposite side a.

The sum of the interior angles of any triangle is 180°.

$$\angle A + \angle B + \angle C = 180°$$

Types of triangles

Triangles are classified according to their specific properties.

Triangles are classified according to their side lengths as follows.

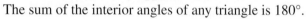

Type of triangle	Diagram	Properties
Equilateral		$AB = BC = AC$ $\angle A = \angle B = \angle C = 60°$
Isosceles		$AB = AC$ $\angle C = \angle B$
Scalene		Sides of different lengths and angles are different sizes.

Triangles are classified according to their angles as follows.

Type of triangle	Diagram	Properties
Right-angled	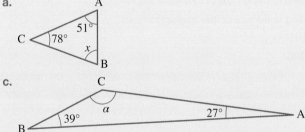	$\angle A = 90°$ The side opposite to the right angle is called the **hypotenuse**.
Acute-angled		All angles are acute angles (less than 90°) $\angle A < 90°$ $\angle B < 90°$ $\angle C < 90°$
Obtuse-angled		One obtuse angle (greater than 90°) $90° < \angle A < 180°$ $\angle B < 90°$ $\angle C < 90°$

WORKED EXAMPLE 7

Classify the following triangles by first calculating the size of the unknown angles. Provide reasons for your answer.

a.

b.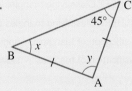

c.

THINK

a. 1. Write an equation connecting the three angles.

2. Substitute any known values.

3. Solve the equation to determine the value of the unknown angle.

4. Name the type of triangle clearly stating the reasons.

b. 1. Look for any sides and angles that are equal, and hence determine the value of x.

WRITE

a. $\angle A + \angle B + \angle C = 180°$
$$51° + x + 78° = 180°$$
$$129° + x = 180°$$
$$129° - 129° + x = 180° - 129°$$
$$x = 51°$$

$\triangle ABC$ is an acute isosceles triangle because
$\angle A = \angle B = 51°$ (isosceles)
$\angle A, \angle B,$ and $\angle C < 90°$ (acute)

b. $AB = AC$
$=>$ the opposite angles to AB and AC are equal in size
$=> \angle C = \angle B = 45°$
$=> x = 45°$

2. Write an equation connecting the three angles.

$\angle A + \angle B + \angle C = 180°$

3. Substitute any known values.

$y + 45° + 45° = 180°$

4. Solve the equation to determine the value of y.

$y + 90° = 180°$
$y + 90° - 90° = 180° - 90°$
$y = 90°$

5. Name the type of triangle, clearly stating the reasons.

$\triangle ABC$ is a right-angled isosceles triangle because
$\angle B = \angle C = 45°$ (isosceles)
$\angle A = 90°$ (right-angled)

c. 1. Write an equation connecting the three angles.

c. $\angle A + \angle B + \angle C = 180°$

2. Substitute any known values.

$27° + 39° + \alpha = 180°$

3. Solve the equation to determine the value of the unknown angle.

$66° + \alpha = 180°$
$66° - 66° + \alpha = 180° - 66°$
$\alpha = 114°$

4. Name the type of triangle clearly stating the reasons.

$\triangle ABC$ is an obtuse scalene triangle because
$\angle A \neq \angle B \neq \angle C$ (scalene)
$\angle C$ is an obtuse angle as
$90° < 114° < 180°$ (obtuse)

---

on **Resources**

Digital documents SkillSHEET Classifying triangles according to the lengths of their sides (doc-6454)

SkillSHEET Classifying triangles according to the size of their angles (doc-6455)

Interactivities Classification of triangles — angles (int-3964)

Classification of triangles — sides (int-3963)

2.3.3 Quadrilaterals

A **quadrilateral** is a four-sided polygon. The name of the quadrilateral in the diagram shown is ABCD. Other notations are possible as long as the letters are read in order in either a clockwise or an anticlockwise direction: BCDA, CDAB or DABC.

The sum of the interior angles of any quadrilateral is 360°.

$$\angle A + \angle B + \angle C + \angle D = 360°$$

Every quadrilateral has two diagonals. A **diagonal** is a line segment which connects two opposite angles of a quadrilateral. In the diagram shown, AC and BD are the diagonals of the quadrilateral ABCD.

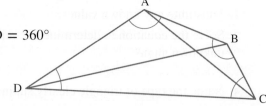

Types of quadrilaterals

Quadrilaterals are classified according to their specific properties as follows.

Quadrilateral	Sides	Angles	Diagonals
Square A B O D C	AB = BC = AD = CD AB ‖ DC and BC ‖ AD	$\angle A = \angle B = \angle C =$ $\angle D = 90°$	$AC \perp BD$ AC = BD AO = OC BO = OD
Rectangle A B O D C	AB‖DC and BC‖AD The symbol ‖ means 'parallel and equal to'.	$\angle A = \angle B = \angle C = \angle D$ $= 90°$	AC = BD AO = OC BO = OD
Parallelogram A B O D C	AB‖DC and BC‖AD	Opposite angles are equal. $\angle A = \angle C$ $\angle B = \angle D$ Consecutive angles are supplementary. $\angle A + \angle B = 180°$ $\angle B + \angle C = 180°$ $\angle C + \angle D = 180°$ $\angle D + \angle A = 180°$	AO = OC BO = OD
Trapezium A B O D C	AB‖CD	All different.	No characteristics.
Isosceles trapezium A B O D C	AB‖CD AD = BC	$\angle A = \angle B$ $\angle C = \angle D$ Opposite angles are supplementary. $\angle B + \angle D = 180°$ $\angle A + \angle C = 180°$	AC = BD
Right-angled trapezium A B O D C	AB‖CD	$\angle A = \angle D = 90°$ $\angle A + \angle D = 180°$ $\angle B + \angle C = 180°$	No characteristics.
Rhombus A D O B C	AB‖DC and BC‖AD AB = BC = CD = AD	Opposite angles are equal. Consecutive angles are supplementary.	$AC \perp BD$ AO = OC BO = OD

Calculate the size of the unknown angles in the following quadrilaterals.

a.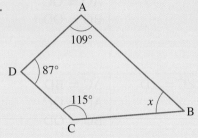

b.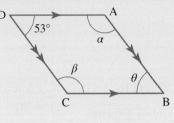

c.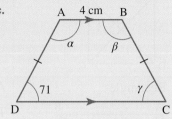

THINK	WRITE
a. 1. Write an equation connecting the four angles.	a. $\angle A + \angle B + \angle C + \angle D = 360°$
2. Substitute any known values.	$109° + x + 115° + 87° = 360°$
3. Solve the equation to determine the size of the unknown angle.	$x + 311° = 360°$
	$x + 311° - 311° = 360° - 311°$
	$x = 49°$
b. 1. State the relationship between angles B and D, and hence calculate the value of θ.	b. $\angle B = \angle D = 53°$
	$=> \theta = 53°$
2. Write an equation stating the relationship between angles A and D and solve the equation to determine the value of α and β.	$\angle A + \angle D = 180°$
	$\alpha + 53° = 180°$
	$\alpha + 53° - 53° = 180° - 53°$
	$\alpha = 127°$
	$=> \beta = \alpha = 127°$
c. 1. State relationship between angles C and D, and hence determine the value of γ.	c. $\angle C = \angle D = 71°$
	$=> \gamma = 71°$
2. Write an equation stating the relationship between angles A and D and solve the equation to determine the value of α and β.	$\angle A + \angle D = 180°$
	$\alpha + 71° = 180°$
	$\alpha + 71° - 71° = 180° - 71°$
	$=> \alpha = 109°$
	$=> \beta = \alpha = 109°$

2.3.4 Circles

A **circle** is a two-dimensional shape represented by a closed curve formed by a set of points equally distanced from a fixed point called the centre of the circle, O. The curve that forms the circle is called the circumference of the circle.

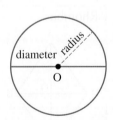

The distance from the centre of the circle to any point on the circumference of the circle is called the radius. The plural of the word radius is **radii**.

The **diameter** of the circle is any straight line connecting two points on the circumference and passing through the centre of the circle.

Exercise 2.3 Two-dimensional geometric shapes

1. Classify the shapes of the mazes in the diagram shown.

2. **WE6** Name the polygon shown in the diagram, with mathematical reasons, and calculate the size of the unknown angle if $\angle A = 97°$, $\angle B = 118°$, $\angle C = 92°$ and $\angle D = 124°$.

3. Calculate the size of the unknown angles of the polygon ABCDEF and state the name of the polygon.

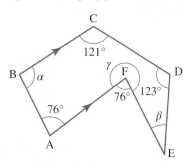

4. Calculate the size of $\angle DOC$ if $\angle AOB = \angle BOC = 75°$ and $\angle AOD = 110°$

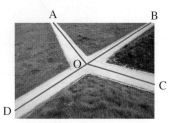

5. **WE7** Classify the following triangles by first calculating the size of the unknown angles. Provide reasons for your answers.

a.
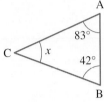

b.
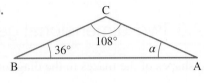

6. Calculate the size of each unknown angle, providing reasons for your answers. Use this information to classify the triangles $\triangle ABC$, $\triangle ABD$ and $\triangle ACD$.

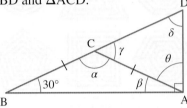

7. Consider the diagram shown.
   a. What type of triangle is $\triangle ABC$?
   b. Determine the value of x.
   c. Calculate the value of y.
   d. Determine the value of z.

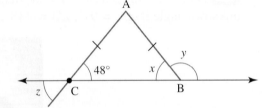

8. **WE8** Calculate the size of the unknown angles in the following quadrilaterals.

a.
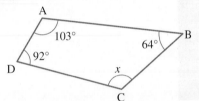

b.
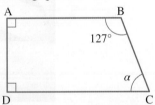

9. The triangles that form the façade of Federation Square in Melbourne are right-angled triangles with the two acute angles of approximate sizes $26.6°$ and $63.4°$. Calculate the internal angles of each polygon and hence state their names.

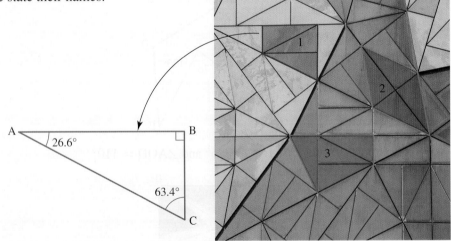

10. **MC** The line that connects the centre of the circle with a point on its circumference is called
    A. the diameter of the circle.
    B. a chord.
    C. the circumference of the circle.
    D. the radius of the circle.

11. Calculate the value of x and the size of $\angle ABC$.

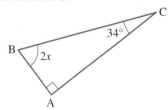

12. Consider the diagram shown. Trapezium ABCF is identical to trapezium CDEF.
   a. Calculate the value of α and the size of $\angle AFE$.
   b. State the type of the quadrilaterals ABCF and FCDE.
   c. State the type of the polygon ABCDEF.
   d. Calculate the value of β.

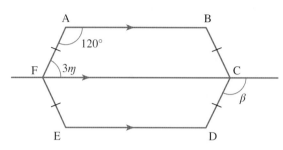

13. Part of the façade of the Perth Arena is displayed in the diagram shown.
   a. State the type of polygons highlighted in the diagram.
   b. Calculate all the interior angles of the polygon ABCDEF if AB ∥ EF, CD ∥ AB, DE ∥ AF, $DE = \frac{1}{2}AF$ and $\angle ABC = 60°$.

14. Consider the polygons shown in the diagram.
   a. State the type of the quadrilateral ABEF with reasons.
   b. State the type of the quadrilateral BCDE with reasons.
   c. Calculate the value of the pronumerals.
   d. Calculate the size of the reflex $\angle FED$.

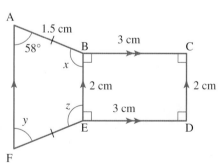

2.4 Three-dimensional objects

Three-dimensional objects are recognised by their property of having length, width and depth (or height). A three-dimensional object with flat surfaces and straight edges enclosing a space is called a **polyhedron** with the plural **polyhedra** or **polyhedrons**. The word polyhedron is made of the Greek words *poly-* ('many') and *-hedron* ('face'). They are also called **geometric solids** or **Platonic solids**. Each flat surface of a polyhedron is called a **face**. Each face is a polygon. Two faces of a polyhedron intersect to form an **edge**. Three faces intersect to form a vertex (plural vertices).

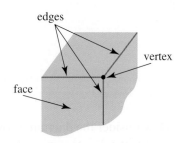

Two basic types of polyhedrons are prisms and pyramids.

2.4.1 Prisms

Prisms are solid shapes with identical opposite polygonal ends which are joined by straight edges. They are three-dimensional objects that can be cut into identical slices, called cross-sections.

Some common prisms and their properties are as follows.

Polyhedron	Shape of the base	Number of faces	Number of edges	Number of vertices
Triangular prism	Triangle	5	9	6
Cube	Cube	6	12	8
Rectangular prism	Rectangle	6	12	8

The following solids are also prisms. This time though the two bases are not basic polygons, they are composite shapes.

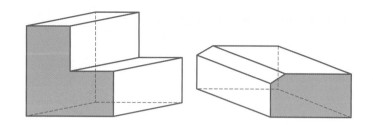

WORKED EXAMPLE 9

State the name, the number of faces, vertices and edges of each of the following prisms.

a.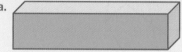

b.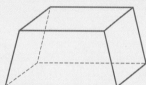

THINK

a. 1. Determine the shape of the base of the prism and state its name.

   2. State the number of faces.

   3. State the number of vertices.

   4. State the number of edges.

b. 1. Determine the shape of the base of the prism and state its name.

   2. State the number of faces.

   3. State the number of vertices.

   4. State the number of edges.

WRITE

a. The base of the prism is a rectangle. This is a rectangular prism.

   6 faces

   8 vertices

   12 edges

b. The base of the prism is a trapezium. This is a trapezoidal prism, or a quadrilateral prism.

   6 faces

   8 vertices

   12 edges

Exercise 2.4 Three-dimensional objects

1. Which of the three-dimensional shapes shown are prisms?

a.

b.

c.

d.

e.

2. **MC** Which three-dimensional object is not a polyhedron?

A.

B.

C.

D.

3. State the number of faces, edges and vertices of each object.

a.

b.

4. The prism shown has edges of equal length. State its name, its number of faces, vertices and edges.

5. **WE9** State the name, the number of faces, vertices and edges of each of the following prisms.

a.

b.

6. Draw a prism using each of the following shapes as the base.

a.

b.

7. Draw a prism using each of the following shapes as the base.

a.

b.

c.

2.5 Two-dimensional representations of three-dimensional objects

2.5.1 Nets of three-dimensional objects

Prisms and pyramids can be drawn as a two-dimensional figure. This shape is called a **net**. In other words, a net is a two-dimensional figure that can be folded to form a three-dimensional object.

Nets of prisms
Net of a cube

The diagram displays the net of a cube. Notice that, when folded, the faces do not overlap.

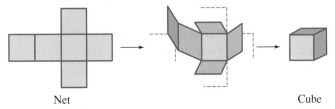

Net   Cube

The following nets are also nets of a cube.

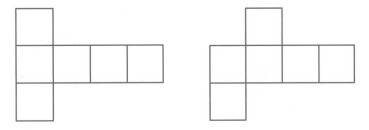

Net of a rectangular prism

The following nets can be folded to form a rectangular prism. Notice that the top and the bottom faces of the rectangular prism are on either side of the net. These two faces cannot be drawn on the same side of the net. The left and right faces alternate with the back and front faces.

Nets of other prisms

A net can be drawn for any type of prism.

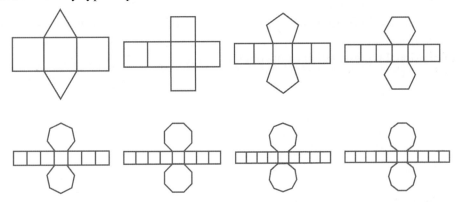

State the name of the three-dimensional object that can be formed by folding the following nets.

a.

b.
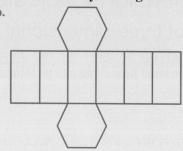

THINK

a. 1. State any features of the two-dimensional shapes.

2. Count the number of faces and check for any possible overlaps or missing faces.

3. State the name of the three-dimensional object.

b. 1. State any features of the two-dimensional shapes.

2. Count the number of faces and check for any possible overlaps or missing faces.

3. State the name of the base and the name of the three-dimensional object.

WRITE

a. All the shapes are equal sized squares.

There are 6 faces.
There are no overlaps.
There are no missing faces.

The object is a cube.

b. There are six identical rectangles and two identical regular hexagons.

There are 8 faces.
There are no overlaps.
There are no missing faces.

The base is a regular hexagon.
The object is a hexagonal prism.

2.5.2 Three-dimensional objects and perspective drawing

Designers, engineers, and architects use two-dimensional drawings to represent three-dimensional objects. It is important to understand how to construct, read and interpret such diagrams. Three-dimensional objects can be drawn using **isometric drawing**.

Plans and views

In the diagram shown there are four views of a car:
- the **plan view** is the diagram of the object as seen looking straight at the object from above
- the **side view,** also called **side elevation,** is the diagram of the object as seen looking straight at the object from one side; left or right
- the **front view,** also called **front elevation,** is the diagram of the object as seen looking straight at the object from the front
- the three-dimensional representation of the object.

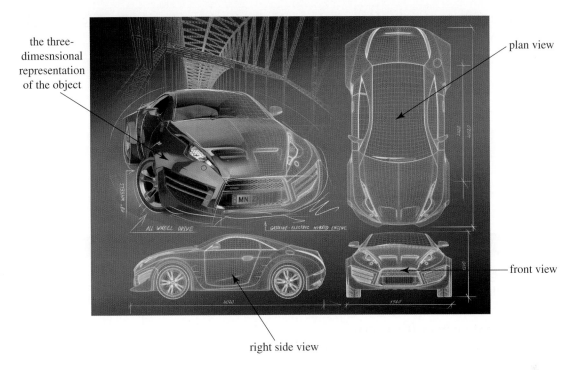

the three-dimesnsional representation of the object

plan view

front view

right side view

Isometric drawing

Isometric drawing is done on **isometric paper**. Isometric paper is a type of graph paper using either dots or lines that form equilateral triangles. These lines allow the parallel lines of the original object to be maintained; however, the right angles are drawn as either acute or obtuse angles.

The simplest three-dimensional object is the cube. Its isometric representation is shown in the diagram at right.

The following diagrams are isometric drawings of the same rectangular prism drawn looking from its left side and from its right side respectively.

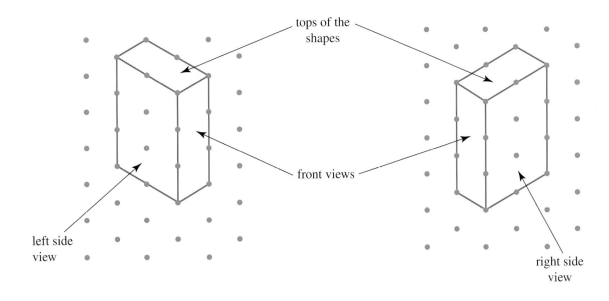

tops of the shapes

front views

left side view

right side view

The three views of the rectangular prism are shown. Notice that the angles are 90° angles although they have different measurements on the isometric drawing.

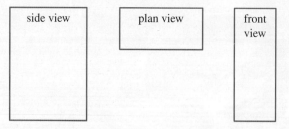

WORKED EXAMPLE 11

The three-dimensional object shown in the diagram is made of three rectangular prisms. Each rectangular prism is 3 cm long and has a square base with side length of 1 cm.
a. Draw the object on isometric paper.
b. Draw the front, side and plan views of the object.

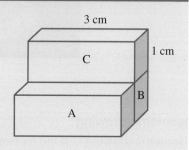

THINK

a. 1. Draw the first rectangular prism by first drawing the corner at the front considering the lengths of its dimensions.

WRITE

a.

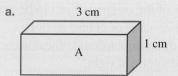

Rectangular prism A has:
- width = 1 unit
- height = 1 unit.
- length = 3 units

Note: the height and the width are equal in length so they are represented by lines of equal lengths.
The length is three times longer than the other dimensions. The proportions between the sides of the original object have to be maintained.

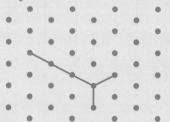

2. Complete the front face of the prism by drawing parallel lines to the existing lines.

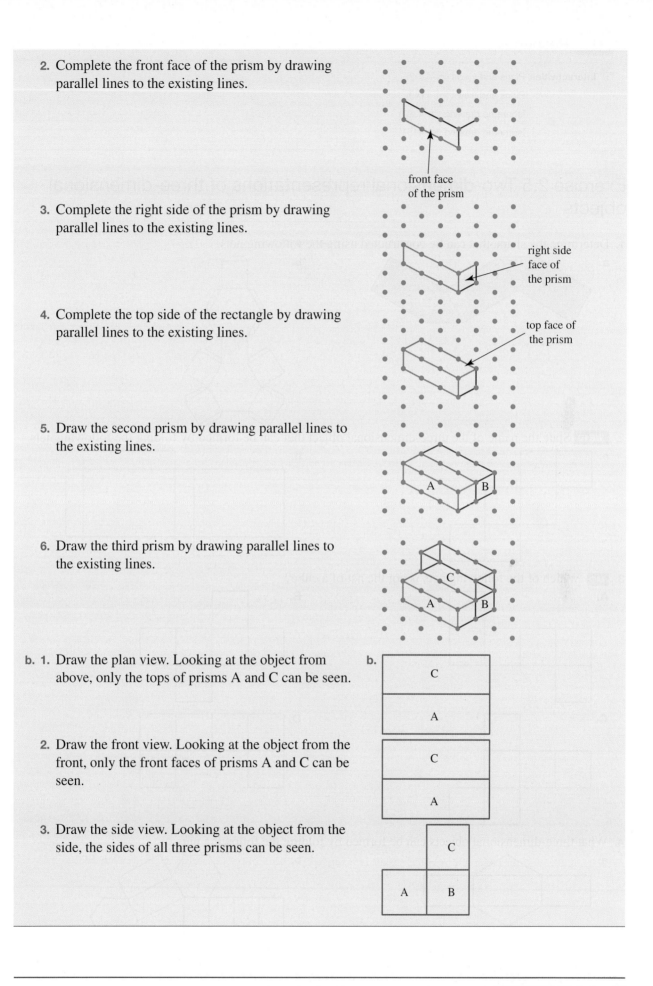

front face of the prism

3. Complete the right side of the prism by drawing parallel lines to the existing lines.

right side face of the prism

4. Complete the top side of the rectangle by drawing parallel lines to the existing lines.

top face of the prism

5. Draw the second prism by drawing parallel lines to the existing lines.

6. Draw the third prism by drawing parallel lines to the existing lines.

b. 1. Draw the plan view. Looking at the object from above, only the tops of prisms A and C can be seen.

b.

2. Draw the front view. Looking at the object from the front, only the front faces of prisms A and C can be seen.

3. Draw the side view. Looking at the object from the side, the sides of all three prisms can be seen.

Exercise 2.5 Two-dimensional representations of three-dimensional objects

1. Determine the shape that can be constructed using the following nets.

a.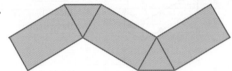

b.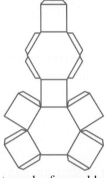

2. **WE10** State the name of the three-dimensional object that can be formed by folding the following nets.

a.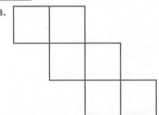

b.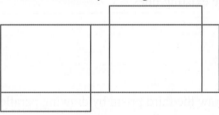

3. **MC** Which of the following nets is not the net of a cube?

A.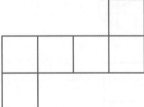

B.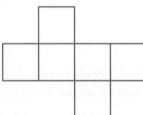

C.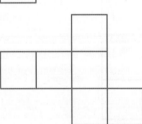

D.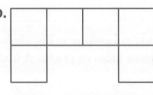

4. What three-dimensional objects can be formed by folding the following nets.

a.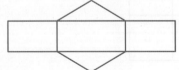

b.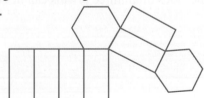

5. **WE11** The three-dimensional object shown in the diagram at right is made of four cubes of side length 1 cm.
   a. Draw the object on isometric paper.
   b. Draw the front, side and plan views of the object.

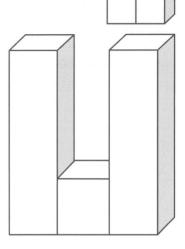

6. Consider the three-dimensional object shown in the diagram.
   a. Draw the object on isometric paper.
   b. Draw the front, side and plan views of the object.

7. Draw an isometric diagram for the following three-dimensional objects.

   a.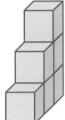

   b.

8. Draw the front view, side view and plan view for the following isometric drawings.

   a.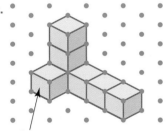

   front

   b.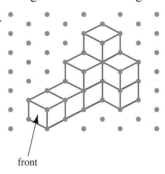

   front

2.6 Review: exam practice

2.6.1 Geometry: summary

Review of angles

- An angle is formed when two lines meet at a point.
- The point at which the two lines meet is called a vertex and the lines themselves are called arms of the angle.
- Angles are named by:
  - writing the letters at the end of each arm and at the vertex (the vertex is the middle letter)
  - writing the letter at the vertex only
  - writing the pronumeral marking the angle inside the arms.
- Acute angles are greater than $0°$ but less than $90°$.

- Right angles are exactly 90°.
- Obtuse angles are greater than 90° but less than 180°.
- Straight angles are exactly 180°.
- Reflex angles are greater than 180° but less than 360°.
- A revolution (or perigon) is exactly 360°.
- Angles are measured in degrees (°).
- Complementary angles add to 90°.

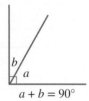

$a + b = 90°$

- Supplementary angles add to 180°.

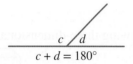

$c + d = 180°$

- Angles with a common vertex that form a circle add to 360°.

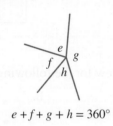

$e + f + g + h = 360°$

- Vertically opposite angles are equal in size.

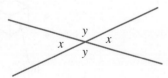

- Corresponding angles are equal in size.

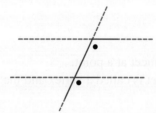

- Alternate angles are equal in size.

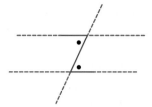

- Co-interior angles add to 180°

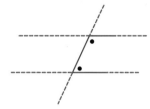

Two-dimensional geometric shapes

- A polygon is a closed two-dimensional shape formed by a set number of straight lines.
- Regular polygons have equal sides and equal angles.
- A triangle is a three-sided polygon.
- Types of triangles:
  - Equilateral triangles have three equal sides and three equal angles.
  - Isosceles triangles have two equal sides and two equal angles.
  - Acute triangles have three acute angles.
  - Obtuse triangles have one obtuse angle.
  - Right-angled triangles have one 90° angle.
  - Scalene triangles have three sides of different length and three differently sized angles.
- A quadrilateral is a four-sided polygon.
- Types of quadrilateral:
  - Squares have four equal sides and four 90° angles.
  - Rectangles have opposite sides parallel and equal and four 90° angles.
  - Parallelograms have opposite sides parallel and equal and opposite angles are equal.
  - Trapeziums have one pair of unequal parallel sides.
  - Isosceles trapeziums have one pair of unequal parallel sides and the consecutive angles of the parallel lines are equal.
  - Right-angled trapeziums have one pair of unequal parallel sides and two consecutive angles equal to 90°.
  - Rhombuses have four equal sides, two pairs of parallel sides and opposite angles are equal.
- A circle is a two-dimensional shape represented by a closed curve formed by all points equally distanced from a point called the centre of the circle, O.
- The curve that forms the circle is called the circumference of the circle.
- The distance from the centre of the circle to any point on the circumference of the circle is called radius. The plural of the word radius is radii.
- The diameter of the circle is any straight line connecting two points on the circumference and passing through the centre of the circle.

Three-dimensional objects

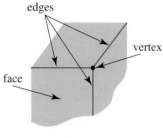

- A three-dimensional object with flat surfaces and straight edges enclosing a space is called a polyhedron with the plural polyhedra or polyhedrons. They are also called geometric solids or Platonic solids.
- Each flat surface of a polyhedron is called a face. Each face is a polygon.
- Two faces of a polyhedron intersect to form an edge.
- Three faces intersect to form a vertex (plural vertices).
- Two basic types of polyhedrons are prisms and pyramids.
- Prisms are solid shapes with identical opposite polygonal ends which are joined by straight edges. They are three-dimensional objects that can be cut into identical slices, called cross-sections.
- A net is a two-dimensional figure that can be folded to form a three-dimensional object.

Two-dimensional representations of three-dimensional objects

- An isometric drawing is a graphical representation of a three-dimensional object in two dimensions using isometric paper.
- Plan view is the diagram of an object as seen looking straight at the object from above.
- Side view also called side elevation is the diagram of an object as seen looking straight at the object from one side; left or right.
- Front view also called front elevation is the diagram of an object as seen looking straight at the object from the front.

Exercise 2.6 Review: exam practice

Simple familiar

1. **MC** The sizes of the angles α and β are respectively
   - **A.** 143° and 53°.
   - **B.** 53° and 143°.
   - **C.** 63° and 143°.
   - **D.** 53° and 153°.

2. **MC** A pair of co-interior angles is
   - **A.** 1 and a.
   - **B.** b and 2.
   - **C.** 4 and 2.
   - **D.** 3 and b.

3. **MC** A quadrilateral is a polygon with the property that
   - **A.** opposite angles are equal.
   - **B.** two opposite sides are parallel.
   - **C.** the sum of all interior angles is 360°.
   - **D.** has five sides and five angles.

4. **MC** Which plan view matches the three-dimensional diagram shown?

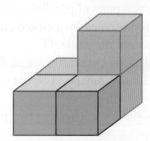

A.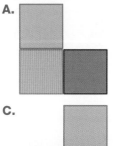

B.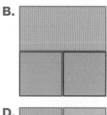

C.

D.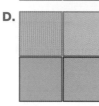

5. Draw the front, right side and top view for the following object.

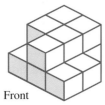

Front

6. Using isometric paper, draw a building with the following right side, front and top views.

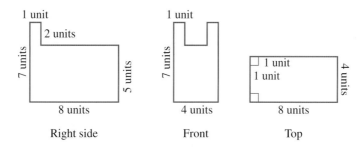

Right side   Front   Top

7. Determine the size of the unknown angles in the diagram.

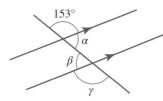

8. Calculate the value of the pronumerals.

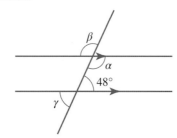

9. Calculate the value of θ.

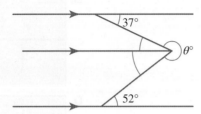

10. Determine the value of the pronumerals in the diagram.

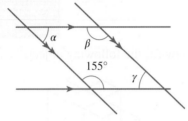

11. Construct a clearly labelled plan view, front view and side view of the letter T shown in the diagram.

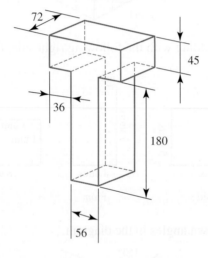

12. Using the properties of angles and circles, determine the value of the following measurements if $CD = 8\,cm$.

   a. α
   b. β
   c. γ
   d. AB
   e. OD

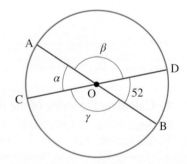

13. Calculate the value of α in the diagram.

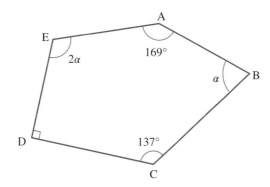

14. Construct an isometric drawing of the following object.

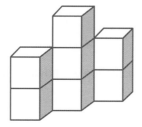

15. Draw four different nets that can make a cube.
16. Determine the value of the pronumerals in the diagram.

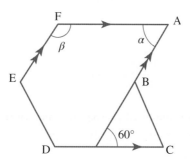

Complex unfamiliar

17. Calculate the values of the unknown angles in the following polygons.

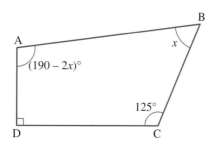

18. Consider the trapezoidal prism shown in the diagram.
    a. State the number of faces.
    b. State the number of edges.
    c. State the type of the quadrilaterals AMQD and MNPQ.
    d. Which lines are parallel and equal?

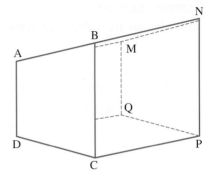

19. Draw the net of the following triangular-based prism.

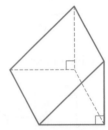

20. A courtyard is in the form of a regular pentagon as shown in the diagram. The side length of the pentagon is 7.5 m.

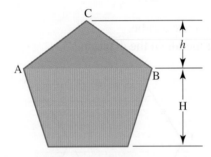

    a. Calculate the size of the interior angles of the pentagonal courtyard.
    A sand path is constructed between points A and B to separate the lawn from the flowerbed.
    b. Calculate the size of the angle ABC.
    A small brick fence is to be built around four sides of the courtyard.
    c. Calculate the length of the fence required.
    d. How many bricks are required if 25 bricks are used to build 1 m of the fence?
    e. Calculate the total cost of the fence if one brick costs $0.95.

Answers

2 Geometry

Exercise 2.2 Review of angles

1. a. i. ∠CAT or ∠TAC
      ∠WAS or ∠SAW
      ∠FAT or ∠TAF
   ii. ∠A
       ∠A
       ∠A
   b. Reflex
      Obtuse
      Acute
2. *See the table at the bottom of the page.
3. a. ∠NMP, ∠MNP, ∠MPN, ∠NOP, ∠OPN, ∠ONP
   b. i. Right angle
      ii. Acute angle
      iii. Acute angle
      iv. Right angle
      v. Acute angle
      vi. Acute angle
   c. b and c, e and f
   d. b and e, c and f
   e. $a = 90°, b = 30°, c = 60°, d = 90°, f = 60°$
4. a. b   b. e   c. g
   d. d   e. e   f. h
5. a. Sample responses can be found in the Worked Solutions
      in the online resources.
   b. ∠XOV and ∠YOW
6. a. $a = 38°$
   b. $b = 110°$
   c. $c = 22°$
   d. $d = 19°$
   e. $e = 39°, g = 141°, f = 141°$
   f. $i = 13°, h = 15°, k = 152°, j = 152°$
7. a. $b = 65°, c = 10°, d = 50°, e = 130°$
   b. $f = 60°, g = 60°, i = 36°, h = 20°$
8. a. $a = 310°$ b. $b = 222°$
   c. $c = 103°$ d. $d = 10°$
   e. $e = 82°$ f. $f = 73.3°$
9. a. $a = 22.5°$
   b. $c = 40°, b = 32.5°, d = 42.5°, e = 42.5°, f = 65°$
10. a. 73°   b. 107°   c. 45°
    d. 89°   e. 80°   f. 60°
11. a. $x = 40°$ b. $q = 45°$
       $y = 80°$ $p = 79°$
    c. $m = 128°$ d. $x = 96°$
       $n = 52°$ $v = 42°$
                               $u = 42°$
12. $x = 38°$

13. a.
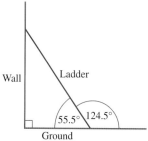
   b. A scalene triangle
   c. 124.5°
14. a. Sample responses can be found in the Worked Solutions
       in the online resources.
    b. Sample responses can be found in the Worked Solutions
       in the online resources.
15. $a = 23.3°$
    $b = 5°$
    $z = 45°$
    $m = 45°$
    $n = 90°$
    $x = 155°$
    $t = 50°$

Exercise 2.3 Two-dimensional geometric shapes

1. Top row: Regular octagon, equilateral triangle, circle
   Second row: Regular hexagon, square, regular pentagon
2. ∠E = 109° ABCDE is an irregular pentagon.
3. $\alpha = 104°$
   $\gamma = 284°$
   $\beta = 12°$ ABCDEF is an irregular hexagon.
4. ∠DOC = 100°
5. a. $x = 55°$
      Scalene triangle
   b. $\alpha = 36°$
      Since two angles are the same it is an isosceles triangle.
6. ΔABC is an isosceles triangle, ΔABD is a right-angled
   triangle, ΔACD is an equilateral triangle, $\beta = 30°$,
   $\alpha = 120°$, $\gamma = 60°$, $\theta = 60°$, $\delta = 60°$.
7. a. Isosceles triangle   b. $x = 48°$
   c. $y = 132°$ d. $z = 48°$
8. a. $x = 101°$
   b. $\alpha = 53°$
9. Polygon 1: Trapezium
   Angle 1: 90°
   Angle 2: 90°
   Angle 3: 63.4°
   Angle 4: 116.6°
   Polygon 2: Rhombus
   Angle 1: 53.2°
   Angle 2: 53.2°
   Angle 3: 126.8°
   Angle 4: 126.8°

*2.

	Acute angles	Obtuse angles	Right angles	Reflex angles
	2°, 5°, 14°, 23°, 45°, 55°, 69°, 78°, 80°, 89°	92°, 92°, 100°, 110°, 122°, 145°, 154°, 160°, 160°, 179°	90°	181°, 190°, 270°, 300°, 359°

Polygon 3: Quadrilateral
Angle 1: 63.4°
Angle 2: 90°
Angle 3: 116.6°
Angle 4: 90°

10. D

11. $x = 28°$
       $\angle ABC = 56°$

12. a. $\alpha = 20°$ b. Isosceles trapezium
         $\angle AFE = 120°$

    c. Hexagon   d. $\beta = 120°$

13. a. Polygon 1 is a concave hexagon, polygon 2 is an
       irregular pentagon, and polygon 3 is a concave hexagon.
    b. $\angle AFE = 120°$, $\angle FED = 60°$, $\angle EDC = 120°$,
       $\angle DCB = 300°$

14. a. ABEF is an isosceles trapezium, AF || BE, AB = EF
    b. BCDE is a rectangle because BC || DE, BE || CD
    c. $x = 122°$, $y = 58°$, $z = 122°$
    d. $\angle FED = 148°$

Exercise 2.4 Three-dimensional objects

1. b and e.
2. C
3. a. 7 faces, 15 edges, 10 vertices,
   b. 10 faces, 24 edges, 16 vertices
4. Cube, 6 faces, 8 vertices, 12 edges
5. a. Triangular prism, 5 faces, 6 vertices, 9 edges
   b. Heptagonal prism, 9 faces, 14 vertices, 21 edges
6. a.   b.

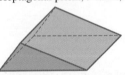

7. a.

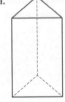

   b.

   c.

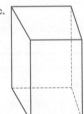

Exercise 2.5 Two-dimensional representations of three-dimensional objects

1. a. Triangular prism
   b. Hexagonal prism
2. a. Cube
   b. Rectangular prism
3. D
4. a. Triangular prism
   b. Hexagonal prism
5. a.

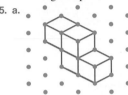

   b.

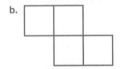

Front view

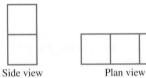

Side view   Plan view

6. a.

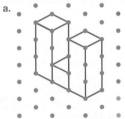

   b.

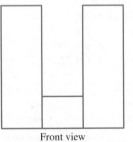

Front view

Side view   Plan view

7. a.

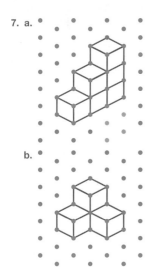

b.

8. a.

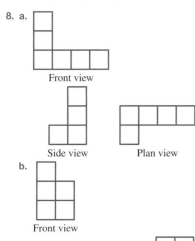

Front view

Side view   Plan view

b.

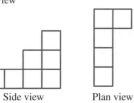

Front view

Side view   Plan view

Exercise 2.6 Review: exam practice

1. B
2. D
3. C
4. D
5.

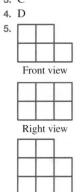

Front view

Right view

Top view

6.

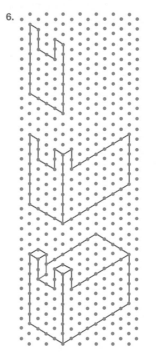

7. $\alpha = 27°, \beta = 27°, \gamma = 153°$
8. $\alpha = 132°, \beta = 132°, \gamma = 48°$
9. $\theta = 271°$
10. $\alpha = 25°, \beta = 155°, \gamma = 25°$
11.

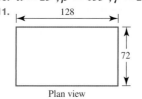

128

72

Plan view

128

45

36

Front view   180

56

72

Side view   45

180

12. a. 52°
    b. 128°
    c. 128°
    d. 8 cm
    e. 4 cm
13. $\alpha = 48°$
14.
15. 6 possible nets are shown

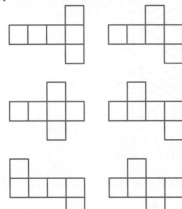

16. $\alpha = 60°$
    $\beta = 120°$
17. $x = 45°$, $\angle A = 100°$, $\angle B = 45°$
18. a. 6
    b. 12
    c. AMQD is a rectangle, MNPQ is a trapezium.
    d. AD and MQ, BC and NP, AM, BN, CP and DQ.
19.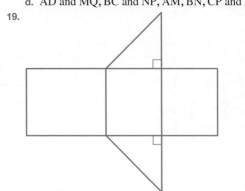

20. a. 108°   b. $\angle ABC = 36°$
    c. 30 m   d. 750 bricks
    e. $712.50

3 Linear and area measure

3.1 Overview

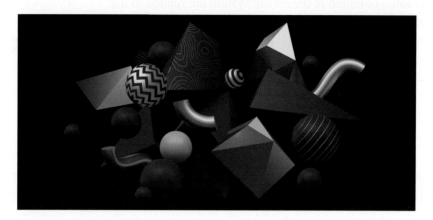

LEARNING SEQUENCE

3.1 Overview
3.2 Units of length and estimation
3.3 Perimeters of familiar and composite shapes
3.4 Units of area and estimation
3.5 Areas of regular shapes
3.6 Areas of composite figures
3.7 Surface areas of familiar prisms
3.8 Surface areas of familiar pyramids and irregular solids
3.9 Review: exam practice

CONTENT

In this chapter, students will learn to:
- use metric units of length (millimetres, centimetres, metres, kilometres), their abbreviations (mm, cm, m, km), conversions between them, and appropriate levels of accuracy and choice of units
- estimate lengths
- calculate perimeters of familiar shapes, including triangles, squares, rectangles, polygons, circles and arc lengths
- calculate perimeters of familiar composite shapes [complex]
- use metric units of area (square millimetres, square centimetres, square metres, square kilometres, hectares), their abbreviations (mm^2, cm^2, m^2, km^2, ha), conversions between them and appropriate choices of units
- estimate the areas of different shapes
- calculate areas of regular shapes, including triangles, squares, rectangles, parallelograms and circles
- calculate areas of regular shapes, including trapeziums and sectors [complex]
- calculate areas of composite figures by decomposing them into regular shapes [complex]
- calculate surface areas of familiar prisms, including cubes, rectangular and triangular prisms, spheres and cylinders [complex]
- calculate surface areas of familiar pyramids, including rectangular-based and triangular-based pyramids [complex]
- calculate surface areas of irregular solids [complex].

Fully worked solutions for this chapter are available in the Resources section of your eBookPLUS at www.jacplus.com.au.

3.2 Units of length and estimation

3.2.1 Units of length

Units of length are used to describe the dimensions of an object, such as its length, width and height, or the distance between two points.

   The metric system we use is a base 10 system with its base unit being the metre. All other units of length in the metric system are related to the metre by powers of 10. The most common of these are the millimetre (mm), centimetre (cm), metre (m) and kilometre (km). They are related in the following way.

$$10 \text{ millimetres} = 1 \text{ centimetre}$$
$$100 \text{ centimetres} = 1 \text{ metre}$$
$$1000 \text{ metres} = 1 \text{ kilometre}$$

Millimetre (mm)	Centimetre (cm)	Metre (m)	Kilometre (km)
1 mm is approximately the width of a computer microchip pin.	1 cm is approximately the width of a person's finger.	1 m is approximately the length of an adult stride.	1 km is approximately the length of Bondi Beach.

   When measuring lengths, you should use an appropriate unit, which is one that gives a reasonable value (not too large and not too small).

3.2.2 Converting between units of length

The relationship between the metric units of length can be used to convert a measurement from one unit to another. The following diagram shows how to convert between metric units.

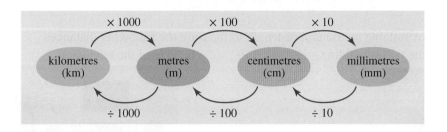

WORKED EXAMPLE 1

Convert the following lengths to the units shown.

a. $0.739 \, \text{km} = \underline{\quad\quad} \, \text{m}$

b. $53\,250 \, \text{mm} = \underline{\quad\quad} \, \text{m}$

THINK	WRITE
a. 1. To convert from km to m, multiply by 1000.	a. 0.739×1000
2. When multiplying by 1000, move the decimal point three places to the right. Make sure the new unit is stated.	$= 739$ Thus: $0.739 \, \text{km} = 739 \, \text{m}$
b. 1. To convert from mm to m, divide by 1000.	b. $53\,250 \div 1000$
2. When dividing by 1000, move the decimal point three places to the left. Make sure the new unit is stated.	$= 53.25$ Thus: $53\,250 \, \text{mm} = 53.25 \, \text{m}$

WORKED EXAMPLE 2

Calculate the value of $7 \, \text{m} \, 36 \, \text{cm} + 5 \, \text{m} \, 45 \, \text{cm}$.

THINK	WRITE
1. Convert both values to metres by dividing the centimetre part of the measurement by 100 and adding it to the metre part of the measurement.	$7 \, \text{m} \, 36 \, \text{cm} = 7.36 \, \text{m}$ $5 \, \text{m} \, 45 \, \text{cm} = 5.45 \, \text{m}$
2. Add the values in metres.	$7.36 + 5.45 = 12.81$
3. Write the answer with the correct units.	$12.81 \, \text{m}$

---

on Resources

Digital documents SkillSHEET Metric units of length (doc-6505)

SkillSHEET Converting units (doc-6509)

SkillSHEET Converting units to compare lengths and distances (doc-6510)

Interactivity Converting units of length (int-4011)

Exercise 3.2 Units of length and estimation

1. Select the most suitable metric unit of length to measure the following distances.

   a. Height of the tallest building

   b. Size of the television

   c. The length of the Brisbane river

   d. The width of the cable connector

2. **WE1** Convert the following lengths to the units shown.

   a. $0.283\,km =$ _____ m

   b. $520\,cm =$ _____ m

   c. $0.418\,m =$ _____ cm

   d. $450\,m =$ _____ km

   e. $5\frac{1}{2}\,km =$ _____ m

   f. $78\,459\,mm =$ _____ m

   g. $0.0378\,km =$ _____ mm

   h. $6500\,cm =$ _____ km

3. **MC** What is the length 375 cm expressed in millimetres?

   A. 0.375 mm   **B.** 3.75 mm   **C.** 37.5 mm   **D.** 3750 mm

4. **MC** When expressed in metres, what is 95.6 kilometres?

   A. 0.956 m   **B.** 956 m   **C.** 9560 m   **D.** 95 600 m

5. **MC** The length 875 mm is equivalent to

   A. 0.875 m.   **B.** 0.0875 m.   **C.** 0.875 km.   **D.** 0.008 75 km.

6. Convert the following lengths into the units shown.

   a. $0.27\,km =$ _____ m

   b. $73\,500\,mm =$ _____ m

   c. $257\,mm =$ _____ cm

   d. $1.3\,km =$ _____ cm

7. **WE2** Calculate the value of the following.

   a. $5\,m\,67\,cm + 9\,m\,51\,cm$

   b. $2\,m\,88\,cm + 3\,m\,20\,cm$

8. Explain why using millimetres is not a good choice to measure the distance between Brisbane and Sydney.

9. The combined height of two students is 3.62 m. If one student is 1.85 m tall, what is the height of the other student in centimetres?

10. Arrange each of the following in ascending order.
    a. 0.15 km, 135 m, 2400 cm
    b. 25 cm, 120 mm, 0.5 m
    c. 9 m, 10 000 mm, 0.45 km
    d. 32 000 cm, 1200 m, 1 km

11. Calculate the value of each of the following in centimetres.
    a. 15 mm + 5 mm
    b. 1.5 m + 40 mm
    c. 995 mm + 1.2 m
    d. 5.67 cm + 1156 mm − 0.25 m

12. Calculate the value of each of the following.
    a. 15 cm 15 mm + 27 cm 86 mm
    b. 3 km 65 m + 79 km 38 m
    c. 66 cm 15 mm − 26 cm 5 mm
    d. 125 m 49 cm − 37 m 79 cm

13. The heights of Melbourne's tallest building, the Eureka Tower, and Queensland's tallest building, the Q1 building, measure 0.297 km and 322.5 m respectively.
    a. Which is the taller building?
    b. Calculate the height difference, in metres, between the buildings.
    c. Calculate the sum of the heights of both buildings, in metres.

14. Given the graph of the average height of the players in each of the AFL clubs in centimetres, determine:
    a. the average height of Collingwood players, in metres
    b. the difference between the average height of Carlton and Sydney players, in centimetres
    c. the difference between the tallest and smallest teams' average height, in centimetres
    d. the League's average height
    e. West Coast's average height in millimetres
    f. two clubs whose difference in average height is 5 mm.

Average height per club

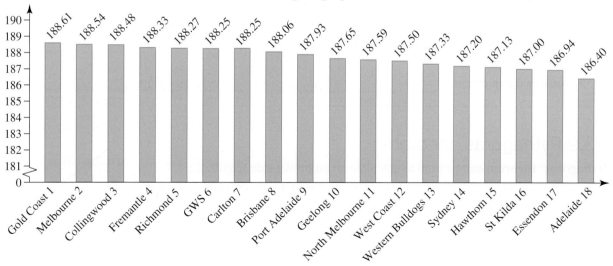

3.3 Perimeters of familiar and composite shapes

3.3.1 Perimeter

The **perimeter** is the distance around the outside of an enclosed shape.
To find the perimeter of a shape, change all lengths to the same units and add them up.

WORKED EXAMPLE 3

Calculate the perimeter of the following shapes.

a.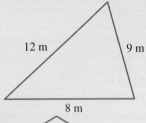

b.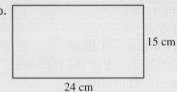

c.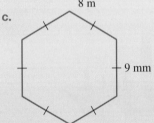

d.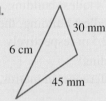

THINK	WRITE
a. 1. Add the distance around the outside of the triangle.	a. $P = 12 + 9 + 8$ $= 29$
2. Answer the question.	The perimeter is 29 m.
b. 1. Add the distance around the outside of the rectangle.	b. $P = 24 + 15 + 24 + 15$ $= 78$
2. Answer the question.	The perimeter is 78 cm.
c. 1. Add the distance around the outside of the hexagon.	c. $P = 9 + 9 + 9 + 9 + 9 + 9$ $= 54$
2. Answer the question.	The perimeter is 54 mm.
d. 1. Convert all lengths into the same units.	d. 6 cm = 60 mm.
2. Add the distance around the outside of the triangle, in millimetres.	$P = 60 + 30 + 45$ $= 135$
3. Answer the question.	Perimeter is 135 mm.

3.3.2 Pythagoras' theorem

Pythagoras' theorem can be used to calculate an unknown side length of a
right-angled triangle if two of the sides are known. Pythagoras' theorem states that:

$$c^2 = a^2 + b^2$$

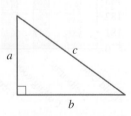

Calculate the perimeter of the following.

THINK

1. To calculate the perimeter we need to know all side lengths. We can use Pythagoras' theorem to determine the unknown side of the right-angled triangle.

2. Substitute the known values into the formula, remembering c represents the hypotenuse, then solve for c. Take the square root of both sides to undo the square.

3. To calculate the perimeter, add the side lengths.

4. Write the answer.

WRITE

$c^2 = a^2 + b^2$

$c^2 = a^2 + b^2$

$c^2 = 6^2 + 8^2$

$c^2 = 36 + 64$

$c^2 = 100$

$c = \sqrt{100}$

$c = 10\,\text{mm}$

$P = 10 + 6 + 8$

$= 24$

The perimeter is 24 mm.

3.3.3 Circumference

The **circumference** (C) is the term used for the perimeter of a circle.

The **diameter** (D) of a circle is the distance from one side of the circle to the other, passing through the centre. The **radius** (r) of a circle is the distance from the centre of the circle to the outside and it is half the distance of the diameter.

> **Circumference of a circle**
> The circumference of a circle is given by the formula
> $$C = 2\pi r$$
> or
> $$C = \pi D$$
> The diameter is twice the length of the radius. That is, $D = 2r$ or $r = \dfrac{D}{2}$.

Calculate the circumference of the following circle to two decimal places.

12 cm

THINK

1. Use the formula for the circumference of a circle in terms of the radius.

WRITE

$C = 2\pi r$

2. Substitute the radius into the equation and solve for C using the π key on your calculator.

$C = 2 \times \pi \times 12$
$= 75.398\,22\,369$
$\approx 75.40\,\text{cm}$

3. Write the answer to two decimal places.

The circumference is 75.40 cm.

3.3.4 Perimeter of a sector

An **arc** is part of the circumference of a circle.

An **arc length** of a circle is calculated by finding the circumference of a circle and multiplying by the fraction of the angle that it forms at the centre of the circle.

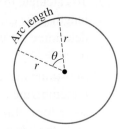

Arc length

$$\text{Arc length}, l = \frac{\theta}{360} \times 2\pi r$$

$$l = \frac{\theta}{180}\pi r$$

where r = radius of the circle and θ = the angle the ends of the arc make with the centre of the circle (in degrees).

A sector is the portion of a circle enclosed by an arc and two radii.

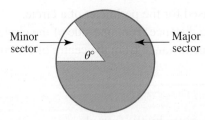

Minor sector

Major sector

$\theta°$

The perimeter of a sector is found by adding the length of the arc and the lengths of two radii.

Perimeter of a sector

$$\text{Perimeter of a sector} = \text{arc length} + 2r = \frac{\theta}{180}\pi r + 2r$$

where r = radius of the circle and θ = the angle the ends of the arc make with the centre of the circle (in degrees).

WORKED EXAMPLE 6

Calculate the perimeter of each of the following shapes to two decimal places.

a.

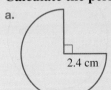

2.4 cm

b.

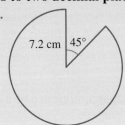

7.2 cm | 45°

THINK	WRITE
a. 1. Determine the angle the ends of the arc make with the centre.	**a.** $\theta = 360° - 90°$ $= 270°$
2. Substitute the values for the radius and the angle into the formula for the perimeter of a sector.	$\text{Perimeter} = \dfrac{\theta}{180} \times \pi r + 2r$ $= \dfrac{270}{180} \times \pi \times 2.4 + 2 \times 2.4$ $= 11.3097... + 4.8$ $\approx 16.11\,\text{cm}$
3. Answer the question.	$\text{Perimeter} = 16.11\,\text{cm}$
b. 1. Determine the angle the ends of the arcs make with the centre.	**b.** $\theta = 360° - 45°$ $= 315°$
2. Substitute the values for the radius and the angle into the formula for the perimeter of a sector.	$\text{Perimeter} = \dfrac{\theta}{180} \times \pi r + 2r$ $= \dfrac{315}{180} \times \pi \times 7.2 + 2 \times 7.2$ $= 39.5841... + 14.4$ $\approx 53.98\,\text{cm}$
3. Answer the question.	$\text{Perimeter} = 53.98\,\text{cm}$

3.3.5 Perimeter of composite shapes

The perimeter of a composite shape is the same as the regular perimeter; it is the total distance around the outside of a shape.

WORKED EXAMPLE 7

Calculate the perimeter of the following shape to two decimal places.

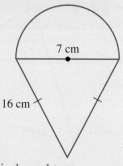

THINK	WRITE
1. Determine the different shapes that make the composite shape.	Shape made up of a semi-circle and two lengths of 16 cm.
2. Calculate the arc length of a semi-circle. Recall that $r = \dfrac{D}{2}$.	$\text{Arc length} = \dfrac{180}{180} \times \pi \times \dfrac{7}{2}$ $= 10.995\,574\,\text{cm}$
3. Calculate the perimeter of the shape by adding all three sides.	$P = 10.995\,574 + 16 + 16$ $= 42.995\,574\,\text{cm}$ $\approx 43.00\,\text{cm}$
4. Answer the question.	The perimeter is 43.00 cm

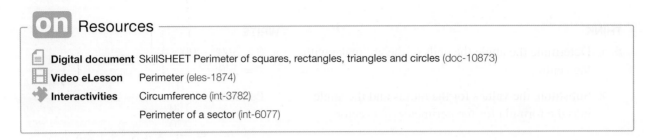
Exercise 3.3 Perimeters of familiar and composite shapes

1. Calculate the perimeter of the following rectangles.

a.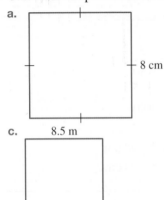

8 cm

b.

10 cm

6 cm

c.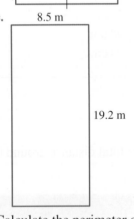

8.5 m

19.2 m

d.

15.38 mm

4.83 mm

2. Calculate the perimeter of the following triangles.

a.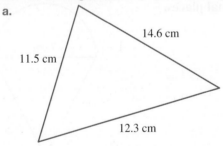

14.6 cm

11.5 cm

12.3 cm

b.

8.6 m   12.16 m

c.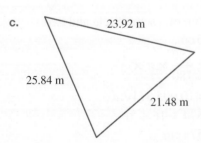

23.92 m

25.84 m

21.48 m

d.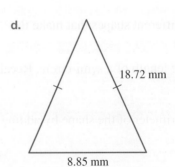

18.72 mm

8.85 mm

3. Calculate the perimeter of the following shapes.

a.

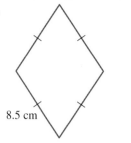

8.5 cm

b.

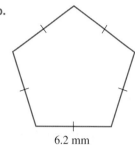

6.2 mm

c.

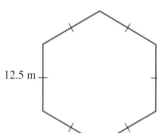

12.5 m

d.

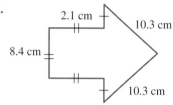

2.1 cm 10.3 cm

8.4 cm

10.3 cm

4. **WE3** Calculate the perimeter of the following shapes.

a.

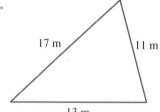

17 m   11 m

13 m

b.

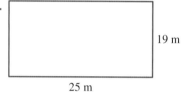

19 m

25 m

c.

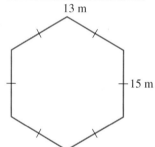

15 m

d.

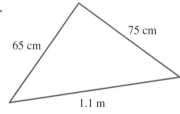

75 cm

65 cm

1.1 m

5. Determine the perimeter of the following.
   a. A square table of length 1.2 metres
   b. A rectangular yard 23 m by 37 m
   c. An octagon of side length 25 cm

6. Calculate the perimeter of the following shapes.

a.

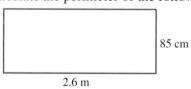

85 cm

2.6 m

b.

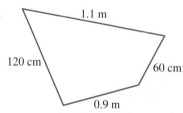

1.1 m

120 cm   60 cm

0.9 m

7. **WE4** Calculate the perimeter of the following shapes, correct to two decimal places where necessary.

a.

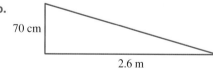

10 cm

b.

70 cm

2.6 m

8. **WE5** Calculate the circumference of the following circles, correct to two decimal places.

a.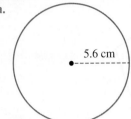
5.6 cm

b.
◄—1055 mm—►

9. **WE6** Calculate the perimeter of the following shapes, correct to two decimal places.

a.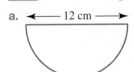
◄—— 12 cm ——►

b.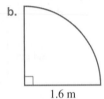
1.6 m

c.
2.2 cm

10. Calculate the perimeter of each of the following shapes to two decimal places.

a.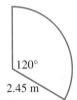
120°
2.45 m

b.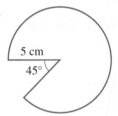
5 cm
45°

11. Calculate the perimeter of each of the following shapes to two decimal places.

a.
14.25 cm

b.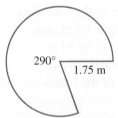
110°
23.48 mm

12. Calculate the perimeter of the following to two decimal places.

a.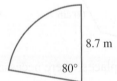
115°
2.3 cm

b.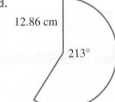
290°
1.75 m

c.
80°
8.7 m

d.
12.86 cm
213°

13. Calculate the perimeter of the following shapes, correct to two decimal places where necessary.

a.
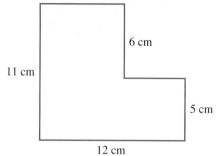
11 cm
6 cm
5 cm
12 cm

b.
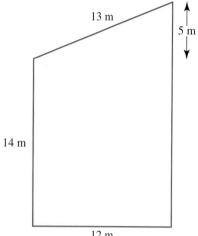
13 m
5 m
14 m
12 m

c.
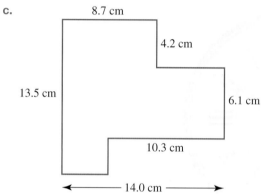
8.7 cm
4.2 cm
13.5 cm
6.1 cm
10.3 cm
14.0 cm

d.
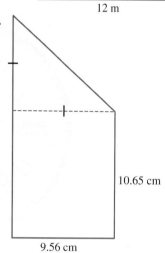
10.65 cm
9.56 cm

e.
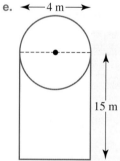
4 m
15 m

f.
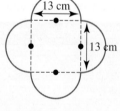
13 cm
13 cm

14. **WE7** Calculate the perimeter of the following shape, to two decimal places.

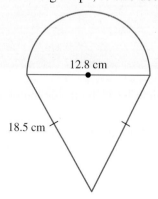
12.8 cm
18.5 cm

15. Calculate the perimeter of the following shape, to one decimal place.

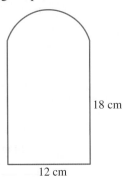

18 cm

12 cm

16. Calculate the radius (in centimetres) of a trundle wheel that travels one metre in one revolution.

17. The minute hand on a clock is 22 centimetres in length. Calculate the distance travelled by the tip of the hand in 3 hours.

18. The straight lengths of a running track are 80 metres. If the half circles at either end of the track are equal in size, calculate the following giving your answers correct to two decimal places.

80 m

a. What is the radius of the end semicircles if the inside length of the track is exactly 400 metres in length?

b. How much further in front of the runner in lane 1 must the runner in lane 2 start to ensure both runners complete 400 metres, if each lane is 1 metre in width? Assume each runner runs in the centre of the lane.

3.4 Units of area and estimation

The **area** of a two-dimensional shape is the amount of space enclosed by the shape.

A square centimetre is the amount of space enclosed by a square with a length 1 cm and width 1 cm.

Some common metric units of area and their abbreviations are: square kilometres (km^2), square metres (m^2), square centimetres (cm^2) and square millimetres (mm^2).

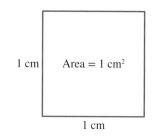

3.4.1 Conversion of area units

When converting area units you need to take into account that the shape is two-dimensional, thus squared units. To convert area units:

- follow the conversion of length
- then square the conversion.

Consider, for example, converting $1\,cm^2$ to mm^2.

To convert from cm to mm, multiply by 10. Since area is two-dimensional we need to square the conversion, so multiply by $10^2 = 100$.

Therefore, $1\,cm^2 = 1 \times 100\,mm^2 = 100\,mm^2$

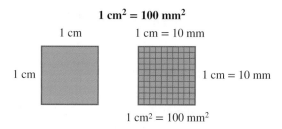

The following diagram shows how to convert between metric units of area.

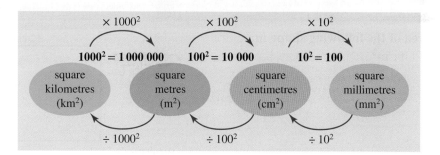

Large parcels of land, such as farmland are often measured in **hectares (ha)**. 1 hectare is the space enclosed by a square with side lengths equal to 100 m.

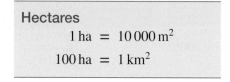

Hectares
$$1\,ha = 10\,000\,m^2$$
$$100\,ha = 1\,km^2$$

WORKED EXAMPLE 8

Convert the following units of area.
a. $56 \, \text{cm}^2 = \underline{\hspace{1.5cm}} \text{mm}^2$
b. $78\,700 \, \text{cm}^2 = \underline{\hspace{1.5cm}} \text{m}^2$
c. $591 \, \text{ha} = \underline{\hspace{1.5cm}} \text{km}^2$

THINK	WRITE
a. 1. To convert from cm^2 to mm^2, multiply by 10^2.	a. $56 \, \text{cm}^2 = 56 \times 10^2 \, \text{mm}^2$
2. Complete the calculation and write the answer.	$56 \, \text{cm}^2 = 56 \times 100 \, \text{mm}^2$
	$56 \, \text{cm}^2 = 5600 \, \text{mm}^2$
b. 1. To convert from cm^2 to m^2, divide by 100^2.	b. $78\,700 \, \text{cm}^2 = 78\,700 \div 100^2 \, \text{m}^2$
2. Complete the calculation and write the answer.	$78\,700 \, \text{cm}^2 = 78\,700 \div 10\,000 \, \text{m}^2$
	$78\,700 \, \text{cm}^2 = 7.87 \, \text{m}^2$
c. 1. To convert from ha to km^2, divide by 100.	c. $591 \, \text{ha} = 591 \div 100 \, \text{km}^2$
2. Complete the calculation and write the answer.	$591 \, \text{ha} = 5.91 \, \text{km}^2$

3.4.2 Area of a rectangle

The area of a rectangle can by calculated by multiplying its length by its width.

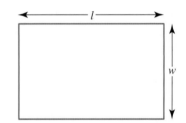

Area of a rectangle

$$A = l \times w$$
where: l = length
w = width

WORKED EXAMPLE 9

Calculate the area of the following shape in:
a. cm^2
b. m^2.

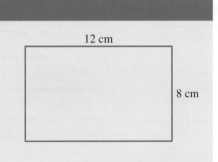

12 cm

8 cm

THINK	WRITE
a. 1. The shape is a rectangle so use the area of a rectangle formula.	a. $A = l \times w$
2. Substitute in $l = 12$ cm and $w = 8$ cm.	$A = 12 \times 8$
	$= 96$
3. Write the answer.	Area $= 96 \, \text{cm}^2$
b. 1. To convert cm^2 to m^2 divide by 100^2.	b. $96 \, \text{cm}^2 = 96 \div 100^2 \, \text{m}^2$
	$= 0.0096 \, \text{m}^2$
2. Write the answer.	Area $= 0.0096 \, \text{m}^2$

3.4.3 Estimation of area

Follow the steps below to estimate the area of a shape where a square is not completely filled by the object.
- If more than half of the square is covered, count it in the area.
- If less than half of the square is covered, don't count it in the area.

WORKED EXAMPLE 10

Estimate the area of the following shape, giving your answer in units of squares.

THINK

1. Place a tick in each square that is more than half full.

2. Count the number of squares that contain a tick. Write the answer.

WRITE

The area is approximately 6 square units.

 Resources

Digital documents SkillSHEET Area units (doc-6512)

SkillSHEET Area of rectangles (doc-6514)

Interactivities Area (int-0005)

Area of a rectangle (int-4017)

Area of rectangles (int-3784)

Conversion chart for area (int-3783)

Metric units of area 1 (int-4015)

Metric units of area 2 (int-4016)

Exercise 3.4 Units of area and estimation

1. Name the most appropriate unit for measuring each of the following.
   a. Area of a floor rug
   b. Area of a fingerprint
   c. Area of a page from a novel
   d. Area of a painted wall in the lounge room
   e. Area of the surface of Uluru
   f. Area of a five-cent coin

2. Which metric unit of area would you use to measure each of the following areas?
   a. The palm of your hand
   b. A suburban house block
   c. A dairy farm
   d. Your fingernail
   e. Australia
   f. Your local football ground

3. Provide two examples (that have not already been mentioned) of area that could be measured in:
   a. mm^2 b. cm^2 c. m^2 d. km^2.

4. **WE8** Convert the following units of area.
   a. $34\,cm^2 = $ _____mm^2 b. $14\,250\,cm^2 = $ _____m^2 c. $592\,ha = $ _____km^2

5. Convert the following units of area.
   a. $0.18\,m^2 = $ _____cm^2 b. $0.0798\,km^2 = $ _____m^2 c. $374\,300\,m^2 = $ _____ha

6. Convert each of the following measurements to the unit shown in brackets.
   a. $2.3\,cm^2\ (mm^2)$ b. $2.57\,m^2\ (cm^2)$ c. $470\,mm^2\ (cm^2)$
   d. $27\,000\,m^2\ (km^2)$ e. $87\,500\,m^2$ (ha)   f. $17\,000\,cm^2\ (m^2)$

7. Convert each of the following measurements to the units shown in the brackets.
   a. $51\,200\,m^2$ (ha)   b. $2.85\,ha\ (m^2)$ c. $8\,380\,ha\ (km^2)$
   d. $12.6\,km^2$ (ha)   e. $0.23\,ha\ (m^2)$ f. $623\,450\,m^2$ (ha)

8. Convert each of the following measurements to the units shown in the brackets.
   a. $0.048\,m^2\ (mm^2)$ b. $0.0012\,km^2\ (cm^2)$ c. $300\,800\,mm^2\ (m^2)$ d. $4478\,cm^2\ (km^2)$

9. If each square is $1\,cm^2$, determine the area of the following shapes.
   a.   b.   c.   d.

10. Calculate the area of the following shapes.
    a. 9 cm / 6 cm
    b. 1.2 mm / 6.5 mm
    c. 24.7 m / 13.1 m
    d. 9.75 cm

11. **WE9** Calculate the area of the following shapes in the specified units.
    i. cm^2
    ii. m^2

    a.
    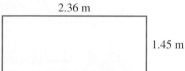
    2.36 m
    1.45 m

    b.
    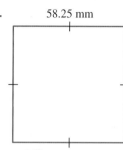
    58.25 mm

    c.

    2.4 m
    85 cm

    d.
    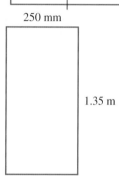
    250 mm
    1.35 m

12. **WE10** Estimate the area of each of the following shapes (giving your answer in units of squares).

    a.
    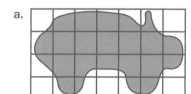

    b.
    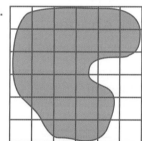

    c.
    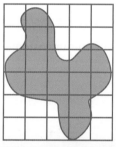

13. Estimate the area of the banner held up by the club members. Use the squares shown in the diagram to help determine the area from one end to the other where each square represents 1 square metre.

14. Estimate the area of the following, by first estimating their dimensions.
    a. A singles tennis court
    b. A basketball court
    c. A netball court
    d. A standard door

15. If a cricket pitch has an area of 60.36 m^2 and it is 3 metres wide, what is the length of the cricket pitch?

16. A rectangular photograph measuring 8 cm by 10 cm is enlarged so that its dimensions are doubled.
    a. Calculate the area of the original photograph.
    b. Calculate the dimensions of the new photograph.
    c. Calculate the area of the new photograph.
    d. Compare the area of the enlarged photo to the original photo and complete the following sentence.
       If the side lengths of the rectangle are doubled its area is _____.

17. A rectangular flower bed measures 25 m by 14 m. A gravel path 2 m wide surrounds it.
    a. Draw a diagram representing the flower bed and path.
    b. Calculate the area of the flower bed.
    c. Calculate the area of the gravel path.
    d. If gravel costs $7.50 per square metre, how much will it cost to cover the path?

18. A wall in a house is to be painted. The wall is rectangular and measures 9.5 metres by 3.2 m. The wall has two rectangular windows, 2.4 metres by 80 cm.
    a. Calculate the area of the wall to be painted in m^2.
    b. If the wall needs three coats, how much coverage of paint is required to paint the wall in m^2?
    c. If 1 litre of paint covers $5\,m^2$, how much paint is required to the nearest litre?
    d. If the paint costs $15.75 per litre, how much will the paint cost?

3.5 Areas of regular shapes

3.5.1 Area of two-dimensional shapes

The area of two-dimensional shapes can be calculated using the following formulae.

Shape	Diagram	Area formula
Square	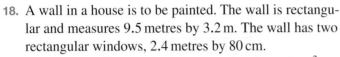	$A_{\text{Square}} = l \times l$ $= l^2$
Rectangle		$A_{\text{Rectangle}} = l \times w$

(Continued)

(Continued)

Shape	Diagram	Area formula
Triangle		$A_{\text{Triangle}} = \dfrac{1}{2} \times b \times h$
Trapezium		$A_{\text{Trapezium}} = \dfrac{1}{2}(a + b) \times h$
Parallelogram		$A_{\text{Parallelogram}} = b \times h$

WORKED EXAMPLE 11

Calculate the area of the following shapes.

a.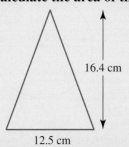
16.4 cm
12.5 cm

b.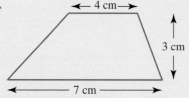
4 cm
3 cm
7 cm

THINK

a. 1. Identify the shape as a triangle and use the area of a triangle formula.

WRITE

a. $A_{\text{Triangle}} = \dfrac{1}{2} \times b \times h$

2. Substitute $b = 12.5$ cm and $h = 16.4$ cm.

$$A_{\text{Triangle}} = \frac{1}{2} \times 12.5 \times 16.4$$
$$= 102.5 \text{ cm}^2$$

3. Write the answer.

The area is 102.5 cm^2.

b. 1. Identify the shape as a trapezium and use the area of a trapezium formula.

b. $A_{\text{Trapezium}} = \frac{1}{2}(a + b) \times h$

2. Substitute $a = 4$ cm, $b = 7$ cm and $h = 3$ cm.

$$A_{\text{Trapezium}} = \frac{1}{2}(4 + 7) \times 3$$
$$= \frac{1}{2}(11) \times 3$$
$$= \frac{33}{2}$$
$$= 16.5 \text{ cm}^2$$

3. Write the answer.

The area is 16.5 cm^2.

3.5.2 Area of a circle

The area of a circle can be calculated using the following formula.

Area of a circle

$$A_{\text{Circle}} = \pi \times r^2$$

where r = radius of the circle.

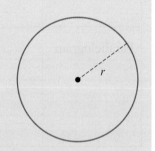

Note that $r = \dfrac{D}{2}$.

WORKED EXAMPLE 12

Calculate the area of the following circle correct to two decimal places.

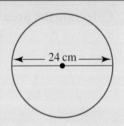

THINK

1. Write the formula for the area of a circle.
2. Determine the radius of the circle.

WRITE

$A_{\text{Circle}} = \pi \times r^2$

The radius is half the length of the diameter, so:
$$r = \frac{24}{2}$$
$$= 12 \text{ cm}$$

3. Substitute $r = 12$ into the formula for the area of a circle and evaluate the expression for the area.

$$A_{\text{Circle}} = \pi \times 12^2$$
$$= 144\pi$$
$$\approx 452.39 \text{ cm}^2$$

WORKED EXAMPLE 13

Calculate the area of the shaded region to two decimal places.

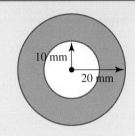

THINK	WRITE
1. The shaded region is the area of the larger circle minus the area of the smaller circle.	$A_{\text{Shaded}} = A_{\text{Large}} - A_{\text{Small}}$
2. Calculate the area of the larger circle with $r = 20$ mm.	$A_{\text{Large}} = \pi \times (20)^2$ $= 400\pi$
3. Calculate the area of the smaller circle with $r = 10$ mm.	$A_{\text{Small}} = \pi \times (10)^2$ $= 100\pi$
4. Calculate the area of the shaded region to two decimal places.	$A_{\text{Shaded}} = 400\pi - 100\pi$ $= 300\pi$ $\approx 942.48 \text{ mm}^2$

3.5.3 Area of a sector

The area of a sector can be calculated by finding the area of the circle and multiplying it by the fraction of the angle that the sector forms at the centre of the circle.

Area of a sector

$$\text{Area of a sector} = \frac{\theta}{360} \times \pi r^2$$

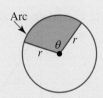

where:
r = radius of the circle
θ = angle that the sector makes with the centre of the circle, in degrees.

Calculate the area of the shaded region to two decimal places.

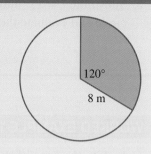

THINK	WRITE
1. Write the formula for the area of a sector.	Area of a sector $= \dfrac{\theta}{360} \times \pi r^2$
2. Substitute $r = 8\,\text{cm}$ and $\theta = 120°$.	Area of a sector $= \dfrac{120}{360} \times \pi \times 8^2$
	$= 67.020\,643$
	$\approx 67.02\,\text{cm}^2$
3. Write the answer.	The shaded area is $67.02\,\text{cm}^2$.

 Resources

  Digital document SkillSHEET Area of squares, rectangles and triangles (doc-6958)

  Interactivities   Area of circles (int-3788)

                    Area of parallelograms (int-3786)

                    Area of trapeziums 1 (int-3789)

                    Area of trapeziums 2 (int-3790)

                    Area of a sector (int-6076)

                    Area of a circle (int-4441)

                    Area of trapeziums (int-4442)

Exercise 3.5 Areas of regular shapes

1. Calculate the area of the following triangles:

  a.

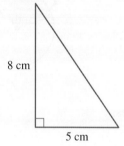

8 cm

5 cm

  b.

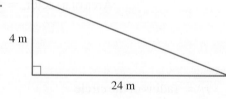

4 m

24 m

c.

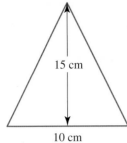

15 cm

10 cm

d.

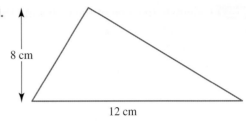

8 cm

12 cm

2. **MC** What is the area of a triangle with base length 12 cm and vertical height of 12 cm?

A. 144 cm^2 **B.** 100 cm^2 **C.** 72 cm^2 **D.** 64 cm^2

3. Calculate the area of the following trapeziums.

a.

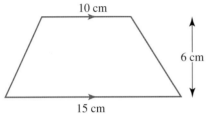

10 cm

6 cm

15 cm

b.

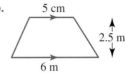

5 cm

2.5 m

6 m

c.

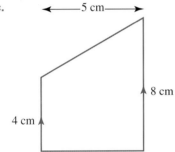

5 cm

8 cm

4 cm

d.

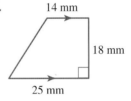

14 mm

18 mm

25 mm

4. **MC** What is the area of a parallelogram of length 25 cm and vertical height of 5 cm?

A. 30 cm^2 **B.** 100 cm^2 **C.** 62.5 cm^2 **D.** 125 cm^2

5. Calculate the area of the following parallelograms.

a.

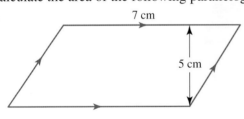

7 cm

5 cm

b.

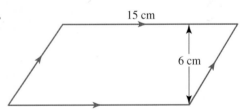

15 cm

6 cm

c.

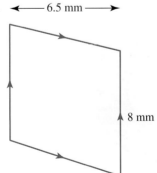

6.5 mm

8 mm

d.

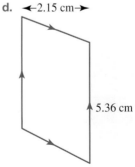

2.15 cm

5.36 cm

6. **WE11** Calculate the area of the following shapes.

a.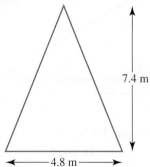
7.4 m
4.8 m

b.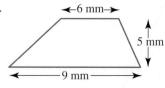
6 mm
5 mm
9 mm

7. **WE12** Calculate the area of the following circles to two decimal places.

a.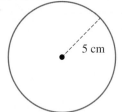
5 cm

b.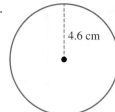
4.6 cm

c.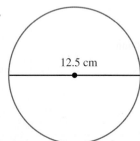
12.5 cm

d.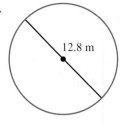
12.8 m

8. **WE13** Calculate the area of the shaded region to two decimal places.

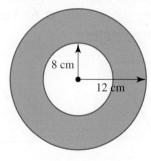
8 cm
12 cm

9. Calculate the area of the shaded region to two decimal places.

a.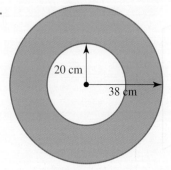
20 cm
38 cm

b.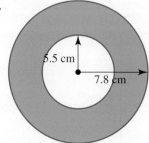
5.5 cm
7.8 cm

10. **WE14** Calculate the area of the shaded region to two decimal places.

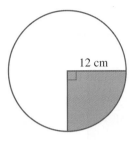

11. Calculate the area of the shaded region to two decimal places.

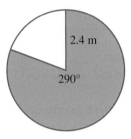

12. Calculate the area of the shaded region to two decimal places.

a.

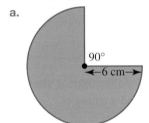

b.

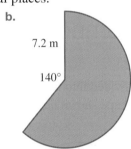

13. **MC** The area of a circle is equal to 113.14 mm^2. The radius of the circle is closest to
    A. 36 mm.
    B. 6 mm.
    C. 2133 mm.
    D. 7 mm.

14. Circular pizza trays come in three different sizes; small, medium and large.
    a. Calculate the area of each tray if the diameters are 20 cm, 30 cm and 40 cm respectively.
    b. How much material must be ordered to make 50 trays of each?
    c. A slice from the large pizza makes an angle from the centre of 45°. What is the area of the pizza slice?

15. A rugby pitch is rectangular and measures 100 m in length and 68 m wide.
    a. What is the area of the pitch?
    b. If the pitch was laid with instant turf, with each sheet measuring 4 m by 50 cm, how many sheets of turf are required to cover the pitch?
    c. If each sheet costs $10.50, how much would it cost to cover the pitch?

16. A school is looking to build four netball courts side by side. A netball court measures 15.25 m wide by 30.5 m long, and they require a 3 m strip between each court and around the outside.

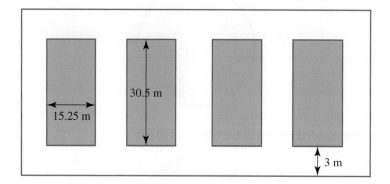

a. Determine the area of each netball court.
b. Calculate the total area, in metres, that the school would need for the four netball courts and its surrounding space.
c. If the four netball courts are painted, determine the area that needs to be painted.
d. If it costs $9.50 per square metre to paint the netball courts, how much would it cost to paint the four courts?

3.6 Areas of composite figures

3.6.1 Sum of areas

A **composite figure** can be divided into two or more sections, each of which is a smaller regular shape with a known area formula. The shape shown below is a combination of a semi-circle and a triangle.

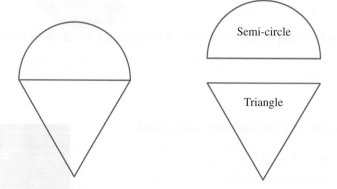

To find the area of a composite figure:
- Divide the shape into smaller figures that have a known area formula.
- Find the area of each smaller figure.
- Calculate the area of the composite figure by *adding* the area of the smaller figures.

WORKED EXAMPLE 15

Calculate the area of the following composite figure.

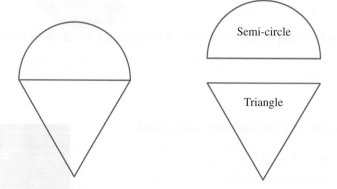

THINK	WRITE
1. This composite figure is made up of a square and a trapezium. Add the areas of both to find the total area.	$A_{Total} = A_{Trapezium} + A_{Square}$
2. Calculate the area of the square.	$A_{Square} = l \times l$ $= 15 \times 15$ $= 225 \text{ mm}^2$
3. Calculate the area of the trapezium, where the height is $23 - 15 = 8$ mm.	$A_{Trapezium} = \dfrac{1}{2}(a+b)h$ $= \dfrac{1}{2}(6+15) \times 8$ $= 84 \text{ mm}^2$
4. Calculate the total area of the figure by adding the two areas.	$A_{Total} = A_{Trapezium} + A_{Square}$ $= 84 + 225$ $= 309 \text{ mm}^2$
5. Write the answer.	The area of the figure is 309 mm^2.

3.6.2 Difference between areas

Not all composite figures are made by adding a number of smaller shapes together; some shapes are made by *subtracting* one shape from another.

WORKED EXAMPLE 16

Calculate the area of the shaded region.

THINK	WRITE
1. This composite shape is a square with a triangle cut out. Find the shaded area by subtracting the area of the triangle from the area of the square.	$A_{Shaded} = A_{Square} - A_{Triangle}$
2. Calculate the area of the square.	$A_{Square} = l \times l$ $= 8 \times 8$ $= 64 \text{ m}^2$
3. Calculate the area of the triangle.	$A_{Triangle} = \dfrac{1}{2} \times b \times h$ $= \dfrac{1}{2} \times 3 \times 4$ $= 6 \text{ m}^2$

4. Calculate the shaded area by subtracting the area of the triangle from the area of the square.

$$A_{\text{Shaded}} = A_{\text{Square}} - A_{\text{Triangle}}$$
$$= 64 - 6$$
$$= 58 \text{ m}^2$$

5. Write the answer.

The shaded area is 58 m².

 Resources

 Digital document WorkSHEET Area (doc-5241)

Video eLesson Composite area (eles-1886)

Exercise 3.6 Areas of composite figures

1. Name the two smaller shapes that make the following composite shapes.

a.

b.

c.

d.

2. **WE15** Calculate the area of the following composite figure.

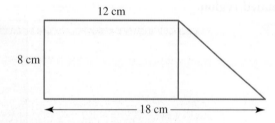

12 cm

8 cm

18 cm

3. Calculate the area of the following shape to two decimal places.

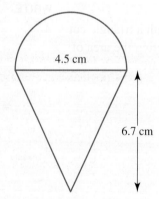

4.5 cm

6.7 cm

4. Calculate the area of the following composite shapes.

a.
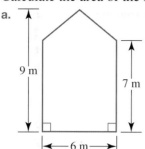

b.
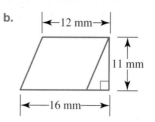

c.
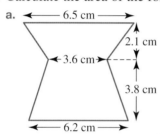

d.
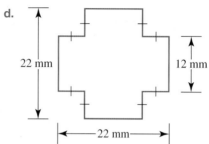

5. Calculate the area of the following composite shapes.

a.
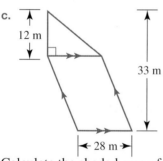

b.
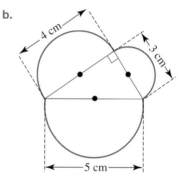

c.
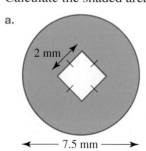

d. 175 mm
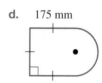

6. Calculate the shaded area of the following shapes.

a.
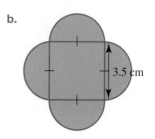

b.

7. **a.** Calculate the area of the hexagon shown by dividing it into two trapeziums.
   b. Calculate the area of the hexagon by dividing it into a rectangle and two triangles.
   c. Compare your answers to part **a** and **b**.

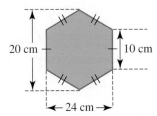

8. A triangular pyramid can be constructed from the net shown. Calculate the total area of the net.

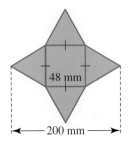

9. Calculate the area of the shaded region.

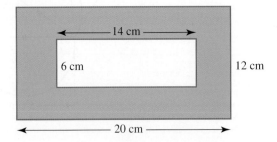

10. **WE16** Calculate the area of the shaded region to two decimal places.

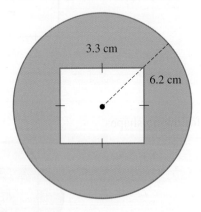

11. The door shown is 1 metre wide, 2.2 metres high and has four identical glass panels, each measuring 76 cm by 15 cm.

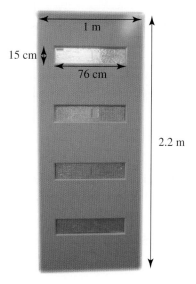

a. Calculate the total area of the glass panels.
b. The door is to be painted on both sides. Calculate the total area to be painted.
c. Two coats of paint are required on each side of the door. If the paint is sold in 1-litre tins at $24.95 per litre and each litre covers 8 m² of the surface, calculate the total cost of painting the door.

12. Calculate the area of the shaded region shown.

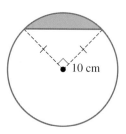

13. The leadlight panel shown depicts a sunrise over the mountains. The mountain is represented by a green triangle 45 cm high. The yellow sun is represented by a section of a circle with an 18 cm radius. There are 10 yellow sunrays in the shape of isosceles triangles with a base of 3 cm and a height of 12 cm, and the sky is blue. Calculate the area of the leadlight panel made of:

a. green glass
b. yellow glass
c. blue glass.

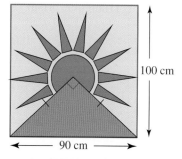

3.7 Surface areas of familiar prisms

The **total surface area** (TSA) of a three-dimensional (3D) object is the total area of each outer face of that object.

The total surface area of a 3D object can be calculated by finding the area of each individual face and adding them together.

Drawing a net of a solid can help you determine the individual shapes of the faces that make up the solid.

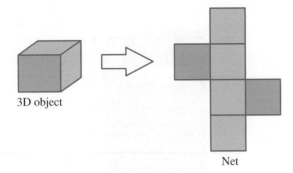

3D object

Net

As with area, total surface area is measured using **squared units** (mm^2, cm^2, m^2, or km^2).
The total surface area of some common 3D objects can be calculated using appropriate formulae.

3.7.1 Total surface area of prisms

A **prism** is a 3D shape with identical opposite ends joined by straight edges. It can be sliced into identical slices called cross-sections.

Prism	Diagram	Total surface area formula
Cube		$TSA_{Cube} = 6l^2$
Rectangular prism		$TSA_{Rectangular\ prism} = 2(lh + lw + wh)$
Triangular prism		$TSA_{Triangular\ prism} = bh + 2ls + bl$

WORKED EXAMPLE 17

Calculate the total surface of the following rectangular prism.

5 cm
4 cm
12 cm

THINK	WRITE
1. Identify the object as a rectangular prism, and write the appropriate total surface area formula.	$TSA_{Rectangular\ prism} = 2(lh + lw + wh)$

2. Identify the dimensions of the object.	$l = 12$, $w = 4$ and $h = 5$
3. Substitute the length, width and height values into the TSA formula and calculate the TSA of the rectangular prism.	$\begin{aligned} \text{TSA}_{\text{Rectangular prism}} &= 2\,(12 \times 5 + 12 \times 4 + 4 \times 5) \\ &= 2\,(60 + 48 + 20) \\ &= 2\,(128) \\ &= 256 \end{aligned}$
4. Write the answer.	The total surface area of the rectangular prism is $256\,\text{cm}^2$.

3.7.2 Total surface area of a cylinder

Drawing the net of a cylinder can help determine the formula for the total surface area.

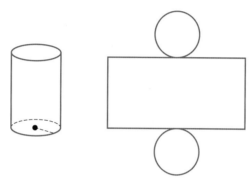

A cylinder is made up of 2 circles and a curved surface (which when rolled out and flattened becomes a rectangle).

Total surface area of a cylinder

The formula for the total surface area of a cylinder is:

$$\text{TSA}_{\text{Cylinder}} = 2\pi rh + 2\pi r^2$$

WORKED EXAMPLE 18

Calculate the total surface area of the following cylinder correct to two decimal places.

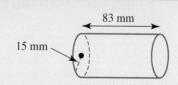

15 mm

83 mm

THINK

1. Identify the object as a cylinder, and write the appropriate total surface area formula.

WRITE

$\text{TSA}_{\text{Cylinder}} = 2\pi rh + 2\pi r^2$

▶

2.	Identify the radius and the height of the cylinder.	$r = 15, h = 83$
3.	Substitute the value of the radius and height into the TSA formula and calculate the TSA of the cylinder.	$\begin{aligned} \text{TSA}_{\text{Cylinder}} &= 2 \times \pi \times 15 \times 83 + 2 \times \pi \times 15^2 \\ &= 2490\pi + 450\pi \\ &= 2940\pi \\ &\approx 9236.28240155 \end{aligned}$
4.	Write the answer.	The total surface area of the cylinder is 9236.28 mm^2.

3.7.3 Total surface area of a sphere

A sphere is a round solid figure, such as a tennis ball, soccer ball or basketball.

Total surface area of a sphere

The formula for the total surface area of a sphere is:

$$\text{TSA}_{\text{Sphere}} = 4\pi r^2$$

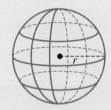

WORKED EXAMPLE 19

Calculate the total surface area of a basketball with a radius of 15 cm. Give your answer correct to two decimal places.

THINK	WRITE
1. Identify the basketball as a sphere, and write the appropriate total surface area formula.	$\text{TSA}_{\text{Sphere}} = 4\pi r^2$
2. Identify the radius of the sphere.	$r = 15$
3. Substitute the value of the radius into the TSA formula and calculate the TSA of the basketball.	$\begin{aligned} \text{TSA}_{\text{Sphere}} &= 4 \times \pi \times 15^2 \\ &= 4 \times \pi \times 225 \\ &= 900\pi \\ &= 2827.433388 \\ &\approx 2827.43 \text{ cm}^2 \end{aligned}$
4. Write the answer.	The total surface area of the basketball is 2 827.43 cm^2.

Exercise 3.7 Surface areas of familiar prisms

1. Which of the shapes below are prisms?

a. 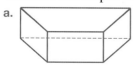   b.    c. 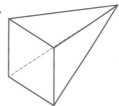   d.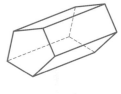

2. Calculate the total surface area of the cubes with the following side lengths.

   a. 4 cm   b. 12.5 mm   c. 0.85 m   d. $7\frac{3}{4}$ cm

3. **WE17** Calculate the total surface area of the following rectangular prism.

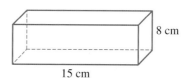

4. Calculate the total surface area of each of the following rectangular prisms.

   a.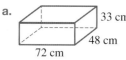

   b.

   c.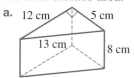

   d.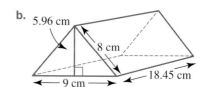

5. For each of the following triangular prisms, calculate:
   i. the area of the triangular face
   ii. the area of each of the three rectangular faces
   iii. the total surface area.

   a. 12 cm, 5 cm, 13 cm, 8 cm

   b. 5.96 cm, 8 cm, 9 cm, 18.45 cm

6. **WE18** Calculate the total surface area of each of the following cylinders, correct to two decimal places.

a.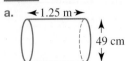
1.25 m
49 cm

b.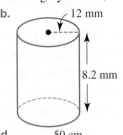
12 mm
8.2 mm

c.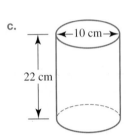
10 cm
22 cm

d.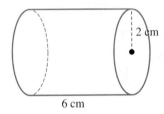
50 cm
65 cm

7. Calculate the total surface area of the following cylinder to two decimal places.

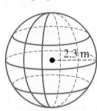
2 cm
6 cm

8. Calculate the side length of a cube with a total surface area of $384 \, \text{m}^2$.

9. Calculate the total surface area of the following sphere, correct to two decimals places.

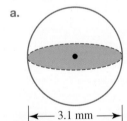
2.3 m

10. Calculate the total surface area of the following shapes, correct to two decimal places.

a.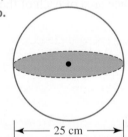
3.1 mm

b.
25 cm

11. **WE19** Calculate the total surface area of a tennis ball with a diameter of 6 cm, correct to two decimal places.

12. **MC** A sphere has a total surface area of 125 cm². What is the radius of the sphere correct to two decimal places?

A. 3.15 cm   B. 9.94 cm   C. 31.5 cm   D. 31.25 cm

13. Calculate the total surface area of the tin of tennis balls shown, correct to two decimal places, in:

a. mm²   b. cm².

14. Calculate the surface area of the following shape, correct to two decimal places.

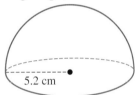

5.2 cm

3.8 Surface areas of familiar pyramids and irregular solids

3.8.1 Total surface area of a cone

The total surface area of a cone, when the radius (r) and the slant-height (s) are known, can be found using the formula shown below.

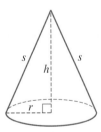

The slant-height is the measurement along the side of the cone, not the vertical height (h) of the cone.

Total surface area of a cone

$$\text{TSA}_{\text{Cone}} = \pi rs + \pi r^2$$

WORKED EXAMPLE 20

Calculate the total surface area of the cone shown, correct to one decimal place.

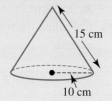
15 cm

10 cm

THINK	WRITE
1. Identify the shape as a cone and write the appropriate total surface area formula.	$\text{TSA}_{\text{Cone}} = \pi r s + \pi r^2$
2. Identify the dimensions of the object.	$r = 10$ and $s = 15$
3. Substitute the values for the radius and the slanting height into the TSA formula and calculate the TSA of the cone.	$\begin{aligned} \text{TSA}_{\text{Cone}} &= \pi \times 10 \times 15 + \pi \times 10^2 \\ &= 150\pi + 100\pi \\ &= 250\pi \\ &= 785.398\,163\,4 \\ &\approx 785.4 \end{aligned}$
4. Write the answer.	The total surface area of the cone is $785.4\,\text{cm}^2$.

3.8.2 Total surface area of pyramids

The total surface area of a pyramid can be calculated by adding the area of the base to the area of each of its sides.

Taking the square-based pyramid as an example, its total surface area can be found as shown.

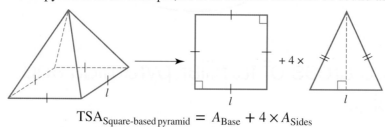

$$\text{TSA}_{\text{Square-based pyramid}} = A_{\text{Base}} + 4 \times A_{\text{Sides}}$$

This method can be adapted to rectangular-based pyramids and triangular-based pyramids.

WORKED EXAMPLE 21

Calculate the total surface area of the pyramid shown correct to two decimal places.

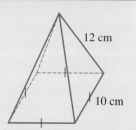
12 cm

10 cm

THINK	WRITE
1. The pyramid is made up of a square base and four triangular sides.	$\text{TSA}_{\text{Square-based pyramid}} = A_{\text{Base}} + 4 \times A_{\text{Sides}}$

2. Calculate the area of the square base with a side length of 10 cm.

$$A_{\text{Base}} = l^2$$
$$= 10^2$$
$$= 100 \text{ cm}^2$$

3. To calculate the area of the triangular side, we need to determine the vertical height using Pythagoras' theorem.
   Hypotenuse is 12 cm and half the width of the base is 5 cm.

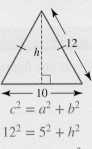

$$c^2 = a^2 + b^2$$
$$12^2 = 5^2 + h^2$$
$$144 = 25 + h^2$$

Subtract 25 from both sides. To undo the square take the square root of both sides.

$$h^2 = 144 - 25$$
$$h^2 = 119$$
$$h = \sqrt{119}$$

4. Calculate the area of the triangular side.

$$A_{\text{Triangle}} = \frac{1}{2}bh$$
$$= \frac{1}{2} \times 10 \times \sqrt{119}$$
$$= 54.5436 \text{ cm}^2$$

5. Calculate the TSA.

$$TSA_{\text{Square-based pyramid}} = A_{\text{Base}} + 4 \times A_{\text{Sides}}$$
$$= 100 + 4 \times 54.5436$$
$$= 318.1744 \text{ cm}^2$$

6. Write the answer.

The total surface area of the pyramid is 318.17 cm^2.

3.8.3 Surface area of combined shapes

Like the area of combined 2D shapes, some 3D shapes are made by combining a number of regular 3D shapes; for example, using a cube and a square-based pyramid as shown.

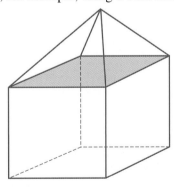

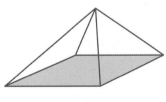

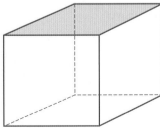

Note: When two 3D shapes are combined together, adding the total surface area of the two individual shapes is generally not correct. Referring to the cube and the square-based pyramid above, the top of the cube (green) and the base of the square based pyramid (pink), sit on top of each other and are therefore no longer an outside surface.

Remember that the total surface area is the area of all the **outside surfaces** (i.e. the surfaces that can be touched).

WORKED EXAMPLE 22

Calculate the total surface area of the following shape.

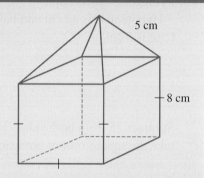

THINK

1. The total surface area of the shape is made up of 5 faces of the cube and the 4 sides of the pyramid.

2. Calculate the area of the cube sides.

3. Calculate the vertical height of the triangular side of the pyramid using Pythagoras' theorem.

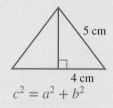

Subtract 16 from both sides.

Take the square root of both sides.

4. Calculate the area of the pyramid sides.

5. Calculate the total surface area of the shape.

6. Write the answer.

WRITE

$TSA = 5 \times A_{\text{Cube sides}} + 4 \times A_{\text{Pyramid sides}}$

$A_{\text{Cube sides}} = 8 \times 8$
$= 64$

$c^2 = a^2 + b^2$
$5^2 = a^2 + 4^2$
$25 = a^2 + 16$
$a^2 = 25 - 16$
$a^2 = 9$
$a = 3$

$A_{\text{Pyramid side}} = \frac{1}{2} \times 8 \times 3$
$= 12$

$TSA = 5 \times A_{\text{Cube sides}} + 4 \times A_{\text{Pyramid sides}}$
$= (5 \times 64) + (4 \times 12)$
$= 320 + 48$
$= 368$

The total surface area of the shape is 368 cm^2.

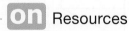

Exercise 3.8 Surface areas of familiar pyramids and irregular solids

1. **WE20** Calculate the total surface area of the cone shown, correct to two decimal places.

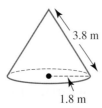

2. Calculate the total surface area of the following shape, correct to two decimal places.

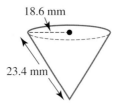

3. Calculate the total surface area of the following shapes, correct to two decimal places.

a.
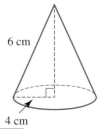

b.
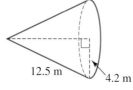

4. **WE21** Calculate the total surface area of the following pyramid, correct to two decimal places.

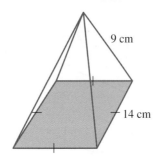

5. Calculate the surface area of the following shapes, correct to two decimal places.

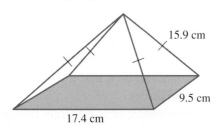

6. Calculate the total surface area of the following shapes, correct to two decimal places.

a.

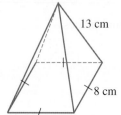

13 cm

8 cm

b.

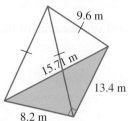

9.6 m

15.7 m

13.4 m

8.2 m

7. Calculate the total surface area of the following triangular-based pyramid, correct to two decimal places.

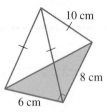

10 cm

8 cm

6 cm

8. **WE22** Calculate the total surface area of the following shape, correct to two decimal places.

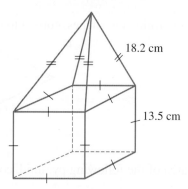

18.2 cm

13.5 cm

9. Calculate the total surface area of the following shape, correct to one decimal place.

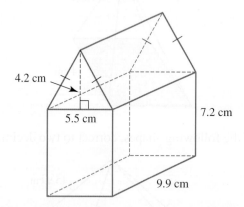

4.2 cm

5.5 cm

7.2 cm

9.9 cm

10. A 22 mm cube has a square-based pyramid sit on top of it with a slant length of 28 mm. Calculate its total surface area correct to two decimal places.

11. Calculate the total surface area of a hemisphere of radius 6 cm that sits on a cylinder of radius 6 cm and height of 12 cm.

12. Calculate the total surface area of two cones joined together at their circular bases if they both have a diameter of 24 cm and a slant length of 18 cm.

13. Calculate the total surface area of the shapes shown to two decimal places.

a.

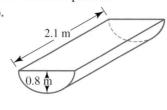

22 cm

|←— 20 cm —→|

b.

2.1 m

0.8 m

14. Calculate the total surface area of the shape shown, correct to two decimal places.

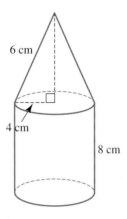

6 cm

4 cm

8 cm

15. The drawing shown is a part of a child's playground toy. Calculate the total surface area, giving the answer to two decimal places.

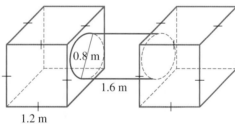

0.8 m

1.6 m

1.2 m

16. Calculate the total surface area of a cube with side length 15 cm that has a hole drilled through it with a diameter of 4 cm. Round the answer correct to two decimal places.

3.9 Review: exam practice

3.9.1 Linear and area measure summary

Units of length

- Units of length can be converted as shown below.

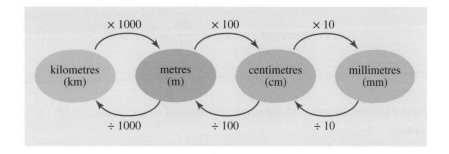

× 1000   × 100   × 10

kilometres (km)   metres (m)   centimetres (cm)   millimetres (mm)

÷ 1000   ÷ 100   ÷ 10

Perimeter

- The perimeter is the distance around the outside of an enclosed shape.
- To find the perimeter of a shape, change all lengths to the same units and add them up.
- Pythagoras' theorem can be used to calculate an unknown side length of a right-angled triangle if two of the sides are known.

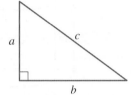

$$c^2 = a^2 + b^2$$

- The diameter (D) of a circle is the distance from one side of a circle to the other, passing through the centre.
- The radius (r) of a circle is the distance from the centre of a circle to the outside and is half the distance of the diameter.
- The circumference (C) is the term used for the perimeter of a circle.

$$C = 2\pi r$$
or
$$C = \pi D$$

- An arc is part of the circumference of a circle.
- An arc length of a circle is calculated by finding the circumference of a circle and multiplying by the fraction of the angle that it forms at the centre of the circle.

$$\text{Arc length, } l = \frac{\theta}{360} \times 2\pi r$$
$$= \frac{\theta}{180}\pi r$$

where θ = the angle the ends of the arc makes with the centre of the circle (in degrees) and r = radius.

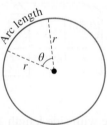

- A sector is the portion of a circle enclosed by an arc and two radii

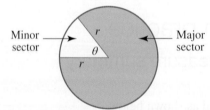

- The perimeter of a sector is found by adding the length of the arc and the lengths of two radii.

$$\text{Perimeter of a sector} = \frac{\theta}{360} \times 2\pi r + 2r$$
$$= \frac{\theta}{180}\pi r + 2r$$

where r = radius of the circle and θ = the angle the ends of the arc makes with the centre of the circle (in degrees).

Area

- The area of a two dimensional shape is the amount of space enclosed by the shape.
- To convert between metric units of area follow the diagram below.

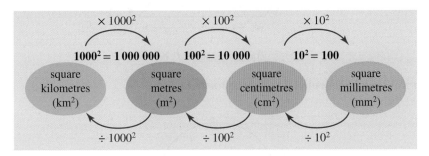

- A common measure of large areas is the hectare (ha) and it can be converted as shown.

$$1\,\text{ha} = 10\,000\,\text{m}^2$$
$$100\,\text{ha} = 1\,\text{km}^2$$

- The area of a rectangle can by calculated by multiplying its length by its width.

$$A = l \times w$$
$$\text{where: } l = \text{length}$$
$$w = \text{width}$$

- The area of a triangle, trapezium and parallelogram can be calculated as shown.

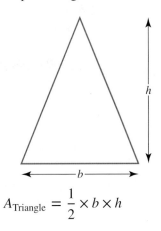

$$A_{\text{Triangle}} = \frac{1}{2} \times b \times h$$

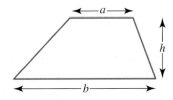

$$A_{\text{Trapezium}} = \frac{1}{2}(a + b) \times h$$

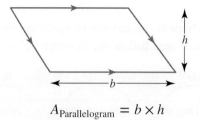

$$A_{\text{Parallelogram}} = b \times h$$

- The area of a circle can be calculated using the formula:

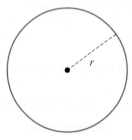

$$A_{\text{Circle}} = \pi \times r^2$$
where r = radius of the circle.

- The area of a sector can be calculated using the formula:

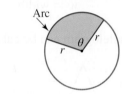

$$\text{Area of a sector } = \frac{\theta}{360} \times \pi r^2$$

where:

r = radius of the circle
θ = angle that the sector makes with the centre of the circle, in degrees.

- Composite shapes can be made by adding regular shapes together.
- Composite shapes can be made by subtracting one shape from another to find its shaded region.

Total surface area

- The total surface area (TSA) of a 3D object is the total area of each outer face of that object.
- The total surface area of a 3D object can be calculated by finding the area of each individual face and add them together.
- As with area, total surface area is measured using squared units (mm^2, cm^2, m^2, or km^2).
TSA of a cube

$$\text{TSA}_{\text{Cube}} = 6l^2$$

TSA of a rectangular prism

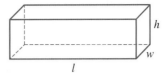

$$\text{TSA}_{\text{Rectangular prism}} = 2\left(lh + lw + wh\right)$$

TSA of a cylinder

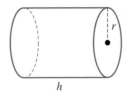

$$\text{TSA}_{\text{Cylinder}} = 2\pi rh + 2\pi r^2$$

TSA of a triangular prism

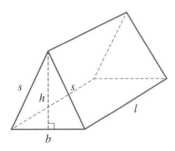

$$\text{TSA}_{\text{Triangular prism}} = bh + 2ls + bl$$

TSA of a sphere

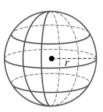

$$\text{TSA}_{\text{Sphere}} = 4\pi r^2$$

TSA of a cone

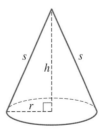

$$\text{TSA}_{\text{Cone}} = \pi rs + \pi r^2$$

The slant-height (s) is the measurement along the side of the cone, not the vertical height (h) of the cone.

TSA of a pyramid

The total surface area of a pyramid can be calculated by adding the area of the base to the area of each of its sides. For example,

$$TSA_{Square\text{-}based\,pyramid} = A_{Base} + 4 \times A_{Sides}$$

This method can be adapted to rectangular-based pyramids and triangular-based pyramids.

- The total surface area of combined 3D shapes is the area of all the outside surfaces (i.e. the surfaces that can touched).
- Be careful when combining shapes that you don't add the surfaces that are no longer an outside surface when the shapes are combined.

Exercise 3.9 Review: exam practice

Simple familiar

1. **MC** The perimeter of the following shape is closest to
   - **A.** 174.35 cm.
   - **B.** 208 cm.
   - **C.** 3.10 m.
   - **D.** 308 cm.

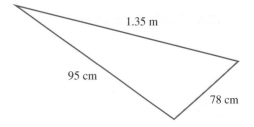

2. **MC** The perimeter of the following, correct to two decimal places, is
   - **A.** 10.80 cm.
   - **B.** 22.62 cm.
   - **C.** 24.16 cm.
   - **D.** 16.96 cm.

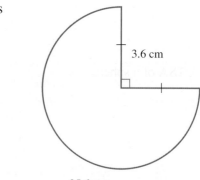

3. **MC** The area of the rectangle shown is
   - **A.** 47.36 cm².
   - **B.** 4.736 cm².
   - **C.** 8.82 cm².
   - **D.** 4.367 mm².

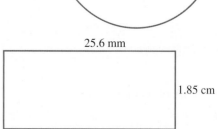

4. **MC** The area of the shape shown is
   - **A.** 33.4 cm².
   - **B.** 69.7 cm².
   - **C.** 139.4 cm².
   - **D.** 48.175 cm².

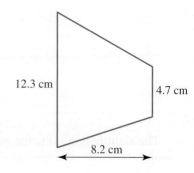

5. **MC** The shaded area of the shape shown to two decimal places is
   A. 162.86 mm^2.
   B. 155.52 mm^2.
   C. 27.14 mm^2.
   D. 135.72 mm^2.

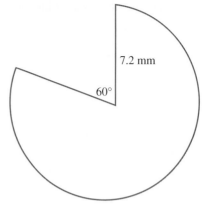

6. **MC** The area of the combined shape shown is closest to
   A. 157.856 cm^2.
   B. 225.543 cm^2.
   C. 218.237 cm^2.
   D. 209.56 cm^2.

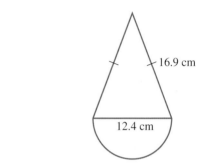

7. **MC** The total surface area of the following rectangular prism is
   A. 221.45 cm^2.
   B. 94.5 cm^2.
   C. 255.55 cm^2.
   D. 271.66 cm^2.

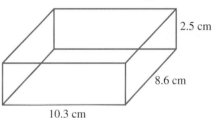

8. **MC** The total surface area of a sphere with a diameter of 5.35 m, to two decimal places is closest to
   A. 359.68 m^2.　　　**B.** 22.48 m^2.　　　**C.** 89.92 m^2.　　　**D.** 33.62 m^2.

9. Calculate the area of the following shapes, correct to two decimal places where appropriate.

   a.

   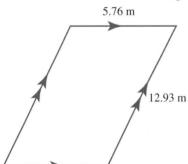

   b.

   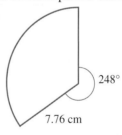

10. Calculate the area of the following shape to three decimal places.

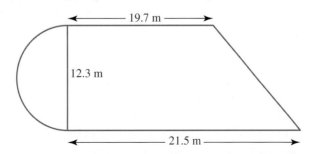

11. Calculate the total surface area of the following shapes to two decimal places.

   a.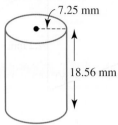
   7.25 mm
   18.56 mm

   b.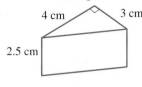
   4 cm   3 cm
   2.5 cm

12. For the given shape:
   a. calculate the perimeter
   b. calculate the area.

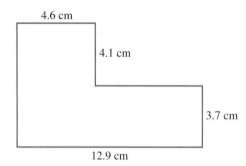
4.6 cm
4.1 cm
3.7 cm
12.9 cm

Complex familiar

13. **MC** The total surface area of the following shape is closest to
   A. $0.78 \, m^2$.
   B. $7804.84 \, cm^2$.
   C. $649.28 \, cm^2$.
   D. $0.65 \, m^2$.

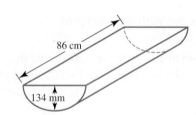
86 cm
134 mm

14. **MC** A cube of length 18 cm has a hole with a 6 cm diameter drilled through its centre. The total surface area of the new shape is closest to
   A. $1866.32 \, cm^2$.
   B. $1548.16 \, cm^2$.
   C. $2226.74 \, cm^2$.
   D. $1944 \, cm^2$.

15. Calculate the total surface area of the following shapes to one decimal place.

   a.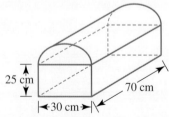
   25 cm
   70 cm
   30 cm

   b.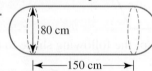
   80 cm
   150 cm

16. A rectangular garden bed has a length of 15.35 m and width 9.52 m. An 85 cm wide path is built on the outside of the garden bed, reducing the size of the garden bed.
   a. Calculate the perimeter of the garden bed before the path was built.
   b. Calculate the perimeter of the garden bed after the path is built.
   c. Determine the area of the garden bed once the path is built.

17. Tennis is played on a rectangular court, with dimensions as shown.

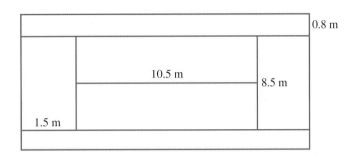

a. Determine the length and width of the tennis court.
b. Calculate the perimeter of the tennis court.
c. To mark all the lines shown with tape on the court, what length of tape would be required?
d. Determine the area of the tennis court.

18. If a rectangular prism can be made from the
    following net, calculate:

    a. the perimeter of the net
    b. the total surface area of the prism.

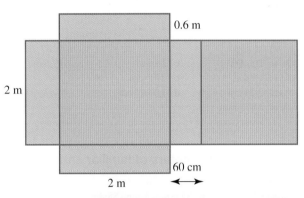

19. An Olympic swimming pool is 50 metres long and
    25 metres wide, by 1.8 metres in depth.
    a. If Haylee swam 25 laps of the pool, how many metres
       did she swim?
    b. If you were to swim the 400-metre medley, how many
       laps would you swim?
    c. If you were to swim the longest Olympic pool event of
       1500 metres, how many laps would you complete?
    d. Determine the surface area of the pool in m^2.
    e. Determine the surface area of the pool in cm^2 .

20. A tin contains four tennis balls, each with a diameter
    of 6.3 cm. The tin has a height of 27.2 cm and a diameter
    of 73 mm.
    a. Determine the circumference of the lid.
    b. Calculate the total surface area of the tin.
    c. Calculate the total surface area of the four tennis balls.
    d. What is the difference between the total surface area of
       the four balls and that of the tin?

Answers

3 Linear and area measure

Exercise 3.2 Units of length and estimation

1. a. Metres
   b. Centimetres
   c. Kilometres
   d. Millimetres
2. a. 283 m   b. 5.2 m   c. 41.8 cm   d. 0.45 km
   e. 5500 m   f. 78.459 m   g. 37 800 mm   h. 0.065 km
3. D
4. D
5. A
6. a. 270 m   b. 73.5 m
   c. 25.7 cm   d. 130 000 cm
7. a. 15.18 m   b. 6.08 m
8. Since Brisbane to Sydney is a long way, you would usually use a large unit such as kilometres.
9. 177 cm
10. a. 2400 cm, 135 m, 0.15 km   b. 120 mm, 25 cm, 0.5 m
    c. 9 m, 10 000 mm, 0.45 km   d. 32 000 cm, 1 km, 1200 m
11. a. 2.0 cm   b. 154 cm   c. 219.5 cm   d. 96.27 cm
12. a. 52 cm 1 mm or 52.1 cm   b. 82 km 103 m or 82.103 km
    c. 41 cm   d. 87 m 70 cm or 87.70 m
13. a. The Q1   b. 25.5 m   c. 619.5 m
14. a. 1.88 m   b. 1.05 cm
    c. 2.21 cm   d. 187.75 cm
    e. 1 875 mm
    f. West Coast (1875 mm) and St. Kilda (1870 mm)

Exercise 3.3 Perimeters of familiar and composite shapes

1. a. 32 cm   b. 32 cm   c. 55.4 m   d. 40.42 mm
2. a. 38.4 cm   b. 29.36 m   c. 71.24 m   d. 46.29 mm
3. a. 34 cm   b. 31 mm   c. 75 m   d. 50 cm
4. a. 41 m   b. 88 m   c. 90 cm   d. 250 cm
5. a. 4.8 m   b. 120 m   c. 200 cm
6. a. 690 cm   b. 3.8 m
7. a. 34.14 cm   b. 5.99 m
8. a. 35.19 cm   b. 3314.38 mm
9. a. 30.85 cm   b. 5.71 m   c. 14.77 cm
10. a. 10.03 m   b. 37.49 cm
11. a. 36.63 cm   b. 149.41 mm
12. a. 9.22 cm   b. 12.36 m   c. 29.55 cm   d. 73.53 cm
13. a. 46 cm   b. 58 m   c. 55 cm
    d. 53.94 cm   e. 40.28 m   f. 81.68 cm
14. 57.11 cm
15. 66.8 cm
16. 15.92 cm
17. 414.69 cm
18. a. 38.20 m   b. 6.30 m

Exercise 3.4 Units of area and estimation

1. a. m^2 b. mm^2 c. cm^2
   d. m^2 e. km^2 f. mm^2
2. a. cm^2 b. m^2 c. ha
   d. mm^2 e. km^2 f. m^2

3. There are many possible answers. Some examples are given.
   a. A computer chip or face of a small diamond
   b. A fridge magnet or dinner plate
   c. A block of land or surface of a concrete slab
   d. Tasmania or Kakadu National Park
4. a. $3400\ mm^2$ b. $1.4250\ m^2$ c. $5.92\ km^2$
5. a. $1800\ cm^2$ b. $79\ 800\ m^2$ c. 37.43 ha
6. a. $230\ mm^2$ b. $25\ 700\ cm^2$ c. $4.7\ cm^2$
   d. $0.027\ km^2$ e. 8.75 ha   f. $1.7\ m^2$
7. a. 5.12 ha   b. $28\ 500\ m^2$ c. $83.8\ km^2$
   d. 1260 ha   e. $2300\ m^2$ f. 62.345 ha
8. a. $48\ 000\ mm^2$ b. $12\ 000\ 000\ cm^2$
   c. $0.3008\ m^2$ d. $0.000\ 000\ 447\ 8\ km^2$
9. a. 5 squares $= 5\ cm^2$ b. 9 squares $= 9\ cm^2$
   c. 10 squares $= 10\ cm^2$ d. 15 squares $= 15\ cm^2$
10. a. $54\ cm^2$ b. $7.8\ mm^2$
    c. $323.57\ m^2$ d. $95.0625\ cm^2$
11. a. i. $34\ 220\ cm^2$ ii. $3.422\ m^2$
    b. i. $33.930\ 625\ cm^2$ ii. $0.003\ 393\ 062\ 5\ m^2$
    c. i. $20\ 400\ cm^2$ ii. $2.04\ m^2$
    d. i. $3375\ cm^2$ ii. $0.3375\ m^2$
12. a. 18 squares   b. 25 squares   c. 15 squares
13. $108\ m^2$
14. Answers will vary. Examples are:
    a. $200\ m^2$ b. $450\ m^2$ c. $450\ m^2$ d. $2\ m^2$
15. 20.12 m
16. a. $80\ cm^2$
    b. Width $= 16$ cm, Length $= 20$ cm
    c. $320\ cm^2$
    d. Multiplied by 4 or quadrupled
17. a.

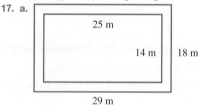

    b. $350\ m^2$
    c. $172\ m^2$
    d. $1290
18. a. $26.56\ m^2$ b. $79.68\ m^2$ c. 16 litres   d. $252

Exercise 3.5 Areas of regular shapes

1. a. $20\ cm^2$ b. $48\ m^2$ c. $75\ cm^2$ d. $48\ cm^2$
2. C
3. a. $75\ cm^2$ b. $13.75\ m^2$ c. $30\ cm^2$ d. $351\ mm^2$
4. D
5. a. $35\ cm^2$ b. $90\ cm^2$ c. $52\ mm^2$ d. $11.524\ cm^2$
6. a. $17.76\ m^2$ b. $37.5\ mm^2$
7. a. $78.54\ cm^2$ b. $66.48\ cm^2$ c. $122.72\ m^2$ d. $128.68\ m^2$
8. $251.33\ cm^2$
9. a. $3279.82\ cm^2$ b. $96.10\ cm^2$
10. $113.10\ cm^2$
11. $14.58\ m^2$
12. a. $84.82\ cm^2$ b. $99.53\ m^2$
13. B
14. a. Small $= 314.16\ cm^2$, Medium $= 706.86\ cm^2$, Large $= 1256.64\ cm^2$
    b. $113\ 883\ cm^2$
    c. $157.08\ cm^2$

15. a. 6800 m^2 b. 3400 sheets of turf
     c. \$35 700
16. a. 465.125 m^2 b. 2774 m^2
     c. 1860.5 m^2 d. \$17 674.75

Exercise 3.6 Areas of composite figures

1. a. Parallelogram and half a circle
   b. Square and four half circles
   c. Trapezium and rectangle
   d. Triangle and half a circle
2. 120 cm^2
3. 23.03 cm^2
4. a. 48 m^2 b. 154 mm^2
   c. 272 cm^2 d. 384 mm^2
5. a. 29.225 cm^2 b. 25.63 cm^2
   c. 756 m^2 d. $42 651.41 \text{ mm}^2$
6. a. 40.18 mm^2 b. 31.49 cm^2
7. a. 360 cm^2
   b. 360 cm^2
   c. They are the same as you would expect.
8. 9600 mm^2
9. 156 cm^2
10. 109.87 cm^2
11. a. 4560 cm^2
    b. 3.488 m^2
    c. Need 1 litre of paint so it will cost \$24.95
12. 28.54 cm^2
13. a. 2025 cm^2 b. 943.41 cm^2 c. 6031.59 cm^2

Exercise 3.7 Surface areas of familiar prisms

1. a. Prism   b. Not a prism
   c. Not a prism   d. Prism
2. a. 96 cm^2 b. 937.5 mm^2
   c. 4.335 m^2 d. 360.375 cm^2
3. 516 cm^2
4. a. $14 832 \text{ cm}^2$ b. 4.16 m^2
   c. 1177 mm^2 d. 28.96 cm^2
5. a.   i. 30 cm^2
       ii. $104 \text{ cm}^2, 40 \text{ cm}^2, 96 \text{ cm}^2$
       iii. 300 cm^2
   b.   i. 26.82 cm^2
       ii. $147.6 \text{ cm}^2, 147.6 \text{ cm}^2, 166.05 \text{ cm}^2$
       iii. 514.89 cm^2
6. a. 2.30 m^2 b. 1523.04 mm^2
   c. 848.23 cm^2 d. $36 128.32 \text{ cm}^2$
7. 100.53 cm^2
8. 8 m
9. 66.48 m^2
10. a. 30.19 mm^2 b. 1963.50 cm^2

11. 113.10 cm^2
12. A
13. a. $72 452.98 \text{ mm}^2$ b. $724.5 298 \text{ cm}^2$
14. 254.85 cm^2

Exercise 3.8 Surface areas of familiar pyramids and irregular solids

1. 31.67 m^2
2. 2454.21 mm^2
3. a. 125.66 cm^2 b. 220.35 m^2
4. 354.39 cm^2
5. 541.01 cm^2
6. a. 261.91 cm^2 b. 179.97 m^2
7. 132.58 cm^2
8. 1367.60 cm^2
9. 344.3 cm^2
10. 3552.95 mm^2
11. 791.68 cm^2
12. 1357.17 cm^2
13. a. 2324.78 cm^2 b. 10.65 m^2
14. 326.73 cm^2
15. 20.30 m^2
16. 1513.36 cm^2

3.9 Review: exam practice

1. D
2. C
3. B
4. B
5. D
6. A
7. D
8. C
9. a. 74.48 m^2 b. 58.86 cm^2
10. 312.791 m^2
11. a. 1175.73 cm^2 b. 42 cm^2
12. a. 41.1 cm b. 66.59 cm^2
13. D
14. C
15. a. $11 105.53 \text{ cm}^2$ b. $57 805.30 \text{ cm}^2$
16. a. 49.74 m b. 42.94 m c. 106.743 m^2
17. a. 13.5 m and 10.1 m   b. 47.2 m
    c. 101.7 m d. 136.35 m^2
18. a. 16.8 m b. 12.8 m^2
19. a. 1250 m b. 8 laps   c. 30 laps
    d. 1520 m^2 e. $15 200 000 \text{ cm}^2$
20. a. 22.93 cm b. 707.50 cm^2
    c. 498.76 cm^2 d. 208.74 cm^2

4 Volume, capacity and mass

4.1 Overview

LEARNING SEQUENCE

4.1 Overview
4.2 Units of volume and capacity
4.3 Volume and capacity of prisms and cylinders
4.4 Volume and capacity of pyramids and spheres
4.5 Problems involving mass
4.6 Review: exam practice

CONTENT

In this chapter, students will learn to:
- use metric units of volume (cubic millimetres, cubic centimetres, cubic metres), their abbreviations $\left(mm^3, cm^3, m^3\right)$, conversions between them and appropriate choices of units
- understand and use the relationship between volume and capacity, recognising that $1\,cm^3 = 1\,mL$ (millilitre), $1000\,cm^3 = 1\,L$ (litre), $1\,m^3 = 1\,kL$ (kilolitre), $1000\,kL = 1ML$ (megalitre)
- estimate volume and capacity of various objects
- calculate the volume and capacity of regular objects, including cubes, rectangular and triangular prisms, and cylinders
- calculate the volume and capacity of right pyramids, including square-based and rectangular-based pyramids, and spheres
- use metric units of mass (milligrams, grams, kilograms, metric tonnes), their abbreviations $\left(mg, g, kg, t\right)$, conversions between them and appropriate choices of units
- estimate the mass of different objects
- recognise the need for milligrams.

Fully worked solutions for this chapter are available in the Resources section of your eBookPLUS at www.jacplus.com.au.

4.2 Units of volume and capacity

The **volume** of a three-dimensional object is the amount of space that is occupied inside it. This can be measured by counting the number of cube-shaped units that will fit into the space that the object occupies. Volume can be used to describe the size of a pool or the space inside a room as examples.

4.2.1 Metric units of volume

The units commonly used to describe the volume of an object are the cubic millimetre (mm^3), cubic centimetre (cm^3) and cubic metre (m^3). One cubic centimetre is the space occupied by a cube with side lengths equal to 1 cm.

One cubic metre is the space occupied by a cube with side lengths equal to 1 m; however, something with a volume of one cubic metre can take many different shapes.

Metric units of volume can be converted according to the diagram below:

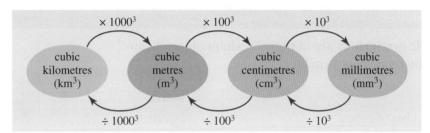

A die has a volume of about $1 \, cm^3$.

The volume of an iPhone is about $78 \, cm^3$.

The spa-bath holds about 1 m³.

WORKED EXAMPLE 1

How many cubic centimetres are in the solid shape shown below?
(Each small cube represents 1 cm³.)

THINK

1. Count the number of cubes.
   There are 12 cubes in each layer, and there are
   2 layers.

2. Each cube has a volume of 1 cm³.
   Express the volume in terms of cm³.

WRITE

$$\text{Number} = 2 \times 12$$
$$= 24 \text{ cubes}$$

$$\text{Volume} = 24 \text{ cm}^3$$

Capacity

If a 3-dimensional object is hollow, it can hold another substance. In this case the
volume is also referred to as **capacity**. Capacity is usually used as a measure of the volume
of liquid a container can hold. The units to describe capacity include the millilitre (mL),
litre (L), kilolitre (kL) and megalitre (ML). For example, 1 litre occupies a space
of 1000 cm³.

Units of capacity are related as shown below.

$$1 \text{ L} = 1\,000 \text{ mL}$$
$$1 \text{ kL} = 1\,000 \text{ L}$$
$$1 \text{ ML} = 1\,000\,000 \text{ L}$$

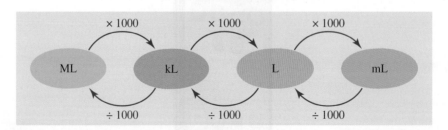

When calculating the capacity of an object, it is sometimes useful to calculate the volume of the object
first, then convert to units of capacity.

The metric unit, $1\,cm^3$, is defined as having a capacity of $1\,mL$. Therefore, volume and capacity units are related as shown.

$$1\,cm^3 = 1\,mL$$
$$1000\,cm^3 = 1000\,mL = 1\,L$$
$$1\,m^3 = 1000\,L = 1\,kL$$

WORKED EXAMPLE 2

Complete the following unit conversions.
a. 70 mL = _____ cm^3   **b. 530 mL = _____ L**   **c. 0.382 L = _____ cm^3**

THINK

a.   There is $1\,cm^3$ in each $1\,mL$.

b.   There are $1000\,mL$ in $1\,L$, so to convert millilitres to litres divide by 1000.

c. 1. There are $1000\,mL$ in $1\,L$, so to convert litres to millilitres multiply by 1000.

   2. Since $1\,mL = 1\,cm^3$.

WRITE

a. $1\,mL = 1\,cm^3$
   Therefore: $70\,mL = 70\,cm^3$

b. $530\,mL = (530 \div 1000)\,L$
         $= 0.53\,L$

c. $0.382\,L = (0.382 \times 1000)\,mL$
            $= 382\,mL$
   $382\,mL = 382\,cm^3$

 **Resources**

📄 **Digital document** SkillSHEET Conversion of volume units (doc-5239)

Exercise 4.2 Units of volume and capacity

1. What does volume measure?
2. Choose the appropriate unit (mm^3, cm^3, m^3) to measure the following volumes.
   a. The volume of a Big M milk carton
   b. The volume a diamond on a ring
   c. The volume of water in a swimming pool
   d. The volume of a basketball
3. **WE1** How many cubic centimetres are in the following solid shapes, given each cube represents $1\,cm^3$?

   a.
   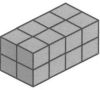

   b.

   c.

   d.

4. How many cubic centimetres are in the following solid shapes?
   a. (Each small cube represents 2 cm^3.)

   b. (Each small cube represents 2 m^3.)

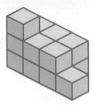

5. Determine the volume of the following shapes, given each cube represents 1 cm^3.
   a.
   b.

   c.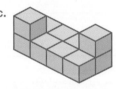
   d.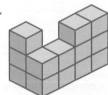

6. Determine the volume of the following shapes, given each cube represents 4 cm^3.
   a.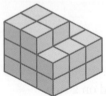
   b.

   c.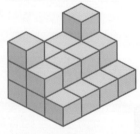
   d.

7. State the difference between capacity and volume.

8. Estimate the volume and capacity of each of the following.
    a. A bath
    b. An Olympic-size swimming pool
    c. Five cans of Coke
    d. Your classroom

9. State whether the following measures are a measure of volume or capacity.
    a. 35 mL
    b. 120 m^3
    c. 1.2 cm^3
    d. 1750 kL
    e. 432 mm^3
    f. 97.37 L

10. Complete the following unit conversions.
    a. 12 L = _____ mL
    b. 3125 L = _____ ML
    c. 397 mL = _____ L
    d. 0.0078 L = _____ mL
    e. 4893 mL = _____ L
    f. 36.97 L = _____ ML

11. **WE2** Complete the following unit conversions.
    a. 372 cm^3 = _____ mL
    b. 1630 L = _____ cm^3
    c. 3.4 L = _____ cm^3
    d. 0.38 mL = _____ cm^3
    e. 163 L = _____ cm^3
    f. 49.28 cm^3 = _____ mL

12. Complete the following unit conversions.
    a. 578 mL = _____ L
    b. 750 mL = _____ L
    c. 0.429 L = _____ mL

13. Complete the following unit conversions.
    a. 47 cm^3 = _____ mL
    b. 3594 cm^3 = _____ L
    c. 0.0042 L = _____ cm^3

14. If the following shapes were broken up into 1-cm cubes:
    i. determine how many 1-cm cubes fit inside the shape
    ii. calculate the volume of the shape in cm^3
    iii. determine the shape's full capacity in millilitres.

    a.
    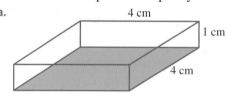
    4 cm, 1 cm, 4 cm

    b.
    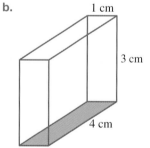
    1 cm, 3 cm, 4 cm

    c.
    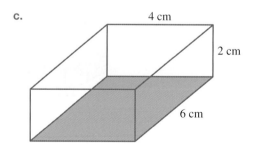
    4 cm, 2 cm, 6 cm

    d.
    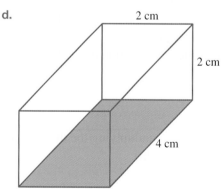
    2 cm, 2 cm, 4 cm

15. A rectangular swimming pool is 12 metres long, 4 metres wide and 1.5 metres deep.
    a. How many $1 \, cm^3$ would fit in the pool?
    b. What is the volume of the pool?
    c. What is the capacity of the pool, in litres, if it is full?
    d. What is the capacity of the pool when it is 80% full?

16. The milk container shown has dimensions of 10 cm by 5 cm by 20 cm.
    a. How many $1 \, cm^3$ would fit in the milk container?
    b. What is the volume of the container?
    c. What is the capacity of the container in litres?
    d. How much milk is in the carton if it is $\dfrac{3}{4}$ full, in mL?

4.3 Volume and capacity of prisms and cylinders

4.3.1 Volume of prisms

A prism has the same cross-sectional area along its length, thus it has the same two end faces. The cross-sectional area is shaded in blue in the diagram shown.

The volume of a prism is calculated by multiplying the cross-sectional area by its height.

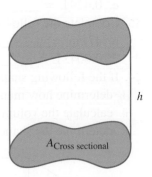

Volume of a prism

$$V_{\text{Prism}} = A_{\text{Cross sectional}} \times h$$

Volume of a rectangular prism

The volume (V) of a rectangular prism follows the general rule of all prisms, hence multiply the cross-sectional area by its height. Taking the bottom face as the cross-sectional area then:

$$V = A_{\text{Cross sectional}} \times h$$
$$V = (l \times w) \times h$$

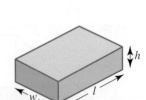

This is the same as multiplying the three dimensions of the prism (length, width and height).

> **Volume of a rectangular prism**
> $$V_{\text{Rectangular prism}} = \text{length} \times \text{width} \times \text{height}$$
> $$V_{\text{Rectangular prism}} = l \times w \times h$$

WORKED EXAMPLE 3

Calculate the volume of the rectangular prism shown.

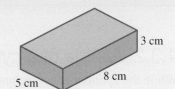

THINK	WRITE
1. Write the volume of a rectangular prism formula.	$V = l \times w \times h$
2. Substitute the values length $= 8\,$cm, width $= 5\,$cm and height $= 3\,$cm into the formula.	$V = 8 \times 5 \times 3$
3. Calculate the volume and state the units.	$V = 120\,\text{cm}^3$

Volume of a triangular prism

The volume of a triangular prism can also be calculated by multiplying the cross-sectional area of the triangular face by its height.

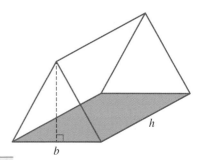

> **Volume of a triangular prism**
> $$V_{\text{Triangular prism}} = A_{\text{Cross sectional}} \times h$$
> $$= \text{area of triangle} \times h$$

WORKED EXAMPLE 4

Calculate the volume of the triangular prism shown.

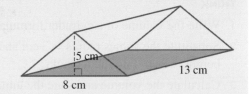

THINK	WRITE
1. Write the volume of a triangular prism formula.	$V = \text{area of triangle} \times h$

▶

2. Recall that the area of a triangle is $A = \dfrac{1}{2} \times$ base $\times$ height. Be careful not to confuse the height of the triangle with the height of the entire object.

$$V = \left(\dfrac{1}{2} \times 8 \times 5\right) \times 13$$

3. Calculate the volume and state the units.

$$V = 260\ \text{cm}^3$$

4.3.2 Volume of a cylinder

While the cylinder is not technically classed as a prism, its volume can be calculated in the same way as a prism since its cross-sectional area (the circle) is along its length. Its volume is calculated by:

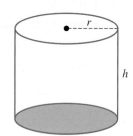

$$V = A_{\text{Cross sectional}} \times \text{height}$$
$$V = A_{\text{Circle}} \times h$$
$$= \pi r^2 \times h$$
$$= \pi r^2 h$$

> **Volume of a cylinder**
> $$V_{\text{Cylinder}} = \pi r^2 h$$
> where: r = radius
> h = height

WORKED EXAMPLE 5

Calculate the volume of the cylinder shown to two decimal places.

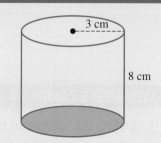

THINK	WRITE
1. Write the volume of a cylinder formula.	$V = \pi r^2 h$
2. Substitute the values radius = 3 cm and height = 8 cm into the formula.	$V = \pi \times 3^2 \times 8$ $= \pi \times 9 \times 8$
3. Calculate the volume and state the units.	$V = 226.19\ \text{cm}^3$

 **Resources**

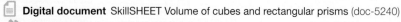

Digital document SkillSHEET Volume of cubes and rectangular prisms (doc-5240)
Interactivity Volume of solids (int-3794)

Exercise 4.3 Volume and capacity of prisms and cylinders

1. Determine the volume of the following prisms.

a.
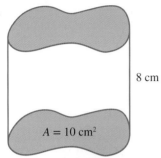
8 cm

$A = 10 \text{ cm}^2$

b.
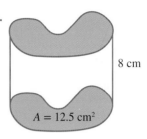
8 cm

$A = 12.5 \text{ cm}^2$

c.
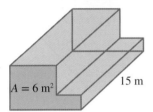
$A = 6 \text{ m}^2$
15 m

d.
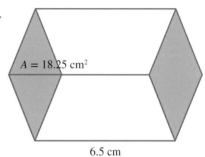
$A = 18.25 \text{ cm}^2$

6.5 cm

2. **WE3** Calculate the volume of the following rectangular prisms.

a.
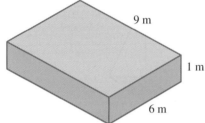
9 m
1 m
6 m

b.
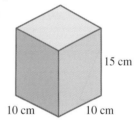
15 cm
10 cm   10 cm

c.
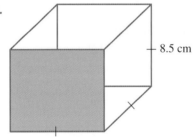
8.5 cm

d.
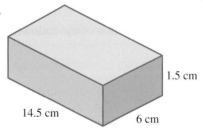
1.5 cm
14.5 cm   6 cm

3. Calculate the volume of the following prisms.

a.
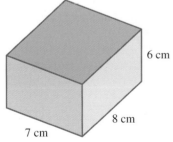
6 cm
8 cm
7 cm

b.
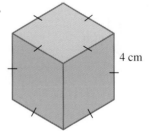
4 cm

4. **WE4** Calculate the volume of the following triangular prisms.

a.

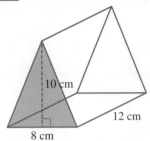

10 cm
12 cm
8 cm

b.

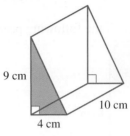

9 cm
10 cm
4 cm

c.

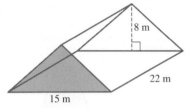

8 m
22 m
15 m

d.

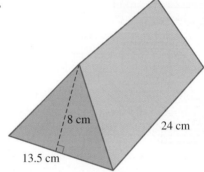

8 cm
24 cm
13.5 cm

5. Calculate the volume of the following triangular prisms.

a.

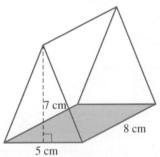

7 cm
8 cm
5 cm

b.

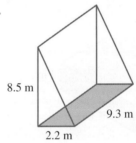

8.5 m
9.3 m
2.2 m

6. **WE5** Calculate the volume of the following cylinders correct to two decimal places.

a.

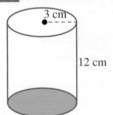

3 cm
12 cm

b.

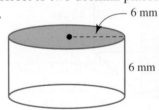

6 mm
6 mm

c.

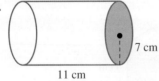

7 cm
11 cm

d.

2.5 m
0.75 m

7. Calculate the volume of the cylinder to two decimal places.

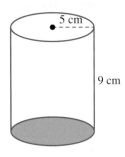

5 cm

9 cm

8. Calculate the volume of the following shape to two decimal places.

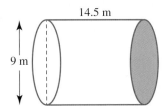

14.5 m

9 m

9. Calculate the volume of the following shapes correct to two decimal places.

a.

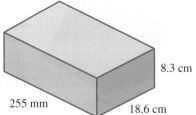

8.3 cm

255 mm

18.6 cm

b.

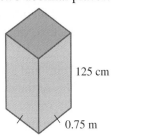

125 cm

0.75 m

c.

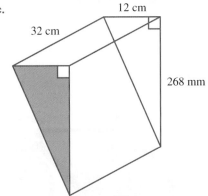

12 cm

32 cm

268 mm

d.

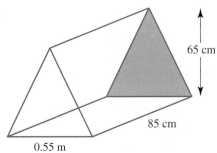

65 cm

85 cm

0.55 m

e.

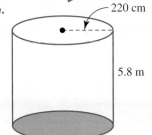

220 cm

5.8 m

f.

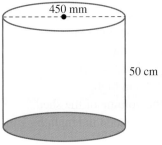

450 mm

50 cm

10. A pool is made in the shape of an open top rectangular prism of length 12 m, width 5 m and depth 1.85 m. What is the total volume of the pool in cubed metres, and how many litres does it hold when full?

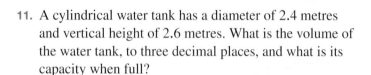

11. A cylindrical water tank has a diameter of 2.4 metres and vertical height of 2.6 metres. What is the volume of the water tank, to three decimal places, and what is its capacity when full?

12. A tent is made in the shape of a triangular prism. If the tent has a vertical height of 2.2 metres, a width of 3.6 metres and a length of 4.2 metres, what is the volume of the inside of the tent?

13. A cylinder has a volume of 245 cm³ and a height of 6 cm. Determine the length of the radius of the cylinder to two decimal places.

14. A garden water feature has a base in the shape of a rectangular prism with dimensions length = 2 metres, width = 1.5 metres and height = 40 cm. Calculate the following.
    a. The volume of the base of the water feature
    b. The volume of water in cm³ in the base if it is filled to a height of 30 cm
    c. The capacity of the water feature base when filled to a height of 25 cm in litres

15. A road bike is packed to be sent interstate in a rectangular prism-shaped box with dimensions 2 m by 85 cm by 40 cm. If the bike takes up 30% of the volume of the box, answer the following.
    a. Determine the volume of the box in m³.
    b. Determine the volume taken up by the road bike.
    c. After the bike was placed inside the box, the rest of the box was filled with packing foam. Determine the volume of packing foam that was used.

16. A rectangular shed has dimensions of 5.5 m long by 5.5 m wide and 2.5 m high.
    a. Determine the volume of the shed.
    b. If the shed is extended in length by 2 metres, what is its new volume?
    c. If the width of the shed is extended so that the volume of the shed is now 100 m³, what is the new width of the shed?

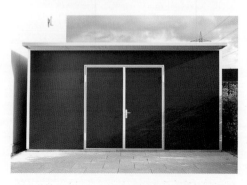

4.4 Volume and capacity of pyramids and spheres

4.4.1 Volume of pyramids

A pyramid has a base and triangular-shaped sides that meet at the top. A pyramid can have any polygon as the base, with the most common bases being squares, rectangles and triangles. These are the pyramids we will look at in this chapter. The general rule for the volume of a pyramid is as follows.

Volume of a pyramid

$$V_{\text{Pyramid}} = \frac{1}{3} \times \text{area of base} \times \text{height}$$

$$V_{\text{Pyramid}} = \frac{1}{3}Ah$$

This general formula can be adapted to square-, rectangular- and triangular-based pyramids.

Rectangular- and square-based pyramids

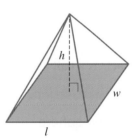

Volume of a rectangular-based pyramid

$$V_{\text{Rectangular pyramid}} = \frac{1}{3} \times l \times w \times h$$

The only difference between a rectangular-based pyramid and a square-based pyramid is that the lengths and widths are equal for the square-based pyramid. Therefore:

Volume of a square-based pyramid

$$V_{\text{Square pyramid}} = \frac{1}{3} \times l^2 \times h$$

Calculate the volume of the following pyramids.

a.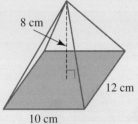

8 cm

12 cm

10 cm

b.

1.2 m

1.4 m

THINK	WRITE
a. 1. Write the formula for the volume of a rectangular-based pyramid.	a. $V_{\text{Rectangular pyramid}} = \frac{1}{3} \times l \times w \times h$
2. Substitute the values length = 12 cm, width = 10 cm and height = 8 cm into the formula.	$V = \frac{1}{3} \times 12 \times 10 \times 8$
3. Calculate the volume and state the units.	$V = 320 \text{ cm}^3$
b. 1. Write the formula of the volume of a square-based pyramid.	b. $V_{\text{Square pyramid}} = \frac{1}{3} \times l^2 \times h$
2. Substitute the values length = 1.4 cm and height = 1.2 cm into the formula.	$V = \frac{1}{3} \times 1.4^2 \times 1.2$
3. Calculate the volume and state the units.	$V = 0.784 \text{ m}^3$

4.4.2 Volume of spheres

A sphere is a 3-dimensional circular shape such as a ball and the volume of a sphere
can be calculated by:

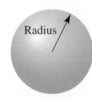

Radius

> **Volume of a sphere**
> $$V_{\text{Sphere}} = \frac{4}{3} \pi r^3$$
> where: r = radius

Calculate the volume of the following sphere.

12 cm

THINK	WRITE
1. Write the volume of a sphere formula.	$V_{\text{Sphere}} = \frac{4}{3} \pi r^3$
2. Substitute $r = 12$ cm into the formula.	$V = \frac{4}{3} \times \pi \times 12^3$
3. Calculate the volume and state the units.	$V = 7238.23 \text{ cm}^3$

Exercise 4.4 Volume and capacity of pyramids and spheres

1. Calculate the volume of the following rectangular-based pyramids, to two decimal places if appropriate.
   a. Base length = 4 cm, base width = 5 cm and height = 6 cm
   b. Base length = 12 cm, base width = 7 cm and height = 14 cm
   c. Base length = base width = 6.5 cm and height = 10.2 cm
   d. Base length = base width = 15.5 cm and height = 20.5 cm
   e. Base length = 2.8 m, base width = 4.2 m and height = 6.9 m
   f. Base length = 1.8 m, base width = 80 cm and height = 1.5 m
2. Calculate the volume of the following spheres to two decimal places.
   a. Radius of 5 cm
   b. Radius of 12 cm
   c. Diameter of 1.8 m
   d. Diameter of 27 cm
   e. Radius of 24.8 cm
   f. Diameter of 32.5 m
3. **WE6a** Calculate the volume of the pyramid shown.

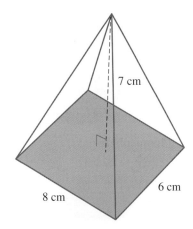

7 cm

8 cm

6 cm

4. **WE6b** Calculate the volume of the pyramid shown.

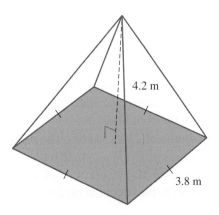

4.2 m

3.8 m

5. Calculate the volume of the following pyramids.

a.

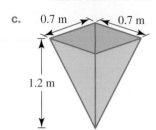

15.3 mm

Area of base = 45 mm²

b.

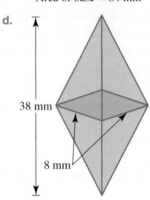

22 mm

Area of base = 84 mm²

c. 0.7 m   0.7 m

1.2 m

d.

38 mm

8 mm

6. **WE7** Calculate the volumes of the following spheres.

a.
17 mm

b.
8.2 mm

c.
38 mm

d.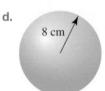
8 cm

e.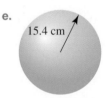
15.4 cm

7. Calculate the volume of the following hemisphere.

32 cm

8. If a square-based pyramid has a height of 12 cm and a volume of 128 cm³, then calculate the length of the square base.

9. A sphere has a volume of 523.81 cm³. What is the radius of the sphere?

10. A spherical container with a radius of 15 cm is filled with water.
    a. Calculate the volume of the sphere, correct to two decimal places.
    b. Determine the total capacity of the sphere, in litres, correct to two decimal places.
    c. If a tap at the bottom of the container allows the water to be emptied at a rate of 80 cm³/s, how long would it take to completely empty the container?

11. **MC** A hemispherical bowl has a volume of 2145.52 cm³. The diameter of the bowl is closest to
    A. 22.63 cm.   **B.** 8.0 cm.   **C.** 20 cm.   **D.** 512 cm.

12. A hemisphere of diameter 24 cm sits perfectly on top of a cylinder of height 46 cm. Calculate the volume of the composite shape correct to two decimal places.

13. Calculate the volume of the shape shown, correct to two decimal places.

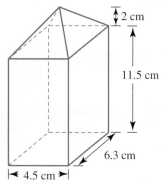

4.5 Problems involving mass

4.5.1 Units of mass

Mass describes how much matter makes up an object, and its standard unit of measurement is the kilogram (kg).

The mass of an object can be measured in milligrams (mg), grams (g), kilograms (kg) and tonnes (t). Just as with length and area, it is important to use the appropriate unit to measure the mass of an object.

Milligrams (mg)

Milligrams are used to measure the mass of a light object such as a strand of hair. The mass of a strand of hair is around 1 mg.

Gram (g)

There are 1000 milligrams in 1 gram. Grams are commonly used when measuring non-liquid cooking ingredients. The mass of 1 cup of flour is about 150 g.

Kilograms (kg)

There are 1000 grams in 1 kilogram. Kilograms are commonly used to measure larger masses such as the mass of a person.

Tonne (t)

There are 1000 kilograms in 1 tonne. Tonnes are used to measure the mass of very large objects such as a truck or boat.

4.5.2 Conversion between units of mass

The relationship between the units of mass can be used to change a measurement from one unit to another.

$$1000 \, \text{mg} = 1 \, \text{g}$$
$$1000 \, \text{g} = 1 \, \text{kg}$$
$$1000 \, \text{kg} = 1 \, \text{t}$$

The diagram shown can be used to assist in the conversion between different units of mass.

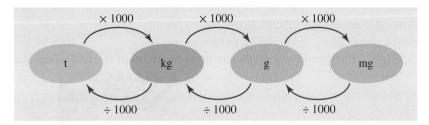

WORKED EXAMPLE 8

Convert the following mass measurements to the units specified.
a. 15.86 kg to grams   **b. 13 650 mg to kilograms**   **c. 0.0071 t to kilograms**

THINK	WRITE
a. 1. To convert from kilograms to grams we need to multiply by 1000.	**a.** $15.86 \text{ kg} = (15.86 \times 1000) \text{ g}$ $= 15\,860 \text{ g}$
b. 1. To convert from milligrams to kilograms we need to divide by 1000 to convert to grams first.	**b.** $13\,650 \text{ mg} = (13\,650 \div 1000) \text{ g}$ $= 13.65 \text{ g}$
2. Once converted to grams then we divide by 1000 again to convert to kilograms.	$13.65 \text{ g} = (13.65 \div 1000) \text{ kg}$ $= 0.013\,65 \text{ kg}$
c. 1. To convert from tonnes to kilograms we need to multiply by 1000.	**c.** $0.0071 \text{ t} = (0.0071 \times 1000) \text{ kg}$ $= 7.1 \text{ kg}$

Exercise 4.5 Problems involving mass

1. Which unit of mass would be best used to describe the mass of each of the following?
   a. A full suitcase checked in for a flight   b. A grass seed
   c. A family car   d. A calculator
   e. A bag of apples   f. A brick
2. Estimate the mass of each of the following objects.
   a. A football   b. A maths textbook

   c. A Blu-ray cover   d. A computer printer

   e. Your pen   f. A can of soft drink
3. Research your answers to question 2 and compare the actual values to your estimated values.

4. **WE8** Convert the following mass measurements to the units specified.
   a. 18.9 kg to grams
   b. 76 490 mg to kilograms
   c. 0.083 t to kilograms

5. Convert the following mass measurements to the units specified.
   a. 717 mg to grams.
   b. 3867 g to tonnes.
   c. 0.0084 tonnes to grams.

6. From your knowledge of household essentials, estimate the mass of the following items.

   a.

   b.

   c.

   d.

7. Change the following mass measurements to the units specified.
   a. 2.4 g to milligrams
   b. 46.7 t to kilograms
   c. 2510 mg to grams
   d. 82 g to kilograms
   e. 0.03 kg to grams
   f. 7893 g to milligrams

8. Change the following mass measurements to the units specified.
   a. 25 781 mg to kilograms
   b. $\frac{7}{8}$ kg to milligrams
   c. 384.2 kg to tonnes
   d. 0.075 kg to milligrams
   e. 0.000 004 5 tonnes to grams
   f. 33 456 milligrams to kilograms

9. For Noah's party, 8 blocks of the chocolate shown were purchased.
   a. What is the total mass of the 8 blocks of chocolate?
   b. The mass of the chocolate is shown on the block. What does the term **net** mean?

10. A family of 5 (2 parents and 3 children) is going on a holiday to Disneyland. The allowable limit for their luggage is 23 kg per person. They pack light with each parent taking 18 kg of luggage and each of the 3 children taking 12 kg. They hope to do some shopping during their holiday. Calculate the maximum mass of their shopping so that they don't exceed the luggage allowance.

11. Christiana decided to go shopping to get some fruit and vegetables for the weekend. Complete the table shown to calculate her total cost to the nearest cent.

Item	Cost per kg	Cost
1.5 kg of bananas	$3.50	$1.5 \times \$3.50 = \$5.25$
3.4 kg of apples	$3.95	
1.2 kg of carrots	$2.80	
2.3 kg of grapes	$7.90	
Total cost		

12. If the mass of Rob is 94.4 kg and his son Jack is 67.8 kg, what is the difference in their masses?

13. A group of contestants entered a weight-loss competition. Given the information in the table, complete the table and answer the following questions.

Name	Original mass (kg)	Loss	New mass (kg)
Matthew	112.3	16.7 kg	
Jack	133.8	24% of his mass	
Sarah	93.6	$\dfrac{1}{6}$ of her mass	
Jane	88.3	Jane now weighs $\dfrac{7}{9}$ of her original mass	
Paul	105.6	13.9 kg	
Kelly	96.1	17.5% of her mass	

a. Who lost the most mass and how much was it?
b. Who lost the least mass and how much was it?
c. Put the contestants in descending order of their new mass.
d. What was the total mass loss of the six contestants?

14. The mass of an object is constant, but its weight changes depending on gravity. To calculate the weight (W) of an object, you multiply its mass (m) by the value of the acceleration due to gravity (g). On the moon the acceleration due to gravity is about $\frac{1}{6}$ of that on Earth (which is $9.8\,\mathrm{m\,s^{-2}}$). The method for calculating the weight of a 100-kg object on the moon, measured in newtons (N), is shown below.

$$W = m \times g$$
$$= 100\,\mathrm{kg} \times \left(\frac{1}{6} \times 9.8\,\mathrm{m\,s^{-2}}\right)$$
$$\approx 163.33\,\mathrm{N}$$

Note: Weight is different to mass as weight describes the gravitational force acting on an object, and is measured in newtons (N).

Use this information to calculate the weight of each of the following objects on the moon.

a. A 4500-kg boat   b. A 0.56-kg bottle   c. A 76.25-kg person   d. A 331-g iPad mini

4.6 Review: exam practice

4.6.1 Volume, capacity and mass summary

Volume and capacity

- The volume of a 3-dimensional object is the amount of space that is occupied inside it. The units commonly used to describe the volume of an object are the cubic millimetre (mm^3), cubic centimetre (cm^3) and the cubic metre (m^3). One cubic centimetre is the space occupied by a cube with side lengths equal to 1 cm.
- Metric units of volume can be converted as shown below.

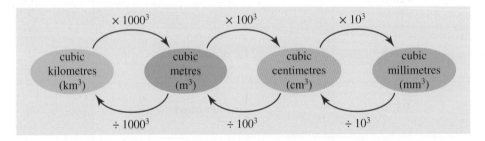

- The capacity of a container refers to the amount of liquid that a container can hold.
- Units of capacity are related as shown below.

$$1\,\mathrm{L} = 1\,000\,\mathrm{mL}$$
$$1\,\mathrm{kL} = 1\,000\,\mathrm{L}$$
$$1\,\mathrm{ML} = 1\,000\,000\,\mathrm{L}$$

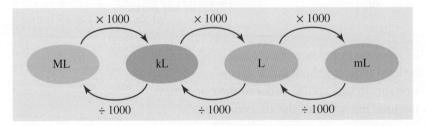

- A container, such as a jar, has both volume and capacity, as it occupies space and can hold, contain or absorb something other than what it is made of.

Volume of a prism

- A prism has the same cross-sectional area along its length, thus it has the same two end faces. The cross-sectional area is shaded in blue in the diagram shown.

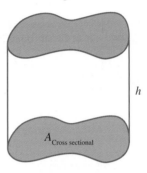

$$V_{\text{Prism}} = A_{\text{Cross sectional}} \times h$$
$$V_{\text{Prism}} = Ah$$

- Volume of a rectangular prism:

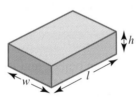

$$V = \text{length} \times \text{width} \times \text{height}$$
$$V = l \times w \times h$$

- Volume of a triangular prism:

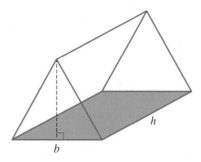

$$V = \text{area of triangle} \times h$$

- Volume of a cylinder:

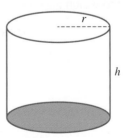

$$V_{Cylinder} = \pi r^2 h$$

where:
$$r = \text{radius}$$
$$h = \text{height}$$

- Volume of a pyramid (general rule):

$$V_{Pyramid} = \frac{1}{3} \times \text{area of base} \times \text{height}$$

$$V_{Pyramid} = \frac{1}{3} Ah$$

- Volume of a sphere:

where $r = $ radius
$$V_{Sphere} = \frac{4}{3} \pi r^3$$

Mass

- Mass describes how much matter makes up an object, and its standard unit is the kilogram (kg).
- The mass of an object can be measured in milligrams (mg), grams (g), kilograms (kg) and tonnes (t). Just like with length and area it is important to use the appropriate unit to measure an object's mass.
- The diagram shown can be used to assist in the conversion between different units of mass.

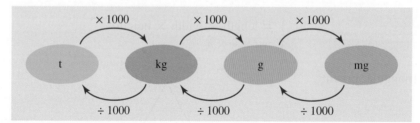

Exercise 4.6 Review: exam practice

Simple familiar

1. **MC** A mass of 259 grams is the same as
   A. 259 kg.   **B.** 0.259 kg.   **C.** 2.59 kg.   **D.** 0.259 tonnes.

2. **MC** If Jack buys 7 250-gram packets of chips for a group of friends who are coming around to watch a movie, then the total mass of chips is
   A. 2.5 kg.
   B. 1.5 kg.
   C. 2.0 kg.
   D. 1.75 kg.

3. **MC** The volume of a jug is 465 cm³. Its capacity is
   A. 564 mL.   B. 46.5 mL.   C. 4.65 L.   D. 0.465 L.

4. **MC** The number of cubes with side length 1 cm that can fit into a larger cube with side length 4 cm is
   A. 64.   B. 16.
   C. 56.   D. 32.

5. **MC** The volume of the shape shown is
   A. 10 m³.
   B. 15 m³.
   C. 20 m³.
   D. 18 m³.

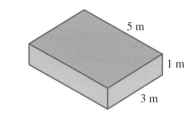

6. **MC** The volume of the triangular prism shown is closest to
   A. 2406 cm³.
   B. 1203 cm³.
   C. 1895 cm³.
   D. 60 cm³.

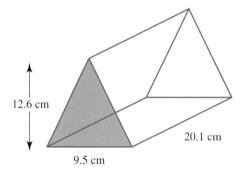

7. **MC** The volume of the cylinder shown is closest to
   A. 285.5 cm³.
   B. 6589 cm³.
   C. 2097.5 cm³.
   D. 1647 cm³.

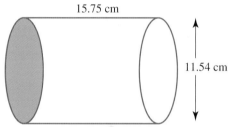

8. **MC** The volume of a rectangular-based pyramid with length 2.5 m, width 1.8 m and vertical height 1.5 m is closest to
   A. 6.75 m³.   B. 5.75 m³.   C. 2.25 m³.   D. 1.25 m³.

9. **MC** If a square-based pyramid has a volume of 12.4 cm³ and a vertical height of 2.2 cm then the length of the pyramid is
   A. 4.1 cm.   B. 2.2 cm.
   C. 2.4 cm.   D. 3.8 cm.

10. **MC** The volume of a disco ball that has a diameter of 55 cm is closest to
    A. 0.32 m³.
    B. 0.07 m³.
    C. 0.03 m³.
    D. 0.09 m³.

11. The Luxor in Las Vegas is known for its square-based pyramid and bright light. The Luxor pyramid is 111 metres high and 90 metres wide. Calculate the volume of the Luxor Pyramid to two decimal places.

12. The volume of a cylinder is $135\,\text{cm}^3$. If it has a radius of 3 cm, calculate the height of the cylinder to two decimal places.

Complex familiar

13. Calculate the volume of water that can completely fill the inside of the washing machine barrel shown in the diagram to four decimal places in:
    a. cm^3
    b. m^3
    c. litres.

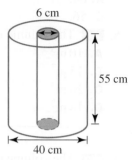

14. Calculate the volume of each of the following objects.

    a.

    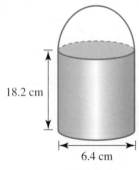

    b.

    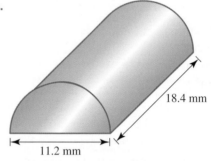

    c.

    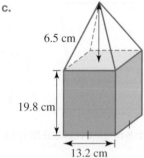

    d.

    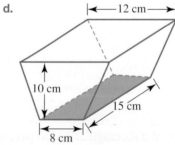

15. A block of cheese is in the shape of a triangular prism. The area of the triangular face is $2.3\,\text{cm}^2$ and the length of the prism is 9.7 cm. If the price of cheese is $0.53 per cm^3, calculate the price of the block.

16. A fish tank has dimensions $90\,\text{cm} \times 40\,\text{cm} \times 60\,\text{cm}$. How much water would be needed to:
    a. three-quarters-fill the tank in cm^3
    b. three-quarters-fill the tank in litres
    c. fill 90% of the tank in cm^3
    d. fill 90% of the tank in litres?

Complex unfamiliar

17. **a.** Calculate the volume of the container shown given that the formula for the volume of a cone is $V_{Cone} = \frac{1}{3}\pi r^2 h$.

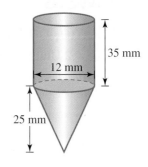

b. The container is completely filled with water. A tap is opened to release the water from the base of the container at a rate of 15 mm³ per minute. If water is added at a rate of 10 mm³ per minute, use your calculations to work out the amount of water left in the container after 20 minutes.

18. The sculpture shown below is to be packaged into a rectangular or cylindrical cardboard box.

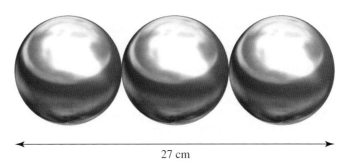

27 cm

a. Assuming each sphere touches the one next to it, calculate the:
   i. volume of the smallest rectangular box that completely contains the spheres
   ii. volume of the smallest cylindrical box that completely contains the spheres
   iii. volume of space in each of the boxes not occupied by the spheres.
b. Use your calculations to justify which box you would choose.

19. A sandpit has the dimensions as shown.
   a. How much sand is required to fill the sandpit?
   b. If the sand came in bags shaped as rectangular prisms with dimensions 45 cm by 30 cm by 10 cm, how many bags would be required to fill the sandpit?
   c. If the bags of sand cost $7.50 each, how much would it cost to fill the sand pit?
   d. The digger's scoop is in the shape of a rectangular-based pyramid with base length of 20 cm, base width of 18 cm and height of 15 cm. Assuming the scoop was full each time, how many scoops would be required to empty the full sandpit?

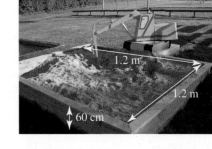

20. Callum's Camping Company uses tents as shown for outback adventures.
   a. Calculate the amount of material required to make each of the tents assuming no base is required.
   b. Calculate the room in each of the tents in m³.
   c. If there was a requirement that a tent provided a minimum of 3 m³ per person, how many people can sleep in the tent shown?
   d. The side wall of the tents can come off so that two tents can join together into one tent. How many people can fit in the double tent given the previous requirement?

Answers

4 Volume, capacity and mass

Exercise 4.2 Units of volume and capacity

1. Volume measures the amount of space that is occupied inside the object.
2. a. cm^3 b. mm^3 c. m^3 d. cm^3
3. a. $16\,cm^3$ b. $12\,cm^3$ c. $6\,cm^3$ d. $27\,cm^3$
4. a. $24\,cm^3$ b. $28\,m^3$
5. a. $9\,cm^3$ b. $7\,cm^3$ c. $10\,cm^3$ d. $13\,cm^3$
6. a. $88\,cm^3$ b. $120\,cm^3$ c. $88\,cm^3$ d. $116\,cm^3$
7. Volume is the amount of space taken up by the object, while capacity is the measure of an object's ability to hold a substance.
8. a. $80\,000\,cm^3$ and $80\,L$ b. $2500\,m^3$ and $2.5\,ML$
      c. $1750\,cm^3$ and $1.75\,L$ d. $160\,m^3$ and $160\,kL$
9. a. Capacity   b. Volume   c. Volume
      d. Capacity   e. Volume   f. Capacity
10. a. $12\,000\,mL$ b. $3.125\,ML$ c. $0.397\,L$
      d. $7.8\,mL$ e. $4.893\,L$ f. $0.036\,97\,ML$
11. a. $372\,ml$ b. $1\,630\,000\,cm^3$ c. $3400\,cm^3$
      d. $0.38\,cm^3$ e. $163\,000\,cm^3$ f. $49.28\,mL$
12. a. $0.578\,L$ b. $0.75\,mL$ c. $429\,mL$
13. a. $47\,mL$ b. $3.594\,L$ c. $4.2\,cm^3$
14. a. i. 16 cubes   ii. $16\,cm^3$ iii. $16\,mL$
      b. i. 12 cubes   ii. $12\,cm^3$ iii. $12\,mL$
      c. i. 48 cubes   ii. $48\,cm^3$ iii. $48\,mL$
      d. i. 16 cubes   ii. $16\,cm^3$ iii. $16\,mL$
15. a. $72\,000\,000$ cubes   b. $72\,000\,000\,cm^3$
      c. $72\,000\,L$ d. $57\,600\,L$
16. a. 1000 cubes   b. $1000\,cm^3$
      c. $1\,L$ d. $750\,mL$

Exercise 4.3 Volume and capacity of prisms and cylinders

1. a. $80\,cm^3$ b. $100\,cm^3$
     c. $90\,m^3$ d. $118.625\,cm^3$
2. a. $54\,m^3$ b. $1500\,cm^3$
     c. $614.125\,cm^3$ d. $130.5\,cm^3$
3. a. $336\,cm^3$ b. $64\,cm^3$
4. a. $480\,cm^3$ b. $180\,cm^3$
     c. $1320\,m^3$ d. $1296\,cm^3$
5. a. $140\,cm^3$ b. $86.955\,m^3$
6. a. $339.29\,cm^3$ b. $678.58\,mm^3$
     c. $1693.32\,cm^3$ d. $4.42\,m^3$
7. $706.86\,cm^3$
8. $922.45\,m^3$
9. a. $3936.69\,cm^3$ b. $0.70\,m^3$ c. $5145.6\,cm^3$
     d. $151\,937.5\,cm^3$ e. $88.19\,m^3$ f. $79\,521.56\,cm^3$
10. $111\,m^3$, $111\,000\,L$
11. $11.762\,m^3$, $11.762\,kL$
12. $16.632\,m^3$
13. $3.61\,cm$
14. a. $1.2\,m^3$ b. $0.9\,m^3$ c. $750\,L$
15. a. $0.68\,m^3$ b. $0.204\,m^3$ c. $0.476\,m^3$
16. a. $75.625\,m^3$ b. $103.125\,m^3$ c. $5.33\,m$

Exercise 4.4 Volume and capacity of pyramids and spheres

1. a. $40\,cm^3$ b. $392\,cm^3$ c. $143.65\,cm^3$
     d. $1641.71\,cm^3$ e. $27.05\,m^3$ f. $0.72\,m^3$
2. a. $523.60\,cm^3$ b. $7238.23\,cm^3$
     c. $3.05\,m^3$ d. $10\,305.99\,cm^3$
     e. $63\,891.58\,cm^3$ f. $17\,974.16\,m^3$
3. $112\,cm^3$
4. $20.22\,m^3$
5. a. $229.5\,mm^3$ b. $616\,mm^3$
     c. $0.196\,m^3$ d. $810.67\,mm^3$
6. a. $20\,579.53\,mm^3$ b. $2309.56\,mm^3$
     c. $28\,730.91\,mm^3$ d. $2144.66\,cm^3$
     e. $15\,298.57\,cm^3$
7. $8578.64\,cm^3$
8. $5.66\,cm$
9. $5.00\,cm$
10. a. $14\,137.17\,cm^3$ b. $14.14\,L$ c. 176.71 seconds
11. C
12. $24\,429.02\,cm^3$
13. $344.93\,cm^3$

Exercise 4.5 Problems involving mass

1. a. kilogram   b. milligram   c. tonne
     d. gram   e. kilogram   f. kilogram
2. a. $200\,g$ b. $300\,g$ c. $10\,g$
     d. $2000\,g$ e. $5\,g$ f. $300\,g$
3. Responses can be found in the Worked Solutions in the online resources.
4. a. $18\,900\,g$ b. $0.076\,490\,mg$ c. $83\,kg$
5. a. $0.717\,g$ b. $0.003\,867\,t$ c. $8400\,g$
6. a. $2\,kg$ b. $250\,g$ c. $750\,g$ d. $380\,g$
7. a. $2400\,mg$ b. $46\,700\,kg$
     c. $2.510\,g$ d. $0.082\,kg$
     e. $30\,g$ f. $7\,893\,000\,mg$
8. a. $0.025\,781\,kg$ b. $\dfrac{7\,000\,000}{8}\,mg$
     c. $0.3842\,t$ d. $75\,000\,mg$
     e. $4.5\,g$ f. $0.033\,456\,kg$
9. a. $2000\,g$
     b. Net means the weight of the chocolate not including the packaging.
10. $43\,kg$
11.

Item	Cost per kg	Cost
1.5 kg of bananas	$3.50	$5.25
3.4 kg of apples	$3.95	$13.43
1.2 kg of carrots	$2.80	$3.36
2.3 kg of grapes	$7.90	$18.17
Total cost		$40.21

12. $26.6\,kg$

13. * See the table at the bottom of the page.
    a. Jack lost 32.112 kg
    b. Paul lost 13.9 kg
    c. Jack, Matthew, Paul, Kelly, Sarah, Jane
    d. 114.752 kg
14. a. 7350 N   b. 0.91 N
    c. 124.54 N   d. 0.541 N

4.6 Review: exam practice

1. B
2. D
3. D
4. A
5. B
6. B
7. D
8. C
9. A
10. D
11. 299 700.00 m^3
12. $h = 4.77$ cm

13. a. 67 559.9500 cm^3 b. 0.0676 m^3 c. 67.5600 L
14. a. 654.1 cm^3 b. 906.39 mm^2
    c. 3827.47 cm^3 d. 1500 cm^3
15. $V = 22.31$ cm^3
    Cost = $ 11.82
16. a. 162 000 cm^3 b. 162 L
    c. 194 400 cm^3 d. 194.4 L
17. a. 4900.88 mm^3
    b. $V_{\text{Left}} = 4800.88$ mm^3
18. a. i. 2187 cm^3
        ii. 1717.67 cm^3
       iii. Rectangular box: 1041.89 cm^3
            Cylindrical box: 572.56 cm^3
    b. Since the cylindrical box only has 572.56 cm^3 of empty
       space compared to the rectangular box having
       1041.89 cm^3, the cylindrical box would be more
       tightly packed and less filling would be required.
19. a. 0.864 m^3 b. 64   c. $480   d. 480
20. a. 21.90 m^2
    b. 10.5 m^3
    c. Three people can fit in the tent under these requirements.
    d. People = 7

*13.

Name	Original mass (kg)	Loss	New mass (kg)
Matthew	112.3	16.7 kg	95.6 kg
Jack	133.8	24% of his mass	101.688 kg
Sarah	93.6	$\frac{1}{6}$ of her mass	78 kg
Jane	88.3	Jane now weighs $\frac{7}{9}$ of her original mass	68.68 kg
Paul	105.6	13.9 kg	91.7 kg
Kelly	96.1	17.5% of her mass	79.28 kg

5 Scales, plans and models

5.1 Overview

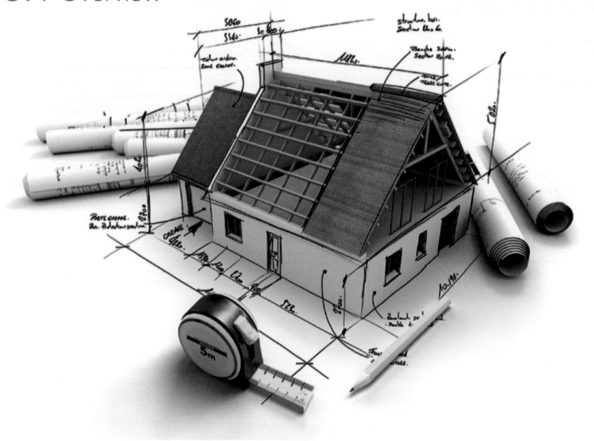

LEARNING SEQUENCE

5.1 Overview
5.2 Reading and interpreting scale drawings
5.3 Calculating measurements from scale drawings
5.4 Creating scale drawings
5.5 Review: exam practice

CONTENT

In this chapter, students will learn to:
- interpret commonly used symbols and abbreviations in scale drawings
- find actual measurements from scale drawings, including lengths, perimeters and areas
- estimate and compare quantities, materials and costs using actual measurements from scale drawings [complex]
- understand and apply drawing conventions of scale drawings, including scales in ratio, clear indications of dimensions and clear labelling [complex]
- construct scale drawings by hand and by using software packages [complex].

Fully worked solutions for this chapter are available in the Resources section of your eBookPLUS at www.jacplus.com.au.

5.2 Reading and interpreting scale drawings

5.2.1 Scales review

A **scale** is a ratio of the length on a drawing to the actual length.

Scale
Scale = length of drawing : actual length

The sign ':' is read as 'to'.

Scales are usually written with no units. If a scale is given in two different units, the larger unit has to be converted into the smaller unit.

A **scale factor** is the ratio of two corresponding lengths in two similar shapes. The scale factor of $\frac{1}{2}$ or a scale (ratio) of $1:2$ means that 1 unit on the drawing represents 2 units in actual size. The unit can be mm, cm, m or km.

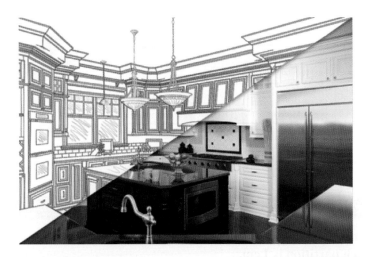

WORKED EXAMPLE 1

Determine the scale and the scale factor of a drawing where 8 cm on the diagram represents 4 km in reality.

THINK	WRITE
1. Convert the larger unit into the smaller unit.	$4 \text{ km} = 4 \times 1000 \times 100$ $= 400\,000 \text{ cm}$
2. Write the scale of the drawing with the same units.	$8 \text{ cm} : 400\,000 \text{ cm}$ $\Rightarrow 8 : 400\,000$
3. Simplify the scale by dividing both sides of the ratio by the highest common factor. We divide both sides of the ratio by 8 as 8 is the highest common factor of both 8 and 400 000. State the scale.	$1 : 50\,000$
4. State the scale factor of the drawing.	The scale factor is $\dfrac{1}{50\,000}$.

5.2.2 Maps and scales

Maps are always drawn at a smaller scale. A map has its scale written or drawn using a diagram.

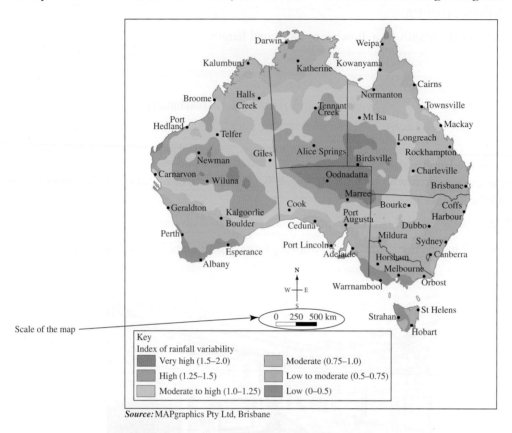

Scale of the map

Key
Index of rainfall variability

- Very high (1.5–2.0)
- High (1.25–1.5)
- Moderate to high (1.0–1.25)
- Moderate (0.75–1.0)
- Low to moderate (0.5–0.75)
- Low (0–0.5)

Source: MAPgraphics Pty Ltd, Brisbane

WORKED EXAMPLE 2

State the scale of the map as a ratio using the graphical scale shown, where the length of each partition is 1 cm.

0   500   1000 km

THINK	WRITE
1. Measure the length of each partition in the diagram.	Each partition is 1 cm.
2. Determine the length that each partition represents in reality.	Each partition represents 500 km.
3. Write the scale as a ratio.	1 cm : 500 km
4. Convert the larger unit into the smaller unit.	500 km = 500 × 1000 × 100 = 50 000 000 cm
5. Write the scale as a ratio in the same units.	1 cm : 50 000 000 cm
6. State the scale of the map in ratio form.	1 : 50 000 000

5.2.3 Lengths

The scale factor $= \dfrac{\text{dimension on the drawing}}{\text{actual dimension}}$. We can rearrange the scale factor formula to determine the actual dimension.

$$\text{Actual dimension} = \frac{\text{dimension on the drawing}}{\text{scale factor}}$$

To calculate the actual dimensions, we measure the dimensions on the diagram and then divide these by the scale factor.

WORKED EXAMPLE 3

A scale of 1 : 200 was used for the diagram of the bedroom shown. Calculate both the length and the width of the master bedroom, given the dimensions on the plan are length = 2.5 cm and width = 1.5 cm.

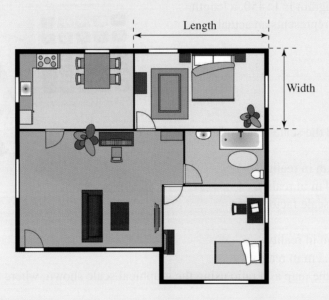

THINK

1. Measure the length and the width of the room in the diagram.

2. Write down the scale of the drawing and state the scale factor.

3. Divide both dimensions by the scale factor.

WRITE

Length = 2.5 cm
Width = 1.5 cm

1 : 200 This means that every 1 cm on the drawing represents 200 cm of the actual dimension. The scale factor is $\dfrac{1}{200}$.

$$\text{Length of the bedroom} = 2.5 \div \frac{1}{200}$$
$$= 2.5 \times 200$$
$$= 500 \, \text{cm}$$
$$= 5 \, \text{m}$$
$$\text{Width of the bedroom} = 1.5 \div \frac{1}{200}$$
$$= 1.5 \times 200$$
$$= 300 \, \text{cm}$$
$$= 3 \, \text{m}$$

Exercise 5.2 Reading and interpreting scale drawings

1. **WE1** Determine the scale and the scale factor of a drawing where 150 mm on the diagram represents 45 m in reality.

2. If the scale of a diagram is 3.9 cm : 520 km, write the scale using same units.

3. **MC** If the scale of a diagram is 1 : 450, a length of 3 cm on the diagram represents an actual length of

   A. 225 cm.

   B. 1350 cm.

   C. 150 cm.

   D. 1.5 m.

4. Determine the scale and the scale factor of a drawing where:

   a. 7 cm represents 3.5 km in reality

   b. 16 mm represents 6.4 m in reality.

5. Calculate the scale and scale factor given the following.

   a. 5 cm represents 185 m in reality

   b. 12 mm represents 19.2 m in reality

6. **WE2** State the scale of the map as a ratio using the graphical scale shown, where the length of each partition is 1 cm.

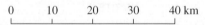

7. State the scale of a map as a ratio using the graphical scale shown, where the length of each partition is 1 cm.

8. State the scale of a map as a ratio using the graphical scale shown.

   a. (each partition is 1.5 cm)

   0   150   300 km

   b. (each partition is 1 cm)

   0   80   160   240 m

9. State the scale of a map as a ratio using the graphical scale shown, given each partition is 2 cm.

10. A house plan is drawn to a scale of 1 : 1000.

    a. What is the actual length of the house if it is represented by 3.2 cm on the plan?

    b. The width of the house is 17 m. What is the width of the house on the plan?

11. A house plan is drawn to a scale of 1 : 500.

    a. What is the actual length of the lounge room if it is represented by 0.6 cm on the plan?

    b. The width of the bedroom is 5.2 m. What is the width of the bedroom on the house plan?

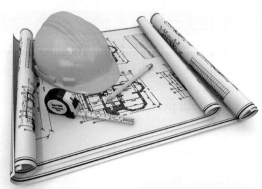

12. Calculate the dimensions of the carpet shown, given that its diagram was drawn using a scale of 1 : 75 and its length and width on the diagram are 3 cm and 2 cm respectively.

13. **WE3** A scale of 1 : 300 was used for the diagram of the house shown. Calculate both the height and width, in metres, of the garage door, given the dimensions on the plan are: height = 0.8 cm and width = 2 cm.

14. **MC** A 15 m long fence is represented by a straight line 4.5 cm long on a drawing. What is the scale of the drawing?

    A. 1000 : 3
    B. 4.5 : 1.5
    C. 45 : 150
    D. 3 : 1000

5.3 Calculating measurements from scale drawings

All diagrams, maps or plans are drawn using a given scale. This scale is used to convert the lengths in the diagram to the actual lengths.

5.3.1 Perimeters

Calculating the actual perimeters of shapes from scaled drawings involves first calculating the lengths required. This procedure was learnt in the previous section.

Once the lengths of the sides of the shape are known, the perimeter can be calculated by adding up all the sides of the shape.

Given a scale of 1 : 1000 for the diagram of the backyard shown, calculate:
a. the actual length and the width of the pool, given the diagram measurements are 2.4 cm by 1.2 cm
b. the actual perimeter of the pool.

THINK	WRITE
a. 1. Measure the length and the width of the pool on the diagram.	**a.** Length = 2.4 cm Width = 1.2 cm
2. Write down the scale of the drawing and state the scale factor.	1 : 1000 This means that every 1 cm on the drawing represents 1000 cm of the actual dimension. The scale factor is $\dfrac{1}{1000}$.
3. Divide both dimensions by the scale factor.	Length of the pool = $2.4 \div \dfrac{1}{1000}$ $= 2.4 \times 1000$ $= 2400$ cm $= 24$ m Width of the pool = $1.2 \div \dfrac{1}{1000}$ $= 1.2 \times 1000$ $= 1200$ cm $= 12$ m
b. 1. Calculate the perimeter of the pool.	**b.** Perimeter of pool = 2 × length + 2 × width $= 2 \times 24 + 2 \times 12$ $= 48 + 24$ $= 72$ m

5.3.2 Areas

Areas of surfaces from scaled drawings can be calculated in a similar way as the perimeter. The dimensions of the shape have to be calculated first. Once the lengths of the shape are known, the area can be calculated by using an appropriate formula.

WORKED EXAMPLE 5

Given a scale of 1 : 15 for the diagram of the tile shown, calculate:
a. the actual length and the width of the tile, given the diagram measurements are 2 cm by 2 cm
b. the actual area of the tile
c. the actual area covered by 20 tiles.

THINK	WRITE
a. 1. Measure the length and the width of the shape.	**a.** Length = 2 cm Width = 2 cm
2. Write down the scale of the drawing and state the scale factor.	1 : 15 This means that every 1 cm on the drawing represents 15 cm of the actual dimension. The scale factor is $\dfrac{1}{15}$.
3. Divide both dimensions by the scale factor.	Length of the tile $= 2 \div \dfrac{1}{15}$ $= 2 \times 15$ $= 30\,\text{cm}$ Width of the tile $= 2 \div \dfrac{1}{15}$ $= 2 \times 15$ $= 30\,\text{cm}$
b. 1. Calculate the area of the tile using the formula for the area of a square. Use $A = l^2$.	**b.** Area of one tile $= \text{length}^2$ $= 30^2$ $= 900\,\text{cm}^2$
c. 1. Calculate the area required by multiplying the area of one tile by 20. *Note:* Recall the conversion from cm² to m² is to divide by 100².	**c.** Total area $= \text{Area of one tile} \times 20$ $= 900 \times 20$ $= 18\ 000\,\text{cm}^2 = 18\ 000 \div 100^2$ $= 1.8\,\text{m}^2$

5.3.3 Quantities, materials and costs

Scaled diagrams are used in many areas of work in order to estimate elements such as production costs and quantities of materials required.

Packaging

There are many products that are usually placed in a box when purchased or delivered. Boxes are cut out of cardboard or other materials using a template.

The dimensions of the package are determined by using a net of the three-dimensional shape as shown in the diagram.

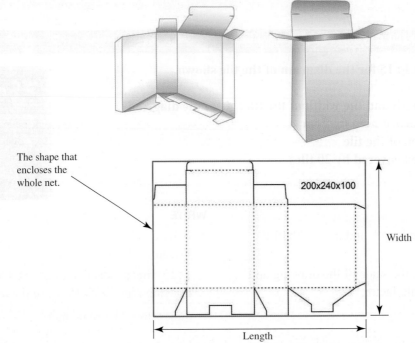

The required area can be estimated by calculating the area of the rectangle that encloses the whole net.

Costs

To calculate the cost of the materials used, we use the cost per m or m², known as the cost per unit.

Cost of materials

$$\text{Cost required} = \frac{\text{area of material}}{\text{area per squared unit}} \times \text{cost per unit}$$

If the area of the packaging is 520 cm² and the cost of the material is $1.20 per square metre, then the cost for packaging is as follows.

$$\text{Cost required} = \frac{\text{area of material}}{\text{area per squared unit}} \times \text{cost per unit}$$

$$= \frac{520\,\text{cm}^2}{1\,\text{m}^2} \times 1.20$$

(*Note:* The two area units have to be the same: 1 m² = 10 000 cm².)

$$= \frac{520}{10\,000} \times 1.20$$

$$= \$0.624$$

Consider the gift box shown and the template required to make it.
The scale of the diagram is 1 : 15, and the diagram measurements are
6 cm by 4 cm.
a. Calculate the width and the height of the packaging template.
b. Estimate the total area of material required.
c. If the cost of materials is $0.25 per square metre, how much would
it cost to make this gift box?

THINK	WRITE
a. 1. Measure the width and the height of the template.	a. Width = 6 cm Height = 4 cm
2. Write down the scale of the drawing and state the scale factor.	1 : 15 This means that every 1 cm on the drawing represents 15 cm of the actual dimension. The scale factor is $\dfrac{1}{15}$.
3. Divide both dimensions by the scale factor.	Width of template $= 6 \div \dfrac{1}{15}$ $\qquad = 6 \times 15$ $\qquad = 90 \text{ cm}$ Height of template $= 4 \div \dfrac{1}{15}$ $\qquad = 4 \times 15$ $\qquad = 60 \text{ cm}$
b. 1. Estimate the area of the box template using the formula for the area of a rectangle.	b. Area of template $=$ width $\times$ height $\qquad = 90 \times 60$ $\qquad = 5400 \text{ cm}^2$
c. 1. Calculate the cost of the material.	c. Cost is $0.25 per m^2 or 10 000 cm^2 Cost for the area required $= \dfrac{\text{Area of material}}{\text{Area per squared unit}}$ $\qquad \times$ Cost per unit $\qquad = \dfrac{5400}{10\,000} \times 0.25$ $\qquad = \$0.135$ $\qquad \approx 0.14$
2. State the answer.	It will cost $0.14 or 14 cents to make the gift box.

5.3.4 Painting

Painting a wall requires some calculations. Three pieces of information are required to calculate the amount
of paint needed:
- area of the surface to be painted
- amount of paint per square metre
- cost of paint per litre.

The following formulae are useful:

Amount of paint required = area to be painted × amount of paint per square metre

Total cost = amount of paint required × cost of paint per litre

WORKED EXAMPLE 7

Indra is going to paint a feature wall in her lounge room. She has a diagram of the room drawn at a scale of 1 : 100 with a length of 4 cm and height 2.5 cm. Calculate:
a. the length and the height of the wall
b. the area of the wall
c. the amount of paint Indra has to buy if the amount of paint needed is 3 L per m²
d. the total cost of the paint if the price is $7.60 per litre.

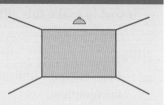

THINK	WRITE
a. 1. Measure the length and the height of the shape.	a. Length = 4 cm Height = 2.5 cm
2. Write down the scale of the drawing and state the scale factor.	1 : 100 This means that every 1 cm on the drawing represents 100 cm of the actual dimension. The scale factor is $\dfrac{1}{100}$.
3. Divide both dimensions by the scale factor.	Length of the wall = $4 \div \dfrac{1}{100}$ $= 4 \times 100$ $= 400$ cm $= 4$ m Height of the wall = $2.5 \div \dfrac{1}{100}$ $= 2.5 \times 100$ $= 250$ cm $= 2.5$ m
b. Calculate the area of the wall using the formula for the area of a rectangle.	b. Area wall = length × width (or height) $= 4 \times 2.5$ $= 10 \text{ m}^2$
c. Calculate the amount of paint required.	c. Amount of paint required = area to be painted × amount of paint per square metre $= 10 \times 3$ L $= 30$ L
d. Calculate the cost of the paint.	d. Total cost = amount of paint required × cost of paint per litre $= 30 \times 7.60$ $= \$228$

5.3.5 Bricklaying

Bricks are frequently used as a basic product in the construction of houses or fences.

The dimensions of a standard brick are shown in the diagram using a scale of $1:10$.

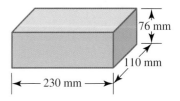

Bricklayers have to calculate the number of bricks required to build a wall by considering the length, width and height of the brick and the dimensions of the wall.

To give the wall strength, bricks have to be laid such that their ends sit on the middle of the layer below as shown in the diagram.

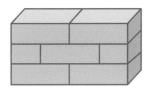

Calculating the number of bricks

Imagine that you want to calculate the number of bricks required to build a $2\,m^2$ wall.

$$\text{Number of bricks} = \frac{\text{area of the wall}}{\text{area of the exposed side of the brick}}$$

The area of the brick is the area of the side that shows on the wall. In the diagram of the wall, using the standard brick dimensions, the exposed side of the brick has length 230 mm and height 76 mm.

The area of the exposed side of the brick is $230 \times 76 = 17\,480\,mm^2$.

The two areas have to be written in the same units. Area of the wall $= 2\,m^2 = 2\,000\,000\,mm^2$.

$$\begin{aligned}\text{Number of bricks} &= \frac{\text{area of the wall}}{\text{area of the exposed side of the brick}}\\ &= \frac{2\,000\,000}{17\,480}\\ &= 114.4\,\text{bricks}\\ &= 115\,\text{bricks}\end{aligned}$$

Note: Notice that although the answer is 114.4 bricks, we round up because we need 115 bricks so we can cut the 0.4 of a brick to finish the whole wall.

This answer is an estimated value as we did not take into account the mortar that has to be added to the structure of the wall.

WORKED EXAMPLE 8

Michael is going to build a brick fence with the dimensions shown in the diagram. The scale factor of the drawing is $\frac{1}{50}$.

Calculate:

a. the length and the height of the fence, given it measures 7 cm by 1 cm on the diagram

b. the area of the front side of the fence

c. the number of bricks required if the dimensions of the exposed side of one brick are 20 cm and 10 cm

d. the cost of building the fence if the price of one brick is $1.20.

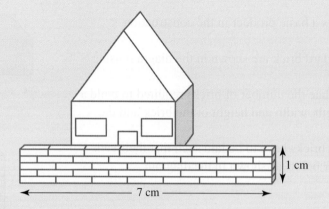

1 cm

7 cm

THINK

WRITE

a. 1. Measure the length and the height of the fence.

a. Length = 7 cm
Height = 1 cm

2. Write down the scale factor.

The scale factor is $\dfrac{1}{50}$.

3. Divide both dimensions by the scale factor.

Length of the fence $= 7 \div \dfrac{1}{50}$

$= 7 \times 50$

$= 350\,\text{cm}$

Height of the fence $= 1 \div \dfrac{1}{50}$

$= 1 \times 50$

$= 50\,\text{cm}$

b. 1. Calculate the area of the front face of the fence using the formula for the area of a rectangle.

b. Area of the front face of the fence = length × width (or height)

$= 350 \times 50$

$= 17\,500\,\text{cm}^2$

c. 1. Calculate the area of the exposed face of one brick.

c. Area of the exposed face of the brick $= 20 \times 10$

$= 200\,\text{cm}^2$

2. Calculate the number of bricks required.

Number of bricks $= \dfrac{17\,500}{200}$

$= 87.5\,\text{bricks}$

$= 88\,\text{bricks}$

d. 1. Calculate the cost.

d. Cost = number of bricks × cost of one brick

$= 88 \times 1.20$

$= \$105.60$

Exercise 5.3 Calculating measurements from scale drawings

1. **WE4** Given a scale of 1 : 250 for the floor plan of the apartment shown, calculate:
   a. the actual length and the width of the apartment, given the diagram measurements are 5 cm by 3 cm
   b. the actual perimeter of the apartment.

2. If the scale of the diagram shown is 1 : 800, calculate, to the nearest metre, the radius and the circumference of the Ferris wheel shown, given the radius measures 1.5 cm on the diagram.

3. **WE5** Given a scale of 1 : 40 for the window shown, calculate:
   a. the actual width and height of each piece of glass needed to cover the window, given that the large piece of glass measures 1.5 cm by 2.5 cm on the diagram, and the width of the small piece of glass measures 0.75 cm
   b. the actual areas of the two glass pieces
   c. the total area covered by glass.

4. If the scale of the diagram shown is 1 : 20, calculate, to the nearest centimetre, the radius and the area of the clock face shown, given the radius measures 1.5 cm on the diagram.

5. **WE6** Consider the DVD mailing box shown and the template required to make it. The scale of the diagram is 1 : 8, and the diagram measurements are 3.5 cm by 1.5 cm.
   a. Calculate the length and the width of the packaging template.
   b. Estimate the total area of material required.
   c. If the cost of materials is $0.85 per square metre, how much would it cost to make this DVD mailing box?

6. Consider the milk box shown and the template required to make it. The scale of the diagram is $1:7$.

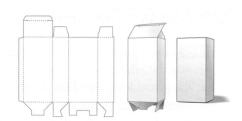

   a. Calculate the length and the width of the packaging template, given the dimensions measure 3 cm by 3 cm on the diagram.
   b. Estimate the total area of material required.
   c. If the cost of materials is $0.45 per square metre, how much would it cost to make this milk box?

7. Jonathan wants to paint a wall with dimensions 5.6 m long and 2.4 m high.

   a. How many litres of paint are required if 1.5 litres of paint covers 1 m²?
   b. What is the cost of the paint if the price per litre is $15.70?

8. **WE7** Ilia was asked by his friend to paint a bedroom wall. His friend gave him a diagram of the bedroom drawn at a scale of $1:100$. If the length and height of the wall are 1.5 cm and 2.5 cm respectively, calculate:

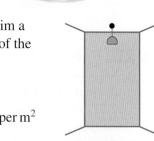

   a. the length and the height of the wall
   b. the area of the wall
   c. the amount of paint Ilia has to buy if the amount of paint needed is 2.6 L per m²
   d. the total cost of the paint if the price is $12.80 per litre.

9. The three identical wooden cubes shown are pictured at a scale of $1:20$. Calculate:

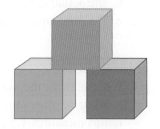

   a. the length of their sides, given they measure 1.3 cm on the diagram
   b. the area of each face
   c. the total area to be painted for the three cubes (State the answer in m² correct to one decimal place.)
   d. the amount of paint required if 2 L of paint are needed per m²
   e. the total cost of the paint if the price is $9.25 per litre.

10. **WE8** Consider a brick fence with the dimensions shown in the diagram. The scale factor of the brick wall shown is $\dfrac{1}{100}$. Calculate:

2.5 cm

6 cm

   a. the length and the height of the fence, given that it measures 6 cm by 2.5 cm on the diagram
   b. the area of the front side of the fence
   c. the number of bricks required if the dimensions of the exposed side of one brick are 20 cm by 5 cm
   d. the cost of the fence if the price of one brick is $0.75.

11. Sky wants to pave a garden path with bricks. The garden path is shown at a scale of $1:150$. Calculate:

   a. the actual dimensions of the garden path, given the diagram measures 8.5 cm by 0.75 cm
   b. the area of the garden path
   c. the number of bricks required if the bricks used are squares with side length 24 cm
   d. the cost of paving the garden path if the price of one brick is $1.65.

12. The scale of the house plan shown is $1:200$. Calculate:

    a. the length of the bathroom, given it measures 1.2 cm by 1 cm on the plan
    b. the width of the hallway, given it is 0.6 cm wide on the plan
    c. the perimeter of the kitchen, given it measures 1 cm by 2 cm on the plan
    d. the floor area of the house, given it measures 4.5 cm by 3.5 cm on the plan.

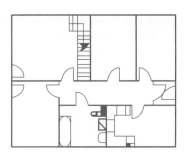

13. A brick has length 215 mm and height 65 mm. A bricklayer is building a wall 6 m long and 7.5 m high. Calculate:

    a. the number of bricks required
    b. the total cost for building the wall if one brick costs $1.95.

14. Consider the gift box shown and the template required to make it. The scale of the diagram is $1:8$.

    a. Calculate the actual length and the width of the packaging template, given it measures 8 cm by 5 cm on the diagram.
    b. Estimate the total area of material required.
    c. If the cost of materials is $2.15 per square metre, how much would it cost to make this gift box?

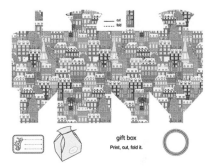

15. If the scale of the diagram shown is $1:12$, calculate, to the nearest centimetre, the radius, the circumference and the area of the circular chopping board shown, given the radius measures 1.5 cm on the diagram.

5.4 Creating scale drawings

Scale drawings are drawings on paper of real-life objects. All dimensions of the drawn object are kept in the same ratio to the actual dimensions of the object using a scale.

5.4.1 Calculating the dimensions of the drawing

Calculating the dimensions of a drawing always requires a scale or a scale factor. The formula for the scale is used to calculate the dimensions of the drawing.

If the scale factor is $\dfrac{\text{dimension on the drawing}}{\text{actual dimension}}$, then

$$\text{dimension on the drawing} = \text{scale factor} \times \text{actual dimension}.$$

Calculate the dimensions required to create a scale drawing of a shipping container 12 m long, 2.5 m wide and 3 m high using a scale of 1 : 100.

THINK	WRITE
1. Convert all dimensions to an appropriate drawing unit.	The dimensions of the diagram are going to be smaller than the actual dimensions of the shipping container. Convert m to cm: $12\,m = 1200\,cm$ $2.5\,m = 250\,cm$ $3\,m = 300\,cm$
2. Write the scale as a scale factor.	$1 : 100 = \dfrac{1}{100}$
3. Calculate the length of the object on the scale drawing.	$\dfrac{\text{Dimension on the drawing}}{\text{Actual dimension}} = \dfrac{1}{100}$ Length on the drawing $=$ scale factor $\times$ actual dimension $= \dfrac{1}{100} \times 1200$ $= \dfrac{1200}{100}$ $= 12\,cm$
4. Calculate the width of the object on the scale drawing.	Width on the drawing $=$ scale factor $\times$ actual dimension $= \dfrac{1}{100} \times 250$ $= \dfrac{250}{100}$ $= 2.5\,cm$
5. Calculate the height of the object on the scale drawing.	Height on the drawing $=$ scale factor $\times$ actual dimension $= \dfrac{1}{100} \times 300$ $= \dfrac{300}{100}$ $= 3\,cm$

5.4.2 Constructing scale drawings

Scale diagrams are drawn on graph paper or plain paper using a pencil and a ruler. Accuracy is very important when constructing these drawings.

Construct a plan view, front view and side view scale drawing of the shipping container from Worked example 9.

THINK	WRITE/DRAW
1. State the drawing dimensions of the plan view of the object.	The drawing dimensions of the plan view of the shipping container are: Length = 12 cm Width = 2.5 cm
2. Draw the plan view of the object. Recall that a plan view is a diagram of the object from above.	To draw the length of the object, construct a horizontal line 12 cm long.

Construct two 90° angles on both sides of the line segment. Ensure the two vertical lines are 2.5 cm long.

Connect the bottom ends of the vertical lines to complete the rectangle.

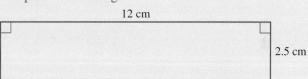

12 cm

2.5 cm

| 3. State the drawing dimensions of the front view of the object. | The drawing dimensions of the front view of the shipping container are:
Length = 12 cm
Height = 3 cm |
| 4. Draw the front view of the object. Recall that the front view is a diagram of the object looking straight at the object from the front. | To draw the length of the object, construct a horizontal line 12 cm long. |

Construct two 90° angles on both sides of the line segment.
Ensure the two vertical lines are 3 cm long.

Connect the bottom ends of the vertical lines to complete the rectangle.

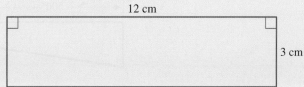

12 cm

3 cm

| 5. State the drawing dimensions of the side view of the object. | The drawing dimensions of the side view of the shipping container are:
Width = 2.5 cm
Height = 3 cm |

6. Draw the side view of the object. Recall that the side view is a diagram of the object looking at the object from one side.

We are going to draw the length of the object which is the width of the shipping container. Draw a horizontal line 2.5 cm long.

Construct two 90° angles on both sides of the line segment. Ensure the two vertical lines are 3 cm long.

Connect the bottom ends of the vertical lines to complete the rectangle.

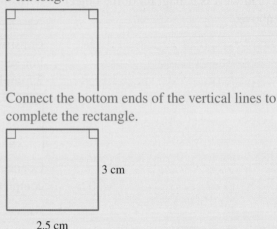

3 cm

2.5 cm

5.4.3 Labelling scale drawings

Scale diagrams are labelled using the actual dimensions of the object. The labels of a scale drawing have to be clear and easily read.

Conventions and line styles

The line style and its thickness indicate different details of the drawing.
- A *thick continuous line* represents a visible line, a contour of a shape or an object.
- A *thick dashed line* represents a hidden line, a hidden contour of a shape or an object.
- *Thin continuous lines* are used to mark the dimensions of the shape or object.

Dimension lines are thin continuous lines showing the dimension of a line. They have arrowheads at both ends.

Projection lines are thin continuous lines drawn perpendicular to the measurement shown. These lines do not touch the object.

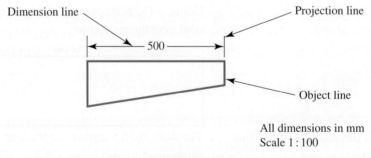

Dimension line

Projection line

500

Object line

All dimensions in mm
Scale 1 : 100

All dimensions on a drawing must be given in millimetres; however, 'mm' is not to be written on the drawing. This can be written near the scale as shown in the diagram above.

Dimensions must be written in the middle of the dimension line.

Label the three scale drawings of the shipping container from Worked example 10.

THINK	WRITE/DRAW
1. Draw the projection lines of the object in the plan view drawing using thin continuous lines.	
2. Draw the dimension lines of the object in the plan view drawing using thin continuous lines with arrowheads at both ends.	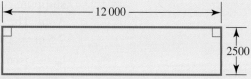
3. Label the drawing.	The actual dimensions of the plan view of the shipping container are 12 m long and 2.5 m wide. *Note:* The dimensions have to be written in mm. 12 m = 12 000 mm and 2.5 m = 2500 mm 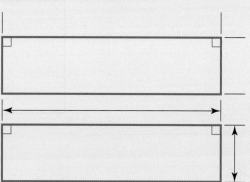
4. Draw the projection lines of the object in the front view drawing using thin continuous lines.	
5. Draw the dimension lines of the object in the front view drawing using thin continuous lines with arrowheads at both ends.	*Note:* The arrowheads of the dimension line for the height are directed from the outside of the projection line. This happens because the space between the projection lines is too small.
6. Label the drawing.	The actual dimensions of the front view of the shipping container are 12 m long and 3 m high. *Note:* The dimensions have to be written in mm. 12 m = 12 000 mm and 3 m = 3000 mm 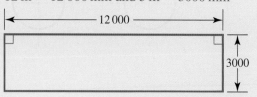

7. Draw the projection lines of the object in the side view drawing using thin continuous lines.

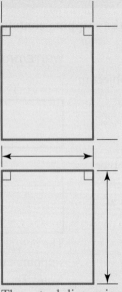

8. Draw the dimension lines of the object in the side view drawing using thin continuous lines with arrowheads at both ends.

9. Label the drawing.

The actual dimensions of the side view of the shipping container are 2.5 m wide and 3 m high. *Note:* The dimensions have to be written in mm.

2.5 m = 2500 mm and 3 m = 3000 mm

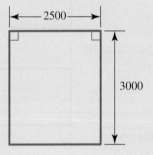

Labelling circles

A circle on a scale diagram is labelled using either the Greek letter Φ (phi) for diameter or R for radius. The diagram shows three correct notations for a 2 cm diameter. *Note:* The dimensions must be written in millimetres.

φ20   φ20   ⊢φ20⊣

The food can shown in the diagram has a diameter of 10 cm. Construct a scale drawing of the plan view of the food can using a scale of 1 : 4.

THINK	WRITE/DRAW
1. State the scale factor and calculate the dimension of the object on the drawing.	Scale factor $= \dfrac{1}{4}$ Diameter on the diagram $= \dfrac{1}{4} \times 10$ $= \dfrac{10}{4}$ $= 2.5\,\text{cm}$
2. Draw the plan view of the object.	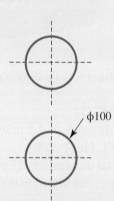
3. Label the drawing.	φ100

Note: The dimensions are written in mm.

Exercise 5.4 Creating scale drawings

1. **MC** A projection line is drawn using
   A. a thin continuous line with two arrowheads.
   B. a thick dashed line.
   C. a thin continuous line.
   D. a thick continuous line

2. **WE9** Calculate the dimensions required to create a scale drawing of a classroom 17 m long and 12 m wide to a scale of 1 : 100.

3. **a.** Calculate the dimensions required to create a scale drawing of a door 1.5 m wide, 2.4 m tall and 9 cm thick to a scale of 1 : 30.
   b. State the measurements that are to be written on a scale drawing.
   c. Are the projection lines parallel or perpendicular to the measurement shown?
   d. Write the symbol used to label the diameter of a circle on a drawing.

4. The drawings described are constructed to a scale of 1 : 50.
   a. Calculate the height, on the drawing, of a tree 3.6 m tall
   b. Calculate the dimensions, on the drawing, of a carpet 2.8 m by 1.4 m.
   c. Calculate the height, on the drawing, of a student 1.8 m tall.

5. a. **WE10** Construct a plan view, front view and side view scale drawing of a skip with drawing dimensions 5 cm long, 2 cm wide and 1.5 cm deep.

   b. **WE11** Label the three scale drawings of the skip from part a, given that the actual dimensions of the skip are 2.5 m long, 1 m wide and 0.75 m deep.

6. Construct a plan view, front view and side view scale drawing of a rectangular lunch box with drawing dimensions 5.3 cm long, 2.7 cm wide and 0.9 cm deep.

7. Calculate the dimensions to construct a scale drawing of the following shapes to the scale stated in brackets.
   a. A square of side length 1.7 cm (2 : 1)
   b. A rectangular pool table with dimensions 4.8 m by 4.0 m (1 : 80)
   c. A circular table with diameter 1.4 m (1 : 200)

8. Draw a house floor plan with actual dimensions 27.6 m by 18.3 m drawn to a scale of 1 : 600.

9. **MC** The dimension on a drawing is calculated using the formula
   A. scale factor × actual length.
   B. scale factor + actual length.
   C. $\dfrac{\text{scale factor}}{\text{actual length.}}$
   D. actual length − scale factor.

10. Construct and label a scale drawing of the floor of a bedroom 3.5 m long and 2.5 m wide drawn to a scale of 1 : 100.

11. Construct and label a scale drawing of a window frame 1.8 m wide and 1.4 m tall drawn to a scale of 1 : 40.

12. Draw and label the projection and dimension lines on the following diagrams.
   a. Square with side length 6.1 m

   b. Rectangle 430 cm long and 210 cm wide

   c. Equilateral triangle with side length 49.7 cm and height 43.0 cm

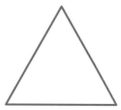

   d. Circle with diameter 7.6 m

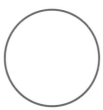

13. **WE12** The pipe shown in the diagram has a diameter of 28 cm. Construct a scale drawing of the plan view of the pipe drawn to a scale of 1 : 10.

14. The hat box shown in the diagram has a diameter of 36 cm. Construct a scale drawing of the plan view of the box drawn to a scale of 1 : 6.

15. A sculptor plans a sculpture by first creating a scale drawing. Determine the height of the sculpture on a drawing if its actual height is intended to be 5.2 m and the scale factor is $\frac{1}{26}$.

16. Measure the dimensions of your bedroom, bed and desk and construct a scale drawing using these measurements and an appropriate scale.

5.5 Review: exam practice

5.5.1 Scale drawings: summary

Reading and interpreting scale drawings

- A scale is a ratio of the length on a drawing to the actual length.

$$\text{Scale} = \text{length of drawing} : \text{actual length}$$

- A scale factor is the ratio of two corresponding lengths in two similar shapes.

Calculating measurements from scale drawings

- Areas and costs:

$$\text{Cost required} = \frac{\text{area of material}}{\text{area per squared unit}} \times \text{cost per unit}$$

- Amount of material and costs:

$$\text{Amount of paint required} = \text{area to be painted} \times \text{amount of paint per square metre}$$

$$\text{Total cost} = \text{amount of paint required} \times \text{cost of paint per litre}$$

- Number of bricks:

$$\text{Number of bricks} = \frac{\text{area of the wall}}{\text{area of the exposed side of the brick}}$$

Creating scale drawings

- Scale drawings are drawings on paper of real-life objects. All dimensions of the drawn object are kept in the same ratio to the actual dimensions of the object using a scale.
  - The dimension on a drawing can be calculated using the formula

$$\text{Dimension on the drawing} = \text{scale factor} \times \text{actual dimension}$$

- Conventions and line styles
  - The line style and its thickness indicate different details of the drawing.
    - A *thick continuous line* represents a visible line, a contour of a shape or an object.
    - A *thick dashed line* represents a hidden line, a hidden contour of a shape or an object.
    - *Thin continuous lines* are used to mark the dimensions of the shape or object.
  - Dimension lines are thin continuous lines showing the dimension of a line. They have arrowheads at both ends.

- Projection lines are thin continuous lines drawn perpendicular to the measurement shown. These lines do not touch the object.

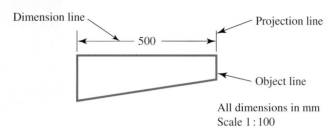

All dimensions in mm
Scale 1 : 100

- A circle on a scale diagram is labelled using either the Greek letter Φ (phi) for diameter or R for radius. The diagram shows three correct notations of the diameter. The diameter should be stated in millimetres.

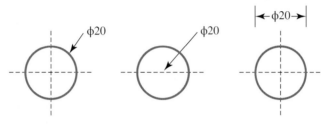

Exercise 5.5 Review: exam practice

Simple familiar

1. If the scale of a diagram is 5 cm : 100 km, convert the scale using same units.
2. **MC** If the scale factor of a diagram is 1 : 250, a length of 7.5 cm on the diagram represents an actual length of
   A. 33.33 m.
   B. 1875 cm.
   C. 7500 cm.
   D. 2.2 m.
3. Calculate the scale and the scale factor of a drawing where:
   a. 7 cm represents 4.9 km in reality
   b. 8 mm represents 6.4 m in reality.
4. The graphical scale on a map is shown below. State the scale of the map as a ratio, if each partition measures 2.5 cm.

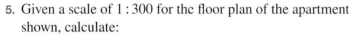

0   150   300 km

5. Given a scale of 1 : 300 for the floor plan of the apartment shown, calculate:
   a. the length and the width of the apartment, given it measures 6 cm by 4 cm on the diagram
   b. the perimeter of the apartment.
6. A sculptor plans a sculpture by first creating a scale drawing. Determine the height of the sculpture on a drawing if its actual height is intended to be 6.3 m and the scale factor is $\dfrac{1}{20}$.

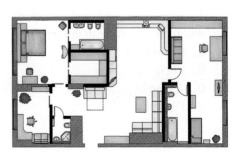

7. **MC** On a map with a scale $1:2\,500\,000$, the distance between two towns is 3.9 cm. The actual distance between the two towns is
   A. 3.9 km.
   B. 97.5 km.
   C. 64.1 km.
   D. 15.6 km.

8. **MC** The total cost of the paint required to paint an area of 68 m² at $16.90 per litre if 2 litres of paint are required per m² is
   A. $1149.20.
   B. $2298.40.
   C. $574.60.
   D. $136.

9. A brick has length 210 mm and height 55 mm. A bricklayer is building a wall 10 m long and 8 m high. Calculate:
   a. the number of bricks required
   b. the total cost for building the wall if one brick costs $2.15.

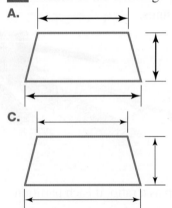

10. **MC** Which of the drawings shown uses the correct conventions?
    A.

    B.

    C.

    D.

11. **MC** A shoe box is 34 cm long, 19 cm wide and 15 cm high, as shown in the diagram. If a scale of $1:2$ is used, a scale diagram of the side view should be drawn with dimensions
    A. 17 cm by 9.5 cm.
    B. 17 cm by 7.5 cm.
    C. 9.5 cm by 7.5 cm.
    D. 19 cm by 15 cm.

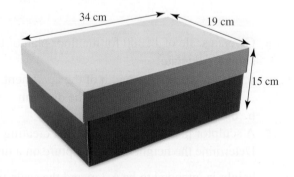

34 cm · 19 cm · 15 cm

12. A scale drawing of a rectangle of length 4.5 cm and width 1.5 cm is to be constructed using a scale $1:3$. Determine the length and width of the rectangle on the drawing.

Complex familiar

13. A house plan is drawn to a scale of $1:500$.
    a. Determine the actual length of the lounge room if it is represented by 0.6 cm on the plan.
    b. The width of the room is 5.2 m. Determine the width of the room on the house plan.

14. Sonia is to paint a feature wall in her lounge room. The plan uses a scale of $1:100$ and it measures 8.5 cm by 24 mm on the plan. Calculate:
    a. the length and the width of the wall
    b. the area of the wall
    c. the amount of paint Sonia has to buy if the amount of paint needed is 2.4 L per m^2
    d. the total cost of the paint if the price is $16.75 per litre.

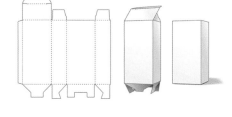

15. Consider the juice container shown and the template required to make it. The scale of the diagram is $1:5$.
    a. Calculate the length and the width of the packaging template, given they both measure 4 cm on the diagram.
    b. Estimate the total area of material required.
    c. If the cost of materials is $0.55 per square metre, calculate how much would it cost to make this juice container, to the nearest cent.

16. The scale factor of the brick wall shown is $\dfrac{1}{250}$.

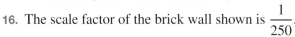

    Calculate:
    a. the length and the width of the wall, given it measures 10 cm by 2.5 cm on the diagram
    b. the area of the front side of the wall
    c. the number of bricks required if the dimensions of the exposed side of one brick are 20 cm by 5 cm
    d. the cost of the wall if the price of one brick is $0.65.

Complex unfamiliar

17. Shae wants to pave a path towards her front door with bricks. The path is shown at a scale of $1:100$. Calculate:
    a. the actual dimensions of the path, given it measures 10.5 cm by 0.75 cm
    b. the area of the path
    c. the number of bricks required if the bricks used are squares with side length 25 cm
    d. the cost of paving the garden path if the price of one brick is $1.85.

18. The scale of the house plan shown is $1:250$. Calculate:
    a. the dimensions of the top left room, given it measures 1.8 cm by 1.7 cm on the plan
    b. the width of the hallway, given it is 0.6 cm wide on the plan
    c. the perimeter of the kitchen, given it measures 1 cm by 1 cm on the plan
    d. the floor area of the house, given it measures 4.5 cm by 3.5 cm on the plan.

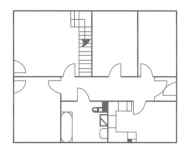

19. A radius of a circular mirror on a plan measures 75 mm. If the scale of the diagram is 1 : 50, calculate, to the nearest centimetre, the radius and the area of the mirror.

20. The objects in the diagram shown are drawing instruments.

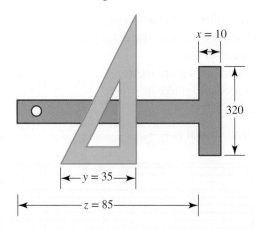

a. Calculate the scale factor of the drawing, given the length that is 32 mm measures 24 mm on the diagram.
b. Calculate the actual measurements of the unknown lengths x, y and z, given they measure 6 mm, 21 mm and 51 mm respectively on the diagram.
c. Determine the lengths of x, y and z on a drawing with a scale of 5 : 4.
d. Construct a scale drawing using the new drawing dimensions from part c. Clearly label all measurements.

Answers

5 Scales, plans and models

Exercise 5.2 Reading and interpreting scale drawings

1. $1:300$, $\dfrac{1}{300}$
2. $3:40\,000\,000$
3. B
4. a. $1:50\,000$, $\dfrac{1}{50\,000}$ b. $1:400$, $\dfrac{1}{400}$
5. a. $1:3700$, $\dfrac{1}{3700}$ b. $1:1600$, $\dfrac{1}{1600}$
6. $1:1\,000\,000$
7. $1:100\,000$
8. a. $1:10\,000\,000$ b. $1:8000$
9. $1:500\,000$
10. a. $32\,m$ b. $1.7\,cm$
11. a. $3\,m$ b. $1.04\,cm$
12. Length $= 2.25\,m$
    Width $= 1.5\,m$
13. Height $= 2.4\,m$
    Width $= 6\,m$
14. D

Exercise 5.3 Calculating measurements from scale drawings

1. a. Width $= 7.5\,m$
     Length $= 12.5\,m$
   b. $40\,m$
2. Radius $= 12\,m$
   Circumference $= 75\,m$
3. a. Height $= 1\,m$
     Big glass width $= 0.6\,m$
     Small glass width $= 0.3\,m$
   b. Area big glass $= 0.6\,m^2$
     Area of small glass $= 0.3\,m^2$
   c. $0.9\,m^2$
4. Radius $= 30\,cm$ and area $= 2827\,cm^2$
5. a. Length $= 28\,cm$
     Width $= 12\,cm$
   b. $336\,cm^2$
   c. 3 cents each
6. a. Length $=$ Width $= 21\,cm$
   b. $441\,cm^2$
   c. 2 cents each
7. a. $20.16\,L$ b. $\$316.51$
8. a. Length $= 1.5\,m$
     Height $= 2.5\,m$
   b. $3.75\,m^2$
   c. $9.75\,L$
   d. $\$124.80$
9. a. Width $=$ Length $= 0.26\,m$
   b. $0.0676\,m^2$
   c. $1.2\,m^2$
   d. $2.4\,L$
   e. $\$22.2$
10. a. $6\,m$, $2.5\,m$ b. $15\,m^2$
    c. 1500 bricks   d. $\$1125$

11. a. $12.75\,m$, $1.125\,m$ b. $14.34\,m^2$
    c. 249 bricks   d. $\$410.85$
12. a. Length $= 2.4\,m$ b. Width $= 1.2\,m$
    c. Perimeter $= 12\,m$ d. Area $= 63\,m^2$
13. a. 3221 bricks   b. $\$6280.95$
14. a. Length $= 64\,cm$
     Width $= 40\,cm$
   b. $0.256\,m^2$
   c. $\$0.55$
15. Radius $= 18\,cm$
    Circumference $= 113.10\,cm$
    Area $= 1017.88\,cm^2$

Exercise 5.4 Creating scale drawings

1. C
2. Length $= 17\,cm$
   Width $= 12\,cm$
3. a. Width $= 5\,cm$, height $= 8\,cm$, depth $= 0.3\,cm$
   b. Actual measurements.
   c. Perpendicular to the measurement shown.
   d. Φ The Greek letter phi
4. a. Height $= 7.2\,cm$
   b. $5.6\,cm$, $2.8\,cm$
   c. Height $= 3.6\,cm$
5. a.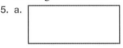
   Plan view

   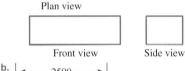
   Front view   Side view

   b.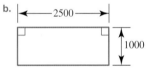

   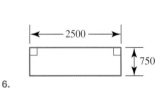 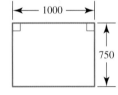

6.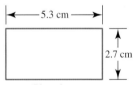
   Plan view

   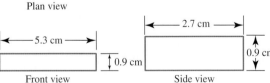
   Front view   Side view

7. a. $3.4\,cm$ b. $6\,cm$, $5\,cm$ c. $0.7\,cm$

8.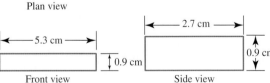

9. A

10.

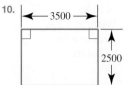

11.

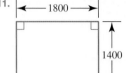

12. a.

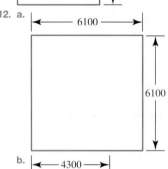

b.

c.

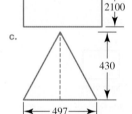

d.

13.

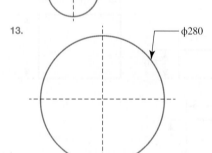

14.

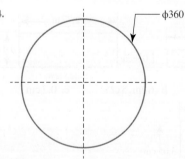

15. 20 cm

16. Sample responses can be found in the Worked Solutions in the online resources.

5.5 Review: exam practice

1. $1 : 2\,000\,000$
2. B
3. a. $1 : 70\,000, \dfrac{1}{70\,000}$ b. $1 : 800, \dfrac{1}{800}$
4. $1 : 60\,00\,000$
5. a. 18 m, 12 m
   b. 60 m
6. 31.5 cm
7. B
8. B
9. a. 6927 bricks b. \$14 893.05
10. C
11. C
12. Length $= 1.5$ cm
    Width $= 0.5$ cm
13. a. 3 m b. 1.04 cm
14. a. Length $= 8.5$ m
    Width $= 2.4$ m
    b. $20.4 \, \text{m}^2$
    c. 48.96 L
    d. \$820.08
15. a. Length $=$ Width $= 20$ cm
    b. $400 \, \text{cm}^2$
    c. 2 cents each
16. a. 25 m, 6.25 m b. $156.25 \, \text{m}^2$
    c. 15 625 bricks d. \$10 156.25
17. a. 10.5 m, 0.75 m b. $7.875 \, \text{m}^2$
    c. 126 bricks d. \$233.10
18. a. Length $= 4.5$ m
    Width $= 4.25$ m
    b. 1.5 m
    c. Length $= 2.5$ m
    Width $= 2.5$ m
    Perimeter $= 10$ m
    d. Length $= 11.25$ m
    Width $= 8.75$ m
    Area $= 98.44 \, \text{m}^2$
19. Radius $= 375$ cm
    Area $= 441\,786 \, \text{cm}^2$
20. a. $3 : 4$
    b. $x = 8$ mm, $y = 28$ mm, $z = 68$ mm
    c. $x = 10$ mm, $y = 35$ mm, $z = 85$ mm
    d.

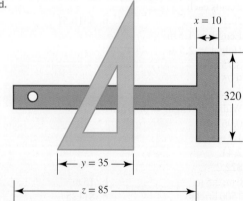

6 Right-angled triangles

6.1 Overview

LEARNING SEQUENCE

6.1 Overview
6.2 Pythagoras' theorem
6.3 Trigonometric ratios: sine, cosine and tangent
6.4 Calculating unknown side lengths of right-angled triangles
6.5 Calculating unknown angles in right-angled triangles
6.6 Angles of elevation and depression
6.7 Review: exam practice

CONTENT

In this chapter, students will learn to:
- apply Pythagoras' theorem to solve problems for all side lengths using $a^2 + b^2 = c^2$
- apply the tangent, sine and cosine ratios to find unknown angles and sides [complex]
- use the concepts of angle of elevation and angle of depression to solve practical problems [complex].

Fully worked solutions for this chapter are available in the Resources section of your eBookPLUS at www.jacplus.com.au.

6.2 Pythagoras' theorem

Pythagoras' theorem allows us to calculate the length of a side of a right-angled triangle if we know the lengths of the other two sides. Consider triangle ABC at right.

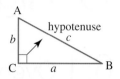

AB is the **hypotenuse** (the longest side). It is opposite the right angle.

Note that the sides of a triangle can be named in either of two ways.

1. A side can be named by the two capital letters given to the vertices at each end. This is what has been done in the figure at right to name the hypotenuse AB.

2. We can also name a side by using the lower-case letter of the opposite vertex. In the figure at right, we could have named the hypotenuse 'c'.

WORKED EXAMPLE 1

Name the hypotenuse in the triangle shown.

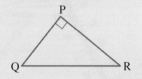

THINK	WRITE
1. Identify the hypotenuse by locating the longest side.	The longest side is opposite the right angle, between vertices Q and R, and opposite vertex P.
2. Name the hypotenuse.	The hypotenuse is QR or p.

Consider the right-angled triangle ABC with sides 3 cm, 4 cm and 5 cm. Squares have been constructed on each of the sides. The area of each square has been calculated by squaring the side length, and indicated inside the square.

Note that the area of the square on the hypotenuse is equal to the sum of the areas of the squares on the other two sides.

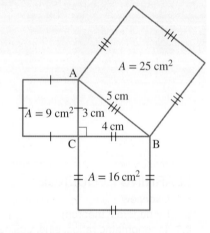

$$25\,\text{cm}^2 = 16\,\text{cm}^2 + 9\,\text{cm}^2$$

Alternatively: $(5\,\text{cm})^2 = (4\,\text{cm})^2 + (3\,\text{cm})^2$

Which means: $\text{hypotenuse}^2 = \text{base}^2 + \text{height}^2$

$$c^2 = a^2 + b^2$$

This result is known as Pythagoras' theorem.

Pythagoras' theorem

In any right-angled triangle, the square of the hypotenuse is equal to the sum of the squares of the two shorter sides. That is,

$$c^2 = a^2 + b^2$$

where a and b are the two shorter sides and c is the hypotenuse.

This is the formula used to find the length of the hypotenuse in a right-angled triangle when we are given the lengths of the two shorter sides.

Determine the length of the hypotenuse in the triangle at right.

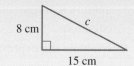

THINK	WRITE
1. This is a right-angled triangle, so we can use Pythagoras' theorem. Write the formula for Pythagoras' theorem.	$c^2 = a^2 + b^2$
2. Substitute the lengths of the shorter sides.	$c^2 = 15^2 + 8^2$
3. Evaluate the expression for c^2 by squaring 15 and 8, then completing the addition.	$= 225 + 64$ $= 289$
4. Calculate the value of c by taking the square root of both sides of the equation.	$\sqrt{c^2} = \sqrt{289}$ $c = \sqrt{289}$ $= 17\,\text{cm}$
5. State the answer.	The length of the hypotenuse is 17 cm.

In Worked example 2, the answer is a whole number because we can find $\sqrt{289}$ exactly. In most examples this will not be possible. In such cases, we are asked to write the answer correct to a given number of decimal places.

By rearranging Pythagoras' theorem, we can write the formula to find the length of a shorter side of a right-angled triangle.

$$c^2 = a^2 + b^2$$

We can solve for a^2 by subtracting b^2 on both sides of the equation.

$$c^2 - b^2 = a^2 + b^2 - b^2$$
$$c^2 - b^2 = a^2$$

Swapping sides gives:

$$a^2 = c^2 - b^2$$

Following a similar method we can rearrange for b^2.

$$b^2 = c^2 - a^2$$

The method of solving this type of question is the same as in the previous example, except that here we use subtraction instead of addition. For this reason, it is important to look at each question carefully to determine whether you are finding the length of the hypotenuse or one of the shorter sides.

Calculate the length of side PQ in triangle PQR, correct to one decimal place.

THINK	WRITE
1. This is a right-angled triangle, so we can use Pythagoras' theorem. Write the formula for Pythagoras' theorem when finding the length of a shorter side.	$b^2 = c^2 - a^2$

2. Substitute the lengths of the known sides. Place the unknown pronumeral on the left-hand side of the equation.

$$r^2 = 16^2 - 9^2$$

3. Evaluate the expression for r^2.

$$= 256 - 81$$
$$= 175$$

4. Calculate the answer by taking the square root of both sides of the equation and rounding the result to one decimal place.

$$\sqrt{r^2} = \sqrt{175}$$
$$r = \sqrt{175}$$
$$\approx 13.2\,\text{m}$$

5. State the answer.

The length of PQ is $13.2\,\text{m}$

Pythagoras' theorem can be used to solve more practical problems. In these cases, it is necessary to draw a diagram that will help you decide the appropriate method for finding a solution. The diagram simply needs to represent the triangle; it does not need to show details of the situation described.

WORKED EXAMPLE 4

The fire brigade attends a blaze in a tall building. They need to rescue a person from the 6th floor of the building, which is 30 metres above ground level. Their ladder is 32 metres long and must be at least 10 metres from the foot of the building. Can the ladder be used to reach the people needing rescue?

THINK

1. Define a pronumeral to represent the unknown distance in this problem.

2. Draw a diagram showing all important information.

3. The triangle is right-angled, so Pythagoras' theorem can be used. Write the rule for Pythagoras' theorem when finding the hypotenuse.

4. Substitute the lengths of the known sides.

5. Evaluate the expression on the right-hand side of the equation.

6. Calculate the answer by taking the square root of both sides of the equation.

WRITE

For the fire brigade's ladder to be long enough, the straight-line distance from a point on the ground 10 m from the base of the building to the window 30 m above the ground must be less than 32 m.
Let c represent the unknown distance.

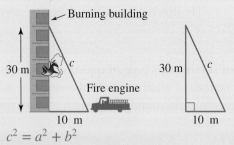

$$c^2 = a^2 + b^2$$

$$c^2 = 10^2 + 30^2$$

$$= 100 + 900$$
$$= 1000$$

$$\sqrt{c^2} = \sqrt{1000}$$
$$c = \sqrt{1000}$$
$$\approx 31.62\,\text{m}$$

7. Answer the question.   The ladder will be long enough to make the rescue, since it is 32 m long.

---

on Resources

Interactivities Finding the hypotenuse (int-3844)

Finding the shorter side (int-3845)

Exercise 6.2 Pythagoras' theorem

1. **WE1** Name the hypotenuse in each of the following triangles.

  a.   **b.** 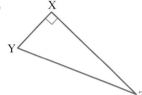   **c.**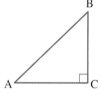

2. **WE2** Determine the length of the hypotenuse in each of the following triangles.

  a. 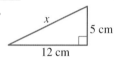   **b.** 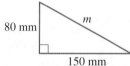   **c.**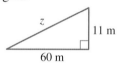

3. In each of the following, determine the length of the hypotenuse, correct to two decimal places.

  a.   **b.**   **c.**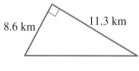

4. **WE3** Calculate the length of the unknown shorter side in each right-angled triangle, correct to one decimal place.

  a.   **b.**   **c.**

5. In each of the following right-angled triangles, determine the length of the side marked with a pronumeral, correct to one decimal place.

  a. 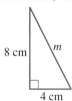 **b.**   **c.** 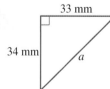 **d.**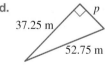

---

6. **MC** The hypotenuse in triangle WXY at right is
   A. WX.
   B. XY.
   C. YZ.
   D. ZW.

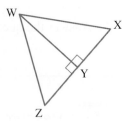

7. **WE4** A television antenna is 12 m high. To support it, wires are attached to the ground 5 m from the foot of the antenna. Determine the length of each wire.

8. Susie needs to clean the guttering on her roof. She places her ladder 1.2 m back from the edge of the guttering that is 3 m above the ground. How long will Susie's ladder need to be (correct to two decimal places)?

9. A rectangular gate is 3.5 m long and 1.3 m wide. The gate is to be strengthened by a diagonal brace as shown. How long should the brace be (correct to two decimal places)?

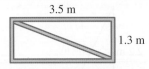

3.5 m

1.3 m

10. A 2.5-m ladder leans against a brick wall. The foot of the ladder is 1.2 m from the foot of the wall. How high up the wall will the ladder reach (correct to one decimal place)?

11. Use the measurements in the diagram to calculate the height of the flagpole, correct to one decimal place.

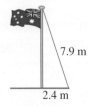

7.9 m

2.4 m

6.3 Trigonometric ratios: sine, cosine and tangent

We have already looked at Pythagoras' theorem, which enabled us to find the length of one side of a right-angled triangle given the lengths of the other two. However, to deal with other relationships in right-angled triangles, we need to turn to **trigonometry**.

Trigonometry allows us to work with the angles also; that is, deal with relationships between angles and sides of right-angled triangles. For example, trigonometry enables us to find the length of a side, given the length of another side and the size of an angle.

So that we are clear about which lines and angles we are describing, we need to identify the given angle, and name the shorter sides with reference to it. For this reason, we label the sides opposite and adjacent — that is, the sides **opposite** and **adjacent** to the given angle. The diagram shows this relationship between the sides and the angle, θ.

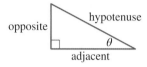

6.3.1 The tangent ratio

Trigonometry uses the ratio of side lengths to calculate the lengths of sides and the size of angles. The ratio of the opposite side to the adjacent side is called the **tangent ratio** (abbreviated 'tan'). This ratio is fixed for any particular angle.

The tangent ratio for any angle, θ, can be found using:

$$\tan \theta = \frac{\text{opposite side}}{\text{adjacent side}}$$

WORKED EXAMPLE 5

Write an expression for the tangent ratio of θ for the triangle shown.

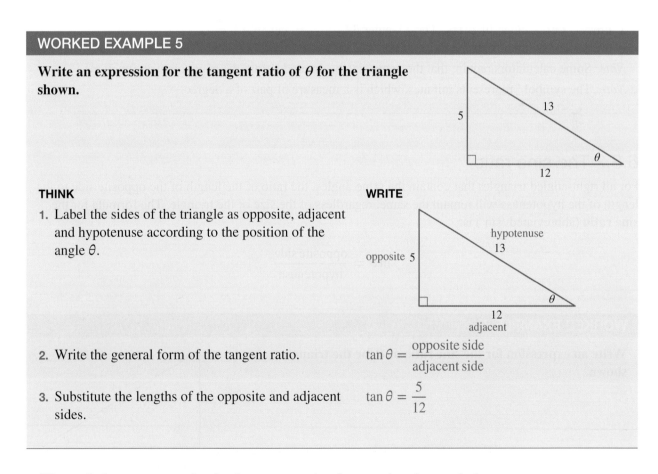

THINK

1. Label the sides of the triangle as opposite, adjacent and hypotenuse according to the position of the angle θ.

2. Write the general form of the tangent ratio.

3. Substitute the lengths of the opposite and adjacent sides.

WRITE

$$\tan \theta = \frac{\text{opposite side}}{\text{adjacent side}}$$

$$\tan \theta = \frac{5}{12}$$

We can find an accurate value for the tangent ratio of any angle using a calculator.

Calculators require a particular sequence of button presses in order to perform this calculation. This sequence may differ for different calculators.

For all calculations in trigonometry you will need to make sure that your calculator is in **DEGREES MODE**. Check the set-up on your calculator to ensure that this is the case.

WORKED EXAMPLE 6

Using your calculator, determine the value of the following, correct to three decimal places.

a. $\tan 60°$

b. $15 \tan 75°$

c. $\dfrac{8}{\tan 69°}$

d. $\tan 49°32'$

THINK

a. With a scientific calculator, press (tan) and enter 60, then press (=).

b. Enter 15, press (×) and (tan), enter 75, then press (=).

c. Enter 8, press (÷) and (tan), enter 69, then press (=).

d. Press (tan), enter 49, press (DMS), enter 32, press (DMS), then press (=).

WRITE

a. $\tan 60° = 1.732$

b. $15 \tan 75° = 55.981$

c. $\dfrac{8}{\tan 69°} = 3.071$

d. $\tan 49°32' = 1.172$

Note: Some calculators require that the angle size be entered before the trigonometric functions.

Note: The symbol $'$ represents minutes, which is a measure of part of a degree.

6.3.2 The sine ratio

For all right-angled triangles that contain the same angles, the ratio of the length of the opposite side to the length of the hypotenuse will remain the same, regardless of the size of the triangle. The formula for the **sine ratio** (abbreviated 'sin') is:

$$\sin \theta = \frac{\text{opposite side}}{\text{hypotenuse}}$$

WORKED EXAMPLE 7

Write an expression for the sine ratio of θ for the triangle shown.

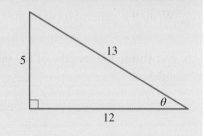

THINK	WRITE
1. Label the sides of the triangle as opposite, adjacent and hypotenuse according to the position of the angle θ.	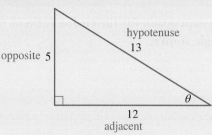
2. Write the general form of the sine ratio.	$\sin \theta = \dfrac{\text{opposite side}}{\text{hypotenuse}}$
3. Substitute the lengths of the opposite side and the hypotenuse.	$\sin \theta = \dfrac{5}{13}$

The value of the sine ratio for any angle can be found using the sin function on the calculator.

WORKED EXAMPLE 8

Using your calculator, determine the value of the following, correct to three decimal places:

a. $\sin 57°$ **b.** $9 \sin 45°$ **c.** $\dfrac{18}{\sin 44°}$ **d.** $9.6 \sin 26°12'$

THINK	WRITE
a. With a scientific calculator, press ⬭sin and enter 57, then press ⬭=.	**a.** $\sin 57° = 0.839$
b. Enter 9, press ⬭× and ⬭sin, enter 45, then press ⬭=.	**b.** $9 \sin 45° = 6.364$
c. Enter 18, press ⬭÷ and ⬭sin, enter 44, then press ⬭=.	**c.** $\dfrac{18}{\sin 44°} = 25.912$
d. Enter 9.6, press ⬭× and ⬭sin, enter 26, press ⬭DMS, enter 12, press ⬭DMS, then press ⬭=.	**d.** $9.6 \sin 26°12' = 4.238$

Note: Check the sequence of button presses required by your calculator.

6.3.3 The cosine ratio

A third trigonometric ratio is the cosine ratio. This ratio compares the length of the adjacent side and the hypotenuse.

The **cosine ratio** (abbreviated 'cos') is found using the formula:

$$\cos \theta = \frac{\text{adjacent side}}{\text{hypotenuse}}$$

WORKED EXAMPLE 9

Write an expression for the cosine ratio of θ for the triangle shown.

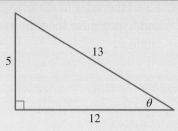

THINK

1. Label the sides of the triangle as opposite, adjacent and hypotenuse according to the position of the angle θ.

2. Write the general form of the cosine ratio.

3. Substitute the lengths of the adjacent side and the hypotenuse.

WRITE

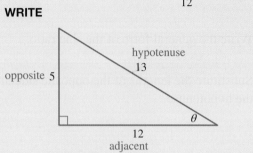

$\cos\theta = \dfrac{\text{adjacent side}}{\text{hypotenuse}}$

$\cos\theta = \dfrac{12}{13}$

To calculate the cosine ratio for a given angle on your calculator, use the cos function.

WORKED EXAMPLE 10

Using your calculator, determine the value of the following, correct to three decimal places:

a. $\cos 27°$ b. $6\cos 55°$ c. $\dfrac{21.3}{\cos 74°}$ d. $\dfrac{4.5}{\cos 82°46'}$.

THINK

a. With a scientific calculator, press ⬭cos⬭ and enter 27, then press ⬭=⬭.

b. Enter 6, press ⬭×⬭ and ⬭cos⬭, enter 55, then press ⬭=⬭.

c. Enter 21.3, press ⬭÷⬭ and ⬭cos⬭, enter 74, then press ⬭=⬭.

d. Enter 4.5, press ⬭÷⬭ and ⬭cos⬭, enter 82, press ⬭DMS⬭, enter 46, press ⬭DMS⬭, then press ⬭=⬭.

Note: Check the sequence of button presses required by your calculator.

WRITE

a. $\cos 27° = 0.891$

b. $6\cos 55° = 3.441$

c. $\dfrac{21.3}{\cos 74°} = 77.275$

d. $\dfrac{4.5}{\cos 82°46'} = 35.740$

6.3.4 Inverse trigonometric functions

If we are given the sine, cosine or tangent of an angle, we can calculate the size of that angle using a calculator. We do this using the inverse trigonometric functions. On most calculators these are the second function of the sin, cos and tan functions and are denoted $\sin^{-1}$, $\cos^{-1}$ and $\tan^{-1}$.

WORKED EXAMPLE 11

Calculate the value of θ, correct to the nearest degree, given that $\sin \theta = 0.738$.

THINK	WRITE
1. With a scientific calculator, press ⟨2nd F⟩[sin⁻¹] (or press SHIFT ⟨ sin ⟩) and enter 0.738, then press ⟨ = ⟩.	$\sin^{-1} 0.738 = 47.56...$
2. Round your answer to the nearest degree.	$\theta = 48°$

Note: Check the sequence of button presses required by your calculator.

Problems sometimes measure angles in degrees (°), minutes (′) and seconds (″), or require answers to be written in the form of degrees, minutes and seconds. On scientific calculators, you will use the **DMS** (Degrees, Minutes, Seconds) function or the ⟨ ° ′ ″ ⟩ function. If you are using a graphics calculator, the 'angle' function provides this facility. To round an angle to the nearest minute, you need to look at the number of seconds. If there are 30 or more seconds, you will need to round the number of minutes up by 1.

WORKED EXAMPLE 12

Given that $\tan \theta = 1.647$, calculate the value of θ correct to the nearest minute.

THINK	WRITE
1. With a scientific calculator, press ⟨2nd F⟩ [tan⁻¹] (or press SHIFT ⟨ tan ⟩) and enter 1.647, then press ⟨ = ⟩.	$\tan^{-1} 1.647 = 58.73...$
2. Convert your answer to degrees and minutes by pressing ⟨ DMS ⟩.	$\theta = 58°44'$

Note: Check the sequence of button presses required by your calculator.

---

on Resources

📄 **Digital document** SkillSHEET Labelling the sides of a right-angled triangle (doc-5226)
🔧 **Interactivity** Trigonometric ratios (int-2577)

Exercise 6.3 Trigonometric ratios: sine, cosine and tangent

1. Label the sides of the following right-angled triangles using the words hypotenuse, adjacent and opposite.

   a.

   b.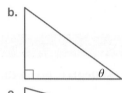

   c.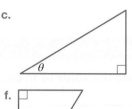

   d.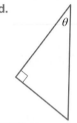

   e.

   f.

2. Write an expression for each of the following ratios for the angle θ.

   a. **WE5** tangent

   b. **WE7** sine

   c. **WE9** cosine

   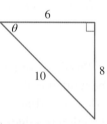

3. **WE6** Using your calculator, determine the value of the following, correct to three decimal places:

   a. $\tan 57°$

   b. $9 \tan 63°$

   c. $\dfrac{8.6}{\tan 12°}$

   d. $\tan 33°19'$

4. **WE8** Using your calculator, determine the value of the following, correct to three decimal places:

   a. $\sin 37°$

   b. $9.3 \sin 13°$

   c. $\dfrac{14.5}{\sin 72°}$

   d. $\dfrac{48}{\sin 67°40'}$

5. **WE10** Using your calculator, determine the value of the following, correct to three decimal places:

   a. $\cos 45°$

   b. $0.25 \cos 9°$

   c. $\dfrac{6}{\cos 24°}$

   d. $5.9 \cos 2°3'$

6. Calculate the value of each of the following, correct to three decimal places, if necessary.

   a. $\sin 30°$

   b. $\cos 15°$

   c. $\tan 45°$

   d. $48 \tan 85°$

   e. $128 \cos 60°$

   f. $9.35 \sin 8°$

7. Calculate the value of each of the following, correct to three decimal places, if necessary.

   a. $\dfrac{4.5}{\cos 32°}$

   b. $\dfrac{0.5}{\tan 20°}$

   c. $\dfrac{15}{\sin 72°}$

8. Calculate the value of each of the following, correct to two decimal places.

   a. $\sin 24°38'$

   b. $\tan 57°21'$

   c. $\cos 84°40'$

   d. $9 \cos 55°30'$

   e. $4.9 \sin 35°50'$

   f. $2.39 \tan 8°59'$

9. Calculate the value of each of the following, correct to two decimal places.

   a. $\dfrac{19}{\tan 67°45'}$

   b. $\dfrac{49.6}{\cos 47°25'}$

   c. $\dfrac{0.84}{\sin 75°5'}$

10. **WE11** Calculate the value of θ, correct to the nearest degree, given that $\sin \theta = 0.167$.

11. Determine θ, correct to the nearest degree, given that:

    a. $\sin \theta = 0.698$

    b. $\cos \theta = 0.173$

    c. $\tan \theta = 1.517$.

12. **WE12** Given that $\cos \theta = 0.058$, calculate the value of θ correct to the nearest minute.

13. Calculate the value of θ, correct to the nearest minute, given that:

    a. $\tan \theta = 0.931$

    b. $\cos \theta = 0.854$

    c. $\sin \theta = 0.277$.

6.4 Calculating unknown side lengths of right-angled triangles

We can use the trigonometric ratios to find the length of one side of a right-angled triangle if we know the length of another side and an angle.

We need to be able to look at a problem and then decide if the solution can be determined using the sine, cosine or tangent ratio. To do this we need to examine the three formulas.

$$\tan \theta = \frac{\text{opposite side}}{\text{adjacent side}}$$

We use the tangent ratio when we are finding either the opposite or adjacent side and are given the length of the other.

$$\sin \theta = \frac{\text{opposite side}}{\text{hypotenuse}}$$

The sine ratio is used when finding the opposite side or the hypotenuse when given the length of the other.

$$\cos \theta = \frac{\text{adjacent side}}{\text{hypotenuse}}$$

The cosine ratio is for problems where we are finding the adjacent side or the hypotenuse and are given the length of the other.

Remembering trigonometric ratios

To remember the trigonometric ratios more easily, we can use this acronym:

$$\text{SOHCAHTOA}$$

We pronounce this acronym as 'Sock ca toe ahh'. The initials of the acronym represent the three trigonometric formulas.

$$\binom{S}{O}{H} \sin \theta = \frac{\text{opposite}}{\text{hypotenuse}} \qquad \binom{C}{A}{H} \cos \theta = \frac{\text{adjacent}}{\text{hypotenuse}} \qquad \binom{T}{O}{A} \tan \theta = \frac{\text{opposite}}{\text{adjacent}}$$

Consider the triangle at right.

In this triangle we are asked to find the length of the opposite side and have been given the length of the adjacent side.

We know that the tangent ratio uses the opposite and adjacent sides:

$\tan \theta = \dfrac{\text{opposite}}{\text{adjacent}}$. In this example, $\tan 30° = \dfrac{x}{14}$.

We can set up an equation that can be solved to determine the value of x.

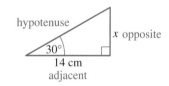

$$\tan \theta = \frac{\text{opposite}}{\text{adjacent}}$$

$$\tan 30° = \frac{x}{14}$$

$$14 \times \tan 30° = \frac{\cancel{14}\,x}{\cancel{14}} \qquad \text{Multiply both sides by 14}$$

$$x = 14 \tan 30° \qquad \text{Enter 14 tan 30° into your calculator}$$

$$= 8.083 \text{ cm}$$

WORKED EXAMPLE 13

Use the tangent ratio to determine the value of *h* in the triangle shown, correct to two decimal places.

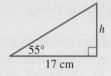

THINK

WRITE

1. Label the sides of the triangle opposite, adjacent and hypotenuse.

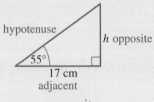

2. Write the tangent ratio.

$$\tan \theta = \frac{\text{opposite}}{\text{adjacent}}$$

3. Substitute the values of the known angle and the known side length: $\theta = 55°$ and adjacent = 17.

$$\tan 55° = \frac{h}{17}$$

4. Make *h* the subject of the equation by multiplying both sides of the equation by 17.

$$17 \times \tan 55° = \frac{17h}{17}$$
$$h = 17 \tan 55°$$

5. Use your calculator to evaluate the expression for *h*.

$$= 24.28 \text{ cm}$$

WORKED EXAMPLE 14

Determine the length of the side marked *x*, correct to two decimal places.

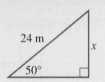

THINK

WRITE

1. Label the sides of the triangle opposite, adjacent and hypotenuse.

2. *x* is the opposite side and 24 m is the hypotenuse; therefore use the sine ratio.

$$\sin \theta = \frac{\text{opposite}}{\text{hypotenuse}}$$

3. Substitute the values of the known angle and the known side length: $\theta = 50°$ and hypotenuse = 24.

$$\sin 50° = \frac{x}{24}$$

4. Make *x* the subject of the equation by multiplying both sides of the equation by 24.

$$24 \times \sin 50° = \frac{24x}{24}$$
$$x = 24 \sin 50°$$

5. Use your calculator to evaluate the expression for *x*.

$$= 18.39 \text{ m}$$

Care needs to be taken at the substitution stage. In the previous two examples, the unknown side was the numerator in the fraction, hence we multiplied to find the answer. If, after substitution, the unknown side is in the denominator, the final step is a division.

WORKED EXAMPLE 15

Calculate the length of the side marked z in the triangle shown, correct to two decimal places.

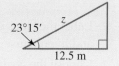

THINK

1. Label the sides of the triangle opposite, adjacent and hypotenuse.

2. Choose the cosine ratio because we are finding the hypotenuse and have been given the adjacent side. Write the cosine ratio.

3. Substitute the values of the known angle and the known side length: $\theta = 23°15'$ and adjacent $= 12.5$.

4. Multiply both sides of the equation by z to remove z from the denominator.

5. Divide both sides of the equation by $\cos 23°15'$ to make z the subject.

6. Use your calculator to evaluate the expression for z.

WRITE

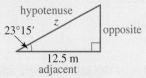

$$\cos \theta = \frac{\text{adjacent}}{\text{hypotenuse}}$$

$$\cos 23°15' = \frac{12.5}{z}$$

$$z \times \cos 23°15' = \frac{\cancel{z} \times 12.5}{\cancel{z}}$$

$$z \cos 23°15' = 12.5$$

$$\frac{z \cancel{\cos 23°15'}}{\cancel{\cos 23°15'}} = \frac{12.5}{\cos 23°15'}$$

$$z = \frac{12.5}{\cos 23°15'}$$

$$z = 13.60 \text{ m}$$

Trigonometry is used to solve many practical problems. In these cases, it is necessary to draw a diagram to represent the problem and then use trigonometry to solve the problem. With written problems that require you to draw the diagram, it is necessary to give the answer in words.

WORKED EXAMPLE 16

A flying fox is used in an army training camp. The flying fox is supported by a cable that runs from the top of a cliff face to a point on the ground 100 m from the base of the cliff. The cable makes a 15° angle with the horizontal. Determine the length of the cable used to support the flying fox, correct to one decimal place.

THINK	WRITE
1. Draw a diagram and show all important information.	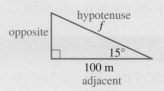

2. Label the sides of the triangle opposite, adjacent and hypotenuse.

3. Choose the cosine ratio to solve this problem because we are finding the hypotenuse and have been given the adjacent side.

4. Write the cosine ratio.

$$\cos \theta = \frac{\text{adjacent}}{\text{hypotenuse}}$$

5. Substitute the values of the known angle and the known side length: $\theta = 15°$ and adjacent $= 100$.

$$\cos 15° = \frac{100}{f}$$

6. Make f the subject of the equation by multiplying both sides by f and then dividing both sides by $\cos 15°$.

$$f \times \cos 15° = \frac{100f}{f}$$
$$f \cos 15° = 100$$
$$\frac{f \cos 15°}{\cos 15°} = \frac{100}{\cos 15°}$$
$$f = \frac{100}{\cos 15°}$$

7. Use your calculator to evaluate the expression for f.

$$f = 103.5 \text{ m}$$

8. Answer the question.

The cable is approximately $= 103.5$ m long.

Exercise 6.4 Calculating unknown side lengths of right-angled triangles

1. Label the sides of each of the following triangles, with respect to the angle marked with the pronumeral.

a.

b.

c.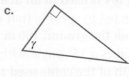

2. **WE13** Use the tangent ratio to determine the length of the side marked x (correct to one decimal place).

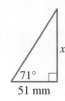

---

3. Use the sine ratio to determine the length of the side marked *a* (correct to two decimal places).

4. Use the cosine ratio to determine the length of the side marked *d* (correct to nearest whole number).

5. **WE14** Determine the length of the side marked with the pronumeral, correct to one decimal place.

a.

b.

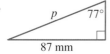

c.

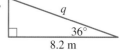

6. **WE15** Calculate the length of the side marked with the pronumeral in each of the following (correct to one decimal place).

a.

b.

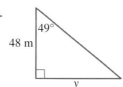

c.

7. Calculate the length of the side marked with the pronumeral in each of the following (correct to one decimal place).

a.

b.

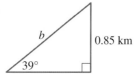

c.

d.

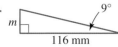

e.

f.

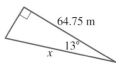

g.

h.

i.

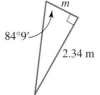

j.

q

$60°32'$

84.6 km

k.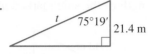

t $75°19'$

21.4 m

l.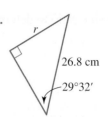

r

26.8 cm

$29°32'$

8. **MC** Look at the diagram and state which of the following is correct.

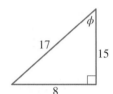

A. $x = 9.2 \sin 69°$

B. $x = \dfrac{9.2}{\sin 69°}$

C. $x = 9.2 \cos 69°$

D. $x = \dfrac{9.2}{\cos 69°}$

x

$69°$

9.2

9. **MC** Study the triangle and state which of the following is correct.

A. $\tan \phi = \dfrac{8}{15}$

B. $\tan \phi = \dfrac{15}{8}$

C. $\sin \phi = \dfrac{15}{17}$

D. $\cos \phi = \dfrac{17}{15}$

ϕ

17

15

8

10. **MC** Study the diagram and state which of the statements is correct.

A. $w = 22 \cos 36°$

B. $w = \dfrac{22}{\sin 36°}$

C. $w = 22 \cos 54°$

D. $w = 22 \sin 54°$

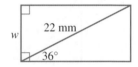

22 mm

w

$36°$

11. **WE16** Calculate the height of the tree in the following diagram, correct to the nearest metre.

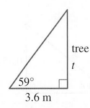

tree

t

$59°$

3.6 m

12. A 10-m ladder just reaches to the top of a wall when it is leaning at $65°$ to the ground. How far from the foot of the wall is the base of the ladder (correct to one decimal place)?

13. The diagram shows the paths of two ships, A and B, after they have left the port. If ship B sends a distress signal, how far must ship A sail to give assistance (to the nearest kilometre).

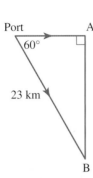

14. A rectangular sign 13.5 m wide has a diagonal brace that makes a 24° angle with the horizontal.
    a. Draw a diagram of this situation.
    b. Calculate the height of the sign, correct to the nearest metre.

15. A wooden gate has a diagonal brace built in for support. The gate stands 1.4 m high and the diagonal makes a 60° angle with the horizontal.
    a. Draw a diagram of the gate.
    b. Calculate the length that the diagonal brace needs to be (correct to one decimal place).

16. The wire support for a flagpole makes a 70° angle with the ground. If the support is 3.3 m from the base of the flagpole, calculate the length of the wire support (correct to two decimal places).

17. A ship drops anchor vertically with an anchor line 60 m long. After one hour the anchor line makes a 15° angle with the vertical.
    a. Draw a diagram of this situation.
    b. Calculate the depth of water, correct to the nearest metre.
    c. Calculate the distance that the ship has drifted, correct to one decimal place.

6.5 Calculating unknown angles in right-angled triangles

So far, we have concerned ourselves with finding side lengths. We are also able to use trigonometry to find the sizes of angles when we have been given two side lengths. We can use the inverse trigonometric ratios $\sin^{-1}$, $\cos^{-1}$ and $\tan^{-1}$ to find an unknown angle in a right-angled triangle.

Inverse trigonometric ratios

$$\text{If } \sin\theta = a, \text{ then } \sin^{-1} a = \theta.$$
$$\text{If } \cos\theta = a, \text{ then } \cos^{-1} a = \theta.$$
$$\text{If } \tan\theta = a, \text{ then } \tan^{-1} a = \theta.$$

Consider the right-angled triangle at right.
We want to find the size of the angle marked θ.

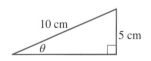

Using the formula $\sin \theta = \dfrac{\text{opposite}}{\text{hypotenuse}}$ we know that in this triangle:

$$\sin \theta = \frac{5}{10}$$
$$= \frac{1}{2}$$
$$= 0.5$$

We then take the inverse sin of both sides of the equation and calculate $\sin^{-1}(0.5)$ to find that $\theta = 30°$. As with all trigonometry, it is important that you have your calculator set to degrees mode for this work.

WORKED EXAMPLE 17

Determine the size of angle θ in the triangle, correct to the nearest degree.

THINK

1. Label the sides of the triangle.

WRITE

2. Choose the tangent ratio to solve the problem as we are given the opposite and adjacent side lengths. Write the tangent ratio.

$\tan \theta = \dfrac{\text{opposite}}{\text{adjacent}}$

3. Substitute the values of the known side lengths: opposite = 4.3 and adjacent = 6.5.

$\tan \theta = \dfrac{4.3}{6.5}$

4. Make θ the subject of the equation by taking the inverse tan of both sides.

$\theta = \tan^{-1}\left(\dfrac{4.3}{6.5}\right)$

5. Use your calculator to evaluate the expression for θ.

$\theta = 33°$

In many cases, we will need to calculate the size of an angle, correct to the nearest minute. The same method for finding the solution is used; however, you will need to use your calculator to convert to degrees and minutes.

WORKED EXAMPLE 18

Calculate the size of the angle θ in the triangle, correct to the nearest minute.

THINK

1. Label the sides of the triangle.

WRITE

2. Choose the sine ratio to solve this problem as we are given the opposite side and the hypotenuse. Write the sine ratio.

$$\sin \theta = \frac{\text{opposite}}{\text{hypotenuse}}$$

3. Substitute the values of the known side lengths: opposite = 4.6 and hypotenuse = 7.1.

$$\sin \theta = \frac{4.6}{7.1}$$

4. Make θ the subject of the equation by taking the inverse sin of both sides.

$$\theta = \sin^{-1}\left(\frac{4.6}{7.1}\right)$$

5. Use your calculator to evaluate the expression for θ and convert your answer to degrees and minutes.

$$\theta = 40°23'$$

The same methods can be used to solve problems requiring an unknown angle to be found. As with finding sides, we set the question up by drawing a diagram of the situation.

WORKED EXAMPLE 19

A ladder is leant against a wall. The foot of the ladder is 4 m from the base of the wall and the ladder reaches 10 m up the wall. Calculate the angle that the ladder makes with the ground, correct to the nearest minute.

THINK

1. Draw a diagram showing all important information and label the sides.

2. Choose the tangent ratio to solve this problem as we are given the opposite and adjacent sides. Write the tangent ratio.

3. Substitute the values of the known side lengths: opposite = 10 and adjacent = 4.

4. Make θ the subject of the equation by taking the inverse tan of both sides.

5. Use your calculator to evaluate the expression for θ.

6. Answer the question.

WRITE

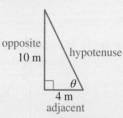

$$\tan \theta = \frac{\text{opposite}}{\text{adjacent}}$$

$$\tan \theta = \frac{10}{4}$$

$$\theta = \tan^{-1}\left(\frac{10}{4}\right)$$

$$= 68°12'$$

The ladder makes an angle of 68°12′ with the ground.

Exercise 6.5 Calculating unknown angles in right-angled triangles

1. **WE17** Use the tangent ratio to determine the size of the angle marked with the pronumeral in each of the following, correct to the nearest degree.

   a.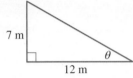

   b.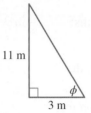

   c.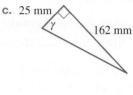

2. **WE18** Use the sine ratio to determine the size of the angle marked with the pronumeral in each of the following, correct to the nearest minute.

   a.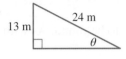

   b.

   c.

3. Use the cosine ratio to determine the size of the angle marked with the pronumeral in each of the following, correct to the nearest minute.

   a.

   b.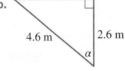

   c.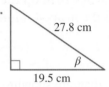

4. Calculate the size of the angle marked θ in the following triangles, correct to the nearest degree.

   a.

   b.

   c.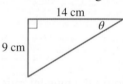

   d.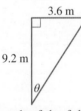

   e.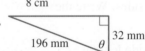

   f.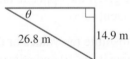

5. In each of the following calculate the size of the angle marked θ, correct to the nearest minute.

   a.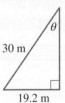

   b.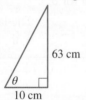

   c.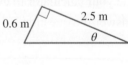

   d.

   e.

   f.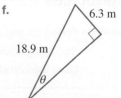

6. **MC** Look at the triangle. Which of the statements below is correct?

   A. ∠ABC = 30°
   B. ∠CAB = 30°
   C. ∠ABC = 45°
   D. ∠CAB = 45°

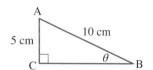

7. **MC** Consider the triangle. The correct value of θ is

   A. 36°39′.
   B. 41°55′.
   C. 41°56′.
   D. 48°4′.

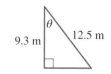

8. **WE19** A 10-m ladder leans against a wall 6 m high. Calculate the angle that the ladder makes with the horizontal, correct to the nearest degree.

9. A kite is flying on a 40-m string. The kite is flying 10 m away from the vertical as shown in the figure. Calculate the angle the string makes with the horizontal, correct to the nearest minute.

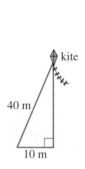

10. A ship's compass shows a course due east of the port from which it sails. After sailing 10 nautical miles, it is found that the ship is 1.5 nautical miles off course as shown in the figure.

Determine the error in the compass reading, correct to the nearest minute.

11. The diagram shows a footballer's shot at goal.

By dividing the isosceles triangle in half to form a right-angled triangle, calculate, to the nearest degree, the angle within which the footballer must kick to get the ball to go between the posts.

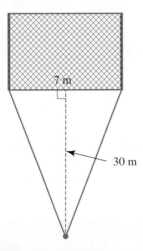

7 m

30 m

12. A golfer hits a ball 250 m, but 20 m off centre. Calculate the angle at which the ball deviated from a straight line, correct to the nearest minute.

6.6 Angles of elevation and depression

The **angle of elevation** is measured upwards from the **horizontal** and refers to the angle at which we need to look up to see an object. Similarly, the **angle of depression** is the angle at which we need to look down from the horizontal to see an object.

We can use the angles of elevation and depression to calculate the heights and distances of objects that would otherwise be difficult to measure.

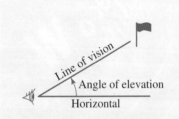

Line of vision
Angle of elevation
Horizontal

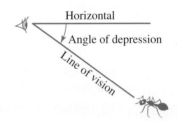

Horizontal
Angle of depression
Line of vision

WORKED EXAMPLE 20

From a point 50 m from the foot of a building, the angle of elevation to the top of the building is measured as 40°. Calculate the height, h, of the building, correct to the nearest metre.

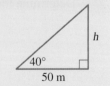

h

40°

50 m

THINK

WRITE

1. Label the sides of the triangle opposite, adjacent and hypotenuse.

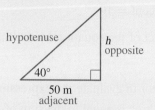

hypotenuse
h
opposite
40°
50 m
adjacent

2. Choose the tangent ratio because we are finding the opposite side and have been given the adjacent side.

3. Write the tangent ratio.

$$\tan\theta = \frac{\text{opposite}}{\text{adjacent}}$$

4. Substitute the values of the known side and angle: $\theta = 40°$ and adjacent = 50.

$$\tan 40° = \frac{h}{50}$$

5. Make h the subject of the equation by multiplying both sides by 50.

$$50 \times \tan 40° = \frac{50h}{50}$$

$$h = 50\tan 40°$$

6. Use your calculator to evaluate the expression for h.

$$h = 42 \text{ m}$$

7. Answer the question.

The height of the building is approximately 42 m.

In practical situations, the angle of elevation is measured using a clinometer (or inclinometer). Therefore, the angle of elevation is measured from a person's height at eye level. For this reason, the height at eye level must be added to the calculated answer.

WORKED EXAMPLE 21

Bryan measures the angle of elevation to the top of a tree as 64°, from a point 10 m from the foot of the tree. If the height of Bryan's eyes is 1.6 m above the ground, calculate the height of the tree, correct to one decimal place.

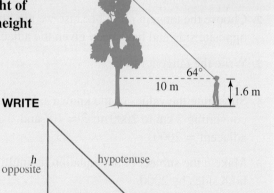

64°
10 m
1.6 m

THINK

WRITE

1. Label the sides opposite, adjacent and hypotenuse.

h
opposite
hypotenuse
64°
10 m
adjacent

2. Choose the tangent ratio because we are finding the opposite side and have been given the adjacent side.

3. Write the tangent ratio.

$$\tan\theta = \frac{\text{opposite}}{\text{adjacent}}$$

4. Substitute the values of the known side and angle: $\theta = 64°$ and adjacent = 10.

$$\tan 64° = \frac{h}{10}$$

5. Make h the subject of the equation by multiplying both sides by 10.

$$10 \times \tan 64° = \frac{\cancel{10}h}{\cancel{10}}$$

$$h = 10 \tan 64°$$

6. Use your calculator to evaluate the expression for h.

$$h = 20.5 \text{ m}$$

7. The height of the tree is equal to the sum of h and the eye height.

Height of tree = 20.5 + 1.6
= 22.1

8. Answer the question.

The height of the tree is approximately 22.1 m.

A similar method for finding the solution is used for problems that involve an angle of depression.

WORKED EXAMPLE 22

When an aeroplane in flight is 2 km from a runway, the angle of depression to the runway is 10°. Calculate the altitude of the aeroplane, correct to the nearest metre.

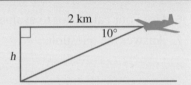

THINK

1. Label the sides of the triangle opposite, adjacent and hypotenuse.

WRITE

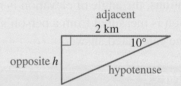

2. Choose the tangent ratio, because we are finding the opposite side and have been given the adjacent side.

3. Write the tangent ratio.

$$\tan \theta = \frac{\text{opposite}}{\text{adjacent}}$$

4. Substitute the values of the known side and angle, converting 2 km to 2000 m: $\theta = 10°$ and adjacent = 2000.

$$\tan 10° = \frac{h}{2000}$$

5. Make h the subject of the equation by multiplying both sides by 2000.

$$2000 \times \tan 10° = \frac{\cancel{2000}h}{\cancel{2000}}$$

$$h = 2000 \tan 10°$$

6. Use your calculator to evaluate the expression for h.

$h = 353 \text{ m}$

7. Answer the question.

The altitude of the aeroplane is approximately 353 m.

Angles of elevation and depression can also be calculated by using known measurements. This is done by drawing a right-angled triangle to represent a situation.

WORKED EXAMPLE 23

A 5.2-m building casts a 3.6-m shadow. Calculate the angle of elevation of the sun, correct to the nearest degree.

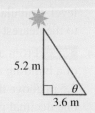

THINK	WRITE
1. Label the sides opposite, adjacent and hypotenuse.	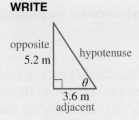
2. Choose the tangent ratio because we are given the opposite and adjacent sides.	
3. Write the tangent.	$\tan\theta = \dfrac{\text{opposite}}{\text{adjacent}}$
4. Substitute the values of the known side lengths: opposite = 5.2 and adjacent = 3.6.	$\tan\theta = \dfrac{5.2}{3.6}$
5. Make θ the subject of the equation by taking the inverse tan of both sides.	$\theta = \tan^{-1}\dfrac{5.2}{3.6}$
6. Use your calculator to evaluate the expression for θ.	$\theta = 55°$
7. Answer the question.	The angle of elevation of the sun is approximately 55°.

 Resources

 Digital document WorkSHEET Trigonometry using elevation and depression (doc-10838)

 Video eLesson Using an inclinometer (eles-0116)

Interactivities Finding the angle of elevation and angle of depression (int-6047)

Angles of elevation and depression (int-4501)

Exercise 6.6 Angles of elevation and depression

1. **WE20** From a point 100 m from the foot of a building, the angle of elevation to the top of the building is 15°.
   Calculate the height of the building, correct to one decimal place.

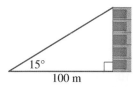

2. From a ship the angle of elevation to an aeroplane is 60°. The aeroplane is located at a horizontal distance of 2300 m away from the ship. Calculate the altitude of the aeroplane, correct to the nearest metre.

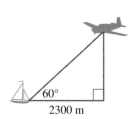

3. From a point out to sea, a ship sights the top of a lighthouse at an angle of elevation of 12°. It is known that the top of the lighthouse is 40 m above sea level. Calculate the horizontal distance of the ship from the lighthouse, correct to the nearest 10 m.

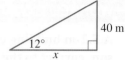

4. **WE21** From a point 50 m from the foot of a building, Rod sights the top of a building at an angle of elevation of 37°. Given that Rod's eyes are 1.5 m above the ground, calculate the height of the building, correct to one decimal place.

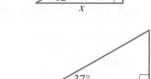

5. Richard is flying a kite and sights the kite at an angle of elevation of 65°. The altitude of the kite is 40 m and Richard's eyes are at a height of 1.8 m. Calculate the length of string the kite is flying on, correct to one decimal place.

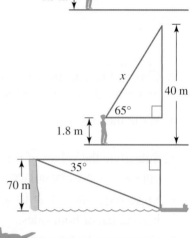

6. From the top of a cliff, 70 m above sea level, a ship is spotted out to sea at an angle of depression of 35°. Calculate the distance of the ship from shore, to the nearest metre.

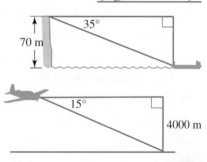

7. **WE22** From an aeroplane flying at an altitude of 4000 m, the runway is sighted at an angle of depression of 15°. Calculate the horizontal distance of the aeroplane from the runway, correct to the nearest kilometre.

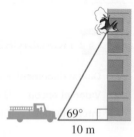

8. There is a fire on the fifth floor of a building. The closest a fire truck can get to the building is 10 m. The angle of elevation from this point to where people need to be rescued is 69°. If the fire truck has a 30-m ladder, can the ladder be used to make the rescue?

9. **WE25** A 12-m high building casts a shadow 15 m long. Calculate the angle of elevation of the sun, to the nearest degree.

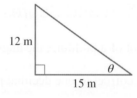

10. An aeroplane, which is at an altitude of 1500 m, is 4000 m from a ship in a horizontal direction, as shown. Calculate the angle of depression from the aeroplane to the ship, to the nearest degree.

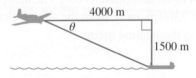

11. The angle of elevation to the top of a tower is $12°$ from a point 400 m from the foot of the tower.

   a. Draw a diagram of this situation.
   b. Calculate the height of the tower, correct to one decimal place.
   c. Calculate the angle of elevation to the top of the tower, from a point 100 m from the foot of the tower.

12. From a navy vessel, a beacon which is 80 m above sea level, is sighted at an angle of elevation of $5°$. The vessel sailed towards the beacon and 30 minutes later the beacon is at an angle of elevation of $60°$.
   Use the diagram to complete the following.

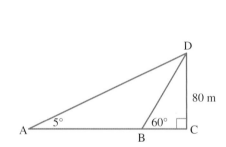

   a. Calculate the distance that the vessel was from the beacon, when the angle of elevation to the beacon was $5°$ (the distance AC).
   b. Calculate the distance that the vessel sailed in the 30 minutes between the two readings.

6.7 Review: exam practice

6.7.1 Right-angled triangles summary

Pythagoras' theorem

• When using Pythagoras' theorem to calculate the length of the hypotenuse of a right-angled triangle, the formula is:

$$c^2 = a^2 + b^2$$

where c is the length of the hypotenuse, and a and b are the two shorter sides.

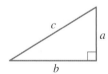

• To find one of the shorter sides of a right-angled triangle, the formula is:

$$a^2 = c^2 - b^2$$
$$\text{or} \quad b^2 = c^2 - a^2$$

Trigonometric ratios: sine, cosine and tangent

- $\sin\theta = \dfrac{\text{opposite}}{\text{hypotenuse}}$
- $\cos\theta = \dfrac{\text{adjacent}}{\text{hypotenuse}}$
- $\tan\theta = \dfrac{\text{opposite}}{\text{adjacent}}$
- SOHCAHTOA — this acronym will help you remember trigonometric ratios.

Steps to find a side of a right-angled triangle

- Label the sides of the triangle opposite, adjacent and hypotenuse.
- Choose the appropriate ratio.
- Substitute given information.
- Make the unknown side the subject of the equation.
- Calculate the answer.

Steps to find an angle in a right-angled triangle

- Label the sides of the triangle opposite, adjacent and hypotenuse.
- Choose the appropriate ratio.
- Substitute given information.
- Make the unknown angle the subject of the equation.
- Calculate the answer by using an inverse trigonometric ratio.

Angles of elevation and depression

- The angle of elevation is the angle we look up from the horizontal to see an object.
- The angle of depression is the angle we look down from the horizontal to see an object.
- Problems are solved using angles of elevation and depression by the same methods as for all right-angled triangles.

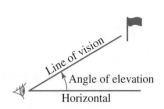

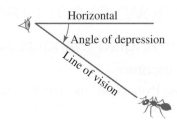

---

on Resources

🗎 **Digital documents** SkillSHEET Selecting an appropriate trigonometric ratio based on the given information (doc-5231)

SkillSHEET Using trigonometry (doc-5233)

Exercise 6.7 Review: exam practice

Simple familiar

1. Determine the length of the side marked with a pronumeral, in each case writing your answer correct to two decimal places.

a.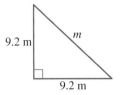
9.2 m
9.2 m
m

b.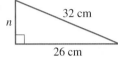
n
32 cm
26 cm

c.
4.8 m
p
3.2 m

d.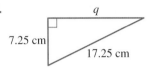
q
7.25 cm
17.25 cm

e.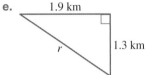
1.9 km
r
1.3 km

f.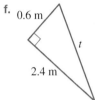
0.6 m
t
2.4 m

2. **MC** What is the length of the third side in this triangle?

A. 48.75 cm
B. 0.698 m
C. 0.926 m
D. 92.6 cm

82 cm
43 cm

3. Which of the following correctly names the sides and angle of the triangle shown?
A. $\angle C = \theta$, AB = adjacent side, AC = hypotenuse, BC = opposite side
B. $\angle C = \theta$, AB = opposite side, AC = hypotenuse, AC = adjacent side
C. $\angle A = \theta$, AB = opposite side, AC = hypotenuse, BC = adjacent side
D. $\angle A = \theta$, AB = adjacent side, AC = hypotenuse, BC = opposite side

A
θ
B
C

4. Calculate each of the following, correct to four decimal places.

a. $\sin 46°$
b. $\tan 76°42'$
c. $4.9 \cos 56°$
d. $8.9 \sin 67°3'$
e. $\dfrac{5.69}{\cos 75°}$
f. $\dfrac{2.5}{\tan 9°55'}$

5. Calculate θ, correct to the nearest degree, given that:
a. $\cos \theta = 0.5874$
b. $\tan \theta = 1.23$
c. $\sin \theta = 0.8.$

6. Calculate θ, correct to the nearest minute, given that:
a. $\cos \theta = 0.199$
b. $\tan \theta = 0.5$
c. $\sin \theta = 0.257.$

7. Determine the length of each side marked with a pronumeral, correct to one decimal place.

a.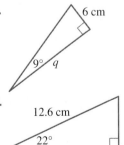
6 cm
9°
q

b.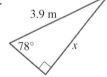
3.9 m
78°
x

c.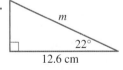
m
22°
12.6 cm

d.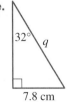
12.6 cm
22°
n

e.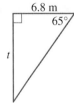
32°
q
7.8 cm

f.
6.8 m
65°
t

8. Determine the length of each side marked with a pronumeral, correct to one decimal place.

a.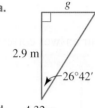
g
2.9 m
26°42′

b.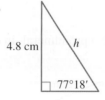
4.8 cm
h
77°18′

c.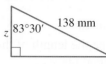
83°30′   138 mm
z

d.
4.32 m
j
29°51′

e.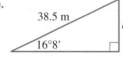
38.5 m
k
16°8′

f.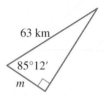
63 km
85°12′
m

9. Determine the size of the angle marked θ in each of the following, giving your answer correct to the nearest degree.

a.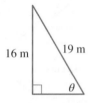
16 m   19 m
θ

b.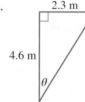
2.3 m
4.6 m
θ

c.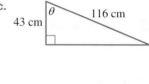
θ 116 cm
43 cm

10. Determine the size of the angle marked θ in each of the following, giving your answer correct to the nearest minute.

a.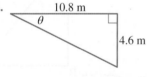
10.8 m
θ
4.6 m

b.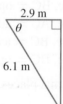
2.9 m
θ
6.1 m

c.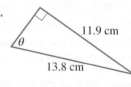
11.9 cm
θ
13.8 cm

11. The top of a building is sighted at an angle of elevation of 40°, when an observer is 27 m back from the base.
    Calculate the height of the building, correct to the nearest metre.

h
40°
27 m

12. A lifesaver standing on his tower 3 m above the ground spots a swimmer experiencing difficulty. The angle of depression of the swimmer from the lifesaver is 12°. How far is the swimmer from the lifesaver's tower? (Give your answer correct to two decimal places.)

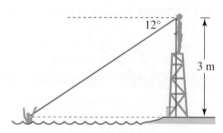
12°
3 m

13. To travel between the towns of Bolong and Molong, you need to travel west along a road for 45 km, then north along another road for another 87 km. Calculate the straight-line distance between the two towns.

14. A rope is 80 m long and runs from a cliff top to the ground, 45 m from the base of the cliff. Calculate the height of the cliff, to the nearest metre.

15. A fire is burning in a building and people need to be rescued. The fire brigade's ladder must reach a height of 60 m and must be angled at 70° to the horizontal. How long must the ladder be to complete the rescue?

16. A rope that is used to support a flagpole makes an angle of 70° with the ground. If the rope is tied down 3.1 m from the foot of the flagpole, calculate the height of the flagpole, correct to one decimal place.

Complex unfamiliar

17. A dirt track runs off a road at an angle of 34° to the road. If I travel for 4.5 km along the dirt track, what is the shortest direct distance back to the road (correct to one decimal place)?

18. A child flies a kite on an 80-m string. He holds the end of the string 1.5 metres above the ground, and the kite reaches a height of 51.5 m above the ground in a strong wind. Calculate the angle the string makes with the horizontal.

19. There is 50 m of line on a fishing reel. When all the line is out, the bait sits on the bed of a lake and has drifted 20 m from the boat. Calculate the angle that the fishing line makes with the vertical.

20. Hakam stands 50 m back from the foot of an 80-m telephone tower. Hakam's eyes are at a height of 1.57 m. Calculate the angle of elevation that Hakam must look to see the top of the tower.

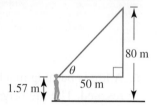

Answers

6 Right-angled triangles

Exercise 6.2 Pythagoras' theorem

1. a. PR or q
   b. YZ or x
   c. AB or c
2. a. 13 cm   b. 170 mm   c. 61 m
3. a. 10.82 cm   b. 6.93 m   c. 14.20 km
4. a. 10.4 cm   b. 1.9 m   c. 3.9 m
5. a. 8.9 cm   b. 22.1 cm   c. 47.4 mm
   d. 37.3 m
6. A
7. 13 m
8. 3.23 m
9. 3.73 m
10. 2.2 m
11. 7.5 m

Exercise 6.3 Trigonometric ratios: sine, cosine and tangent

1. a.
b.
c.
d.
e.
f.

2. a. $\tan\theta = \dfrac{8}{6}$ b. $\sin\theta = \dfrac{8}{10}$ c. $\cos\theta = \dfrac{6}{10}$
3. a. 1.540   b. 17.663   c. 40.460   d. 0.657
4. a. 0.602   b. 2.092   c. 15.246   d. 51.893
5. a. 0.707   b. 0.247   c. 6.568   d. 5.896
6. a. 0.5   b. 0.966   c. 1
   d. 548.643   e. 64   f. 1.301
7. a. 5.306   b. 1.374   c. 15.772
8. a. 0.42   b. 1.56   c. 0.09
   d. 5.10   e. 2.87   f. 0.38
9. a. 7.77   b. 73.30   c. 0.87
10. 10°
11. a. 44°   b. 80°   c. 57°
12. 86°40′
13. a. 42°57′   b. 31°21′   c. 16°5′

Exercise 6.4 Calculating unknown side lengths of right-angled triangles

1. a.
b.

c.

2. 148.1 mm
3. 5.08 m
4. 30 cm
5. a. 12.5 m   b. 89.3 mm   c. 10.1 m
6. a. 12.1 cm   b. 55.2 m   c. 9.4 km
7. a. 5.4 m   b. 1.4 km   c. 2.1 km
   d. 18.4 mm   e. 3.2 cm   f. 66.5 m
   g. 5.4 m   h. 5.4 km   i. 0.2 m
   j. 41.6 km   k. 84.4 m   l. 13.2 cm
8. D
9. A
10. C
11. 6 m
12. 4.2 m
13. 20 km
14. a.
b. 6 m
15. a.
b. 1.6 m
16. 9.65 m
17. a.
b. 58 m   c. 15.5 m

Exercise 6.5 Calculating unknown angles in right-angled triangles

1. a. 30°   b. 75°   c. 81°
2. a. 32°48′   b. 45°3   c. 35°16′
3. a. 53°8′   b. 55°35′   c. 45°27′
4. a. 50°   b. 32°   c. 33°
   d. 21°   e. 81°   f. 34°
5. a. 39°48′   b. 80°59′   c. 13°30′
   d. 79°6′   e. 63°1′   f. 19°28′
6. A
7. C
8. 37°
9. 75°31′
10. 8°38′
11. 13°
12. 4°35′

Exercise 6.6 Angles of elevation and depression

1. 26.8 m
2. 3984 m
3. 190 m
4. 39.2 m
5. 42.1 m
6. 100 m
7. 15 km

8. 28 m, so a 30-m ladder can be used.
9. 39°
10. 21°
11. a.

b. 85 m   c. 40°

12. a. 914 m   b. 868 m

6.7 Review: exam practice

1. a. 13.01 m   b. 18.65 cm   c. 3.58 m
   d. 15.65 cm   e. 2.30 km   f. 2.47 m
2. B
3. D
4. a. 0.7193   b. 4.2303   c. 2.7400
   d. 8.1955   e. 21.9845   f. 14.2998
5. a. 54°   b. 51°   c. 53°

6. a. 78°31′   b. 26°34′   c. 14°54′
7. a. 37.9 cm   b. 3.8 m   c. 13.6 cm
   d. 11.7 cm   e. 14.7 cm   f. 14.6 m
8. a. 1.5 m   b. 4.9 cm   c. 15.6 mm
   d. 7.5 m   e. 10.7 m   f. 5.3 km
9. a. 57°   b. 27°   c. 68°
10. a. 23°4′   b. 61°37′   c. 59°35′
11. 23 m
12. 14.11 m
13. 98 km
14. 66 m
15. 63.9 m
16. 8.5 m
17. 2.5 km
18. 39°
19. 24°
20. 57°

7 Summarising and interpreting data

7.1 Overview

LEARNING SEQUENCE

7.1 Overview
7.2 Measures of central tendency and mode
7.3 Measures of spread
7.4 Outliers
7.5 Applications of measures of centre and spread
7.6 Review: exam practice

CONTENT

In this chapter, students will learn to:
- identify the mode from a data set
- calculate measures of central tendency, the mean and the median from a data set
- investigate the suitability of measures of central tendency in various real-world contexts [complex]
- investigate the effect of outliers on the mean and the median [complex]
- calculate quartiles from a dataset [complex]
- interpret quartiles, deciles and percentiles from a graph [complex]
- use everyday language to describe spread, including spread out, dispersed, tightly packed, clusters, gaps, more/less dense regions and outliers
- calculate and interpret statistical measures of spread, such as the range, interquartile range and standard deviation [complex]
- investigate real-world examples from the media illustrating inappropriate uses of measures of central tendency and spread [complex].

Fully worked solutions for this chapter are available in the Resources section of your eBookPLUS at www.jacplus.com.au.

7.2 Measures of central tendency and mode

7.2.1 Mean and median

In statistics, a population is the entire set of subjects or objects being studied or investigated.

A sample is a smaller selection of subjects or objects taken from the population.

Statistical analysis on a sample can often be used to make generalisations about the population, as long as the subjects or objects in the sample are selected at random.

Values associated with populations are called **parameters**, whereas those associated with samples are called **summary statistics**.

The mean

The **mean** of a population is a theoretical measure of the centre of the entire population. The mean is referred to as the *average* in everyday language. It is not always a value in a data set; for example, the mean number of dogs in an Australian household is 1.5.

Obtaining the mean of a population is not always practicable, so the mean of a sample is often calculated instead.

The mean of a population, μ, and the mean of a sample, $\bar{x}$, are calculated using the same formula.

Population and sample means

$$\mu = \frac{\text{sum of all the values in the population}}{\text{number of values in the population}} = \frac{\sum x}{N}$$

and

$$\bar{x} = \frac{\text{sum of all the values in the sample}}{\text{number of values in the sample}} = \frac{\sum x}{n},$$

where x is a data value, N is the total number of data values in the population and n is the total number of data values in the sample.

$\sum$ is the summation symbol. The summation symbol, $\sum$ (the Greek capital letter sigma), means to add all the occurrences of the formula to the right of the summation symbol. Since the data sets in this chapter are samples, we will use the formula $x = \dfrac{\sum x}{N}$ to calculate the mean.

The median

The **median** is the middle value of the data set by position. It is the value that half the observations are less than and half are greater than.

The median is also a *theoretical measure* of the centre of the set of data. It is not always a value in the set of data.

To calculate the median, arrange the values in the set of data in *ascending* order (smallest to largest).

- For an *odd* number of values, the median is the *middle* value. For example, if the data is 6, 7, 9, 10 and 15, then the median is 9.

$$6, 7, 9, 10, 15$$

$$\uparrow$$

$$\text{Median} = 9$$

- For an *even* number of values, the median is the *mean* of the two middle values. In this case the median may not be an actual data value. For example, if the data is 3, 5, 7, 9, 11 and 17, the median is 8.

$$3, 5, 7, | 9, 11, 17$$
$$\uparrow$$
$$\text{Median} = \frac{7+9}{2} = 8$$

If the set of data contains n values, the position of the median when the values are arranged in numerical (ascending) order can be calculated using the following formula.

$$\text{Median position} = \frac{n+1}{2}$$

If, $n = 6$ then $\frac{n+1}{2} = \frac{6+1}{2}$

$$= \frac{7}{2}$$

$$= 3.5$$

The median is half-way between the third and fourth values.

If, $n = 5$ then $\frac{n+1}{2} = \frac{5+1}{2}$

$$= \frac{6}{2}$$

$$= 3$$

The median is the third value.

7.2.2 The mode

The **mode** is the value that occurs most often. It is often described as the typical value of the observations. The mode is not considered a measure of central tendency; it is simply one of the features of a data set.

The mode can be found for both numerical and categorical data. (Recall that **categorical data** are data that can be grouped or classified, and **numerical data** are data that can be counted or measured.)

Some sets of data have no mode at all. If all of the values occur only once, there is no mode.

```
                    Data
          /                      \
  Categorical data          Numerical data
     /        \               /          \
Ordinal   Nominal       Discrete    Continuous
 data      data          data         data
```

It is possible for a data set to have more than one mode. A data set with two modes is referred to as 'bimodal'. A data set with three modes is referred to as 'trimodal'. A data set with more than three modes is referred to as 'multimodal'.

WORKED EXAMPLE 1

Ten Year 11 students were asked how many hours they spent competing in sport out of school hours. The results of the survey are: 2, 4, 5, 6, 7, 2, 2, 0, 5, 1. Calculate the values of:

a. the mean

b. the median

c. the mode

and explain what they tell us about the data.

THINK	WRITE
a. 1. Calculate the sum of all the values in the data set.	**a.** Sum of all the values $= 2 + 4 + 5 + 6 + 7 + 2 + 2 + 0 + 5 + 1$ $= 34$
2. Count the number of values in the data set (the 0 must be included).	Number of values $= 10$
3. The mean is the sum of all the values divided by the number of values.	Mean $(\bar{x}) = \dfrac{\text{sum of all the values}}{\text{number of values}}$ $= \dfrac{\sum x}{n}$ $= \dfrac{34}{10}$ $= 3.4$
4. Explain what the mean tells us about the data.	If the total number of hours spent competing in sport was shared equally between all ten students, each student would spend on average 3.4 hours competing in sport.
b. 1. Arrange the values in the data set in ascending order (smallest to largest).	**b.** 0, 1, 2, 2, 2, 4, 5, 5, 6, 7
2. The median is the middle value. Locate the position of the median.	There are 10 values in the set of data so the median is the $\dfrac{10 + 1}{2} = 5.5$th value (mean of the 5th and 6th values). 0, 1, 2, 2, 2, | 4, 5, 5, 6, 7
3. Determine the median value by calculating the mean of 2 and 4.	Median $= \dfrac{2 + 4}{2}$ $= \dfrac{6}{2}$ $= 3$
4. Explain what the median tells us about the data.	Half of the students compete in more than 3 hours of sport and half of the students compete in less than 3 hours.
c. 1. The mode is the most common value in the set of data.	**c.** 2, 4, 5, 6, 7, 2, 2, 0, 5, 1 Mode $= 2$
2. Explain what the mode tells us about the data.	The most common amount of time that students spend competing in sport is 2 hours.

7.2.3 Measures of central tendency in dot plots and stem plots

Recall that a list of data may be represented in different formats, such as a dot plot or stem (stem-and-leaf) plot.

In a dot plot, each data value is represented as a dot on a number line.

In a stem-and-leaf plot, each data value is split into two components, the stem and the leaf. Data is then grouped according to its stem.

Calculate the mean of each of the following sets of data. (Give your answer to one decimal place.)

a.

b. Key : 12|1 = 121

Stem	Leaf
12	1 5 9
13	4 7 9
14	8 8
15	2 8
16	3 8
17	2

THINK

a. 1. Write the list of data represented in the dot plot.

2. Calculate the sum of all the values in the data set.

3. Count the number of values in the data set.

4. The mean is the sum of all the values divided by the number of values.

b. 1. Write the list of data represented in the stem plot.

2. Calculate the sum of all the values in the data set.

3. Count the number of values in the data set.

4. The mean is the sum of all the values divided by the number of values.

WRITE

a. 8, 10, 11, 11, 11, 12, 12, 12, 12, 12, 13, 13, 14, 18

Sum of all the values $= 8 + 10 + 11 + 11 + 11 + 12 + 12$
$$+ 12 + 12 + 12 + 13 + 13 + 14 + 18$$
$$= 169$$

Number of values $= 14$

$$\text{Mean } (\bar{x}) = \frac{\text{sum of all the values}}{\text{number of values}}$$
$$= \frac{\sum x}{n}$$
$$= \frac{169}{14}$$
$$\approx 12.1$$

b. 121, 125, 129, 134, 137, 139, 148, 148, 152, 158, 163, 168, 172

Sum of all the values $= 121 + 125 + 129 + 134 + 137 + 139$
$$+ 148 + 148 + 152 + 158 + 163$$
$$+ 168 + 172$$
$$= 1894$$

Number of values $= 13$

$$\text{Mean } (\bar{x}) = \frac{\text{sum of all the values}}{\text{number of values}}$$
$$= \frac{\sum x}{n}$$
$$= \frac{1894}{13}$$
$$\approx 145.7$$

7.2.4 Measures of central tendency and mode in frequency distribution tables

Cumulative frequency

Cumulative frequency of a data value is the number of observations that are above or below the particular value. Cumulative frequency is recorded as a **cumulative frequency table**.

The final value in a cumulative frequency table will always equal the total number of observations in the data set. This cumulative frequency table shows the number of movies watched in the last month by a group of 30 Year 9 students.

Data (x)	Frequency (f)	Cumulative frequency (cf)
3	3	3
4	5	$3 + 5 = 8$
5	7	$8 + 7 = 15$
6	10	$15 + 10 = 25$
7	0	$25 + 0 = 25$
8	5	$25 + 5 = 30$

Ogives

Data from a cumulative frequency table can be plotted to form a **cumulative frequency curve**, which is also called an **ogive** (pronounced '*oh-jive*').

To plot an ogive for data that is in class intervals, the maximum value for the class interval is used as the value against which the cumulative frequency is plotted.

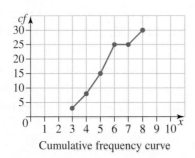

Cumulative frequency curve

Calculating the mean, median and mode from a frequency distribution table

If data are presented in a frequency distribution table, the formula used to calculate the mean is

$$\bar{x} = \frac{\text{sum of (frequency} \times \text{data values)}}{\text{sum of frequencies}} = \frac{\sum (f \times x)}{n}$$

Here, each data value (score) in the table is multiplied by its corresponding frequency; then all the $f \times x$ products are added together and the total sum is divided by the number of observations in the set. To find the median, find the position of each score from the cumulative frequency column. The mode is the score with the highest **frequency**.

For the data shown in the frequency distribution table, calculate the:

a. the mean   b. the median   c. the mode.

Score (x)	Frequency (f)
4	1
5	2
6	5
7	4
8	3
Total	15

THINK

a. 1. Rule up a table with four columns titled Score (x), Frequency (f), Frequency × score ($f \times x$) and Cumulative frequency (cf).

WRITE

a.

Score (x)	Frequency (f)	Frequency × score ($f \times x$)	Cumulative frequency (cf)
4	1		
5	2		
6	5		
7	4		
8	3		
Total	$n = 15$		

2. • Calculate the values of the $f \times x$ column by multiplying the frequency by the score in each row.
   In the first row:
   $f \times x = 4 \times 1$
   $ = 4$
   • Calculate the total sum of all the data values by adding all values in the $f \times x$ column. The total of the $f \times x$ column is 96.

Score (x)	Frequency (f)	Frequency × score ($f \times x$)	Cumulative frequency (cf)
4	1	④	
5	2	10	
6	5	30	
7	4	28	
8	3	24	
Total	$n = 15$	$\sum (f \times x) = ⑨⑥$	

3. Write the rule for the mean.

$$\bar{x} = \frac{\sum (f \times x)}{n}$$

4. Substitute the known values into the rule and evaluate.

$$\bar{x} = \frac{96}{15}$$

$$= 6.4$$

5. Answer the question.

The mean of the data set is 6.4.

b. 1. Complete the *cf* column by calculating the running total of frequencies.

b.

Score (*x*)	Frequency (*f*)	Frequency × score (*f* × *x*)	Cumulative frequency (*cf*)
4	1	4	1
5	2	10	$1 + 2 = 3$
6	5	30	$3 + 5 = 8$
7	4	28	$8 + 4 = 12$
8	3	24	$12 + 3 = 15$
	$n = 15$	$\sum (f \times x) = 96$	

2. Locate the position of the median using the rule $\frac{n+1}{2}$, where $n = 15$.
This places the median as the 8th score.

The median is the $\frac{15+1}{2}$th or 8th score.

3. Use the cumulative frequency column to locate the 8th score and answer the question.

The median of the data set is 6.

c. 1. The mode is the score with the highest frequency.

c. The score with the highest frequency is 6.

2. Answer the question.

The mode of the data set is 6.

7.2.5 Measures of central tendency and mode in real-world contexts

The mean is often called the average in real-world contexts. It is used in scientific research to describe many different phenomena, from weather patterns to sports statistics. The mean is also used to analyse the results of students in schools and universities. Examples include: average number of children per household in Australia, average daily rainfall, average number of goals scored in a game, and average mark on a Mathematics exam.

The median is the middle value of a data set by position, and is often used when the information being investigated has some extremely high or extremely low values in the data set. The median is not greatly affected by these extreme values, as it looks at the number of data points, not their values. For this reason, the median is often used in economics. Examples include median household income and median house price.

The mode is less commonly used in real-world contexts, but it can still be a practical and useful way of describing information. For example, clothing shop owners will look at the highest-selling clothing size to make sure they have the most floor stock in that particular size.

Always remember to relate summary statistics back to real-world contexts, ensuring they have relevance and meaning. For example, a shoe-shop owner might determine that she sells a mean of 26.7 pairs of shoes a day. This can be interpreted as selling, on average, between 26 and 27 pairs of shoes per day.

7.2.6 Grouped data

For large data sets, it may be necessary to group data into smaller sets of data called **class intervals**. These class intervals must be the same size and must be set so that each value belongs to one interval only.

Class intervals are written differently for different types of data. Recall that **discrete data** are numerical data that can only take certain values (usually whole numbers), and **continuous data** are numerical data that can take any value within an interval.

A class interval for:
- discrete data is written showing the minimum and maximum values for the interval (e.g. $10 - 20$ includes the values between and including 10 and 20, for example 12 or 16)
- continuous data is written like an inequation (e.g. for values of 10 up to values less than $20 \, (10 \leq x < 20)$, the class interval is written $10- < 20$.

A frequency table shows the frequency of values appearing in each class interval.

The class interval with the highest frequency is the mode of the grouped data and is called the **modal class**.

The **midpoint** of the class interval is the average of the maximum and minimum values for the class interval.

The midpoint is used as the representative value for the class interval. It is assumed that half of the data values will be greater than the midpoint and half will be less than the midpoint.

The mean of grouped data is calculated using the midpoint of each class interval to represent the data values in the interval.

Calculating the mean from grouped data

$$\bar{x} = \frac{\text{sum of (midpoints} \times \text{frequency)}}{\text{sum of frequencies}} = \frac{\sum (x_{\text{mid}} \times f)}{n}$$

where:

x_{mid} is the midpoint of the class interval

f is the frequency of the class interval

n is the total number of data values

$\sum$ is the summation symbol.

A takeaway food shop is trying to improve its ordering system. The manager takes a survey of the number of hamburgers made each day for a month and collects the data shown in the frequency table. Calculate the mean number of hamburgers made each day.

Class interval	Frequency (f)
0–4	6
5–9	8
10–14	2
15–19	3
20–24	5

THINK

1. To calculate the mean of grouped data, a single value is needed to represent the class interval.
   - To calculate the midpoint of the class intervals, calculate the mean of the maximum and minimum values for each class interval.
     – For the first class interval 0–4,
       $$x_{mid} = \frac{0+4}{2} = 2$$
     – For the second class interval 5–9, $x_{mid} = \frac{5+9}{2} = 7$

2. The midpoint will be the representative value for the interval.
   - Add a new column to the table, $x_{mid} \times f$, and multiply the midpoint by the frequency of each class interval to calculate a value to represent the sum of all values in that class interval.
     – For the first class interval 0–4, $x_{mid} \times f = 2 \times 6 = 12$

WRITE

Class interval	Frequency (f)	Midpoint (x_{mid})
0–4	6	②
5–9	8	⑦
10–14	2	12
15–19	3	17
20–24	5	22

Class interval	Frequency (f)	Midpoint (x_{mid})	$x_{mid} \times f$
0–4	6	2	⑫
5–9	8	7	56
10–14	2	12	24
15–19	3	17	51
20–24	5	22	110
Totals	㉔		㉕㉓

- Calculate the total sum of all the data values by adding all values in the $x_{mid} \times f$ column. The total of the $x_{mid} \times f$ column is 253.
- Calculate the total number of data values by adding all values in the frequency column. The total of the frequency column is 24.

3. Substitute the total of the $x \times f$ column and the total of the frequency column into the formula for the mean of grouped data.

$$\bar{x} = \frac{\sum (x_{mid} \times f)}{n}$$
$$= \frac{253}{24}$$
$$\approx 10.54$$

4. Answer the question.

The mean number of hamburgers made per day is approximately 10.54. This means that, on average, between 10 and 11 burgers are made every day.

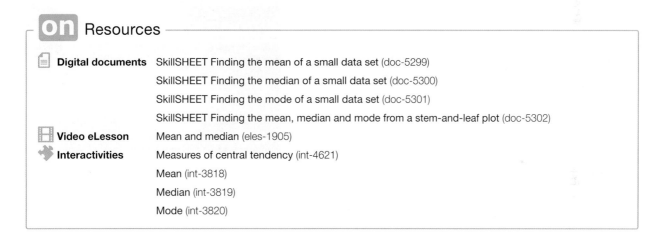

on Resources

Digital documents	SkillSHEET Finding the mean of a small data set (doc-5299)
	SkillSHEET Finding the median of a small data set (doc-5300)
	SkillSHEET Finding the mode of a small data set (doc-5301)
	SkillSHEET Finding the mean, median and mode from a stem-and-leaf plot (doc-5302)
Video eLesson	Mean and median (eles-1905)
Interactivities	Measures of central tendency (int-4621)
	Mean (int-3818)
	Median (int-3819)
	Mode (int-3820)

Exercise 7.2 Measures of central tendency and mode

Note: Where necessary give answers correct to one decimal place.

1. **WE1** For each of the following data sets, calculate the:
   i. mean
   ii. median
   iii. mode.
   a. 3, 3, 4, 5, 5, 6, 6, 7, 8, 8, 8, 9
   b. 12, 18, 4, 17, 5, 12, 0, 10, 12
   c. 42, 29, 11, 28, 21
   d. 8, 2, 5, 6, 9, 9, 7, 3, 2, 9, 3, 7, 6, 8
   e. 5, 5, 6, 4, 8, 3, 4
   f. 3.7, 3.5, 3.8, 3.8, 3.5

2. For each of the following data sets, calculate the mean, median and mode.
   a. 10, 12, 21, 23, 23, 25, 44
   b. 7, 8, 10, 6, 9, 11, 4, 12, 2
   c. 50, 44, 50, 46, 50, 48
   d. 2.5, 1.4, 1.7, 2.1, 1.4, 1.8, 1.6, 1.7, 2.9

3. **MC** Which of the following are the mean, median and mode, respectively, of this data set?
   1024, 1032, 1067, 1112, 1112, 1178, 1236, 1269, 1290, 1301, 1345, 1357, 1365, 1377, 1400
   A. 1269, 1112, 1231   **B.** 1231, 1269, 1112   **C.** 1112, 1231, 1269   **D.** 1231, 1112, 1269

4. The number of students standing in line at the tuck shop 5 minutes after the start of lunch was recorded over a 2-week period. The results were as follows: 52, 45, 41, 42, 53, 45, 47, 32, 52, 56. What was the mean number of students standing in line at this time for this 2-week period? Round your answer to the nearest whole number.

5. The police conducted a survey of the speed of cars down a highway. The lowest and highest speeds recorded were 91 km/h and 154 km/h respectively. The average speed was 104 km/h, and the police found that the speed most commonly recorded was 101 km/h. Half of the cars were also found to be travelling under 102 km/h. State the value of the mean, median and mode.

6. **WE2a** Calculate the mean of the data set displayed below.

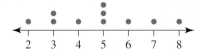

7. **WE2b** Calculate the mean of the data set displayed below.
   Key: $1|6 = 16$ years

Stem	Leaf
1	5 6 7 7 8 9 9
2	1 2 4 8 8
3	0 1 1 1 5
4	2 3
5	3

8. **WE3** Calculate the mean, median and mode in each of the following frequency tables.

a.

Score (x)	Frequency (f)
1	12
2	10
3	8
4	7
5	2

b.

Score (x)	Frequency (f)
25	1
26	15
27	11
28	7
29	3

c.

Score (x)	Frequency (f)
1.5	2
2.0	9
2.5	7
3.0	11
3.5	4

9. **WE4** Calculate the mean number of DVDs presented in this frequency table.

Number of DVDs (x)	Frequency (f)
1–15	3
16–30	9
31–45	8
46–60	11
61–75	10
76–90	14
91–105	15
106–120	18

10. Calculate the mean number of calls made on mobile phones in the month shown in the graph.

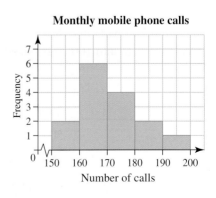

Monthly mobile phone calls

11. The number of goals a netballer scored in the 12 games of a season was as follows.

$$1, 1, 1, 1, 2, 2, 2, 3, 3, 3, 8, 12$$

A local newspaper reporter asked the netballer what his average was for the season.

a. Which measure of centre (mean or median) should the netballer give the reporter as his 'average' so that the value of the average is as high as possible?

b. Which measure of centre would *you* choose to best describe the 'average' number of goals the netballer scored each game? Why?

12. Create a data set that fits each of the following descriptions.

a. Five data values with a mean of 3 and a mode of 3

b. Five data values with a mean of 3 and a mode of 4

c. Five data values with a mean of 3 and a median of 2

13. The mean length of three pieces of string is 145 cm. If they are joined together from end to end, what will their total length be?

14. The median mark for a Science test was 45. No student actually achieved this result. Explain how this is possible.

7.3 Measures of spread

The **spread** of a set of data indicates how far the data values are spread from the centre or from each other. This is also known as the distribution. There are many statistical measures of spread, including the range, interquartile range and the standard deviation. We can use every day language to interpret the spread, such as 'spread out' 'dispersed' or 'tightly packed'.

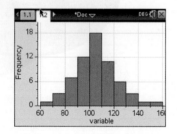
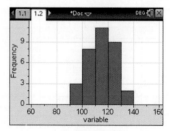

In comparing the two histograms, the histogram on the left has a larger spread of values compared to the histogram on the right.

7.3.1 Quartiles, deciles and percentiles

Data sets can be split up into any given number of equal parts called **quantiles**. Quantiles are named after the number of parts that the data is divided into. **Deciles** divide the data into 10 equal-sized parts, percentiles divide the data into 100 equal-sized parts, and **quartiles** divide the data into 4 equal-sized parts.

Percentile	Quartile and symbol	Common name
25th percentile	First quartile, Q_1	Lower quartile
50th percentile	Second quartile, Q_2	Median
75th percentile	Third quartile, Q_3	Upper quartile
100th percentile	Fourth quartile, Q_4	Maximum

A percentile is named after the percentage of data that lies at or below that value. For example, 60% of the data values lie at or below the 60th percentile.

The following cumulative frequency graph shows the cumulative frequency versus the number of skips with a skipping rope per minute for Year 11 students at a school. Calculate:

a. the median

b. Q_1

c. Q_3

d. the number of skips per minute for the 40th percentile.

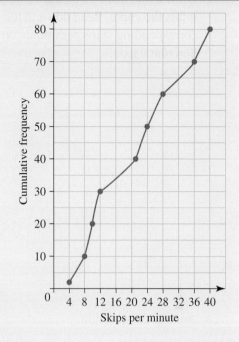

THINK

a. 1. Determine the total number of subjects. This is the highest vertical value on the cumulative frequency curve.

2. The median occurs at the middle of the total cumulative frequency, so divide the total frequency by 2.

3. Draw a line from 40 on the vertical axis until it hits the curve. Read down to the corresponding skips per minute on the horizontal axis.

WRITE

a. 80 students

The median occurs at the $\dfrac{80}{2} = $ 40th person.

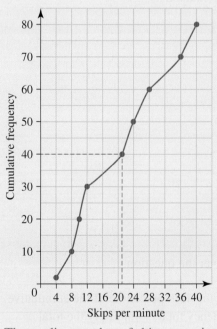

4. Answer the question.

The median number of skips per minute is approximately 21.

b. 1. Q_1 occurs at a quarter of the total cumulative frequency, so divide the total frequency by 4.

b. Q_1 occurs at the $\dfrac{80}{4} = 20$th person.

2. Draw a line from 20 on the vertical axis until it hits the curve. Read down to the corresponding skips per minute on the horizontal axis.

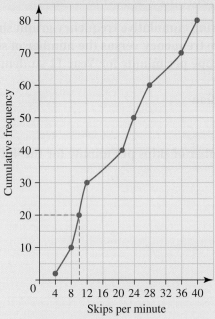

3. Answer the question.

Q_1 is at approximately 10 skips per minute.

c. 1. Q_3 occurs at three-quarters of the total cumulative frequency, so divide the total frequency by 4 and multiply the result by 3.

c. Q_3 occurs at the $\dfrac{80}{4} \times 3 = 60$th person.

2. Draw a line from 60 on the vertical axis until it hits the curve. Read down to the corresponding skips per minute on the horizontal axis.

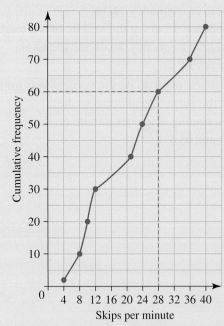

3. Answer the question.

Q_3 is at approximately 28 skips per minute.

d. 1. Find the 40th percentile of the cumulative frequency total, so multiply the total frequency by $\dfrac{40}{100}$.

d. $\dfrac{40}{100} \times 80 = 32$nd person.

2. Draw a line from 32 on the vertical axis until it hits the curve. Read down to the corresponding skips per minute on the horizontal axis.

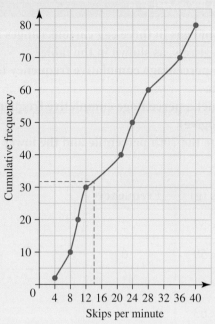

3. Answer the question.

The 40th percentile is at approximately 14 skips per minute.

Percentiles can be read off a percentage cumulative frequency curve. A percentage cumulative frequency curve is created by:
- writing the cumulative frequencies as a percentage of the total number of data values
- plotting the percentage cumulative frequencies against the maximum value for each interval.

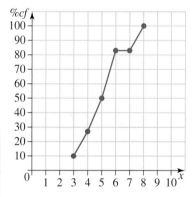

Data (x)	Frequency (f)	Cumulative frequency (cf)	Percentage cumulative frequency (%cf)
3	3	3	$\frac{3}{30} \times \frac{100}{1}\% = 10\%$
4	5	8	$\frac{8}{30} \times \frac{100}{1}\% = 27\%$
5	7	15	$\frac{15}{30} \times \frac{100}{1}\% = 50\%$
6	10	25	$\frac{25}{30} \times \frac{100}{1}\% = 83\%$
7	0	25	$\frac{25}{30} \times \frac{100}{1}\% = 83\%$
8	5	30	$\frac{30}{30} \times \frac{100}{1}\% = 100\%$

The mass of eggs in three egg cartons ranges between
55 and 65 grams, as shown in the table.

a. Draw a percentage cumulative frequency table for
the data.

b. Construct a percentage cumulative frequency
curve (ogive).

c. Evaluate the 70th percentile and the lower quartile.

Mass (g)	Frequency
55–<57	2
57–<59	6
59–<61	12
61–<63	11
63–<65	5

THINK

a. 1. Construct the cumulative
frequency table by calculating
the cumulative frequency for
each class interval, as shown in
blue.

2. Calculate the percentage
cumulative frequency for each
interval by dividing the
cumulative frequency for each
interval by the total cumulative
frequency, as shown in red.

WRITE

a.

Mass (g)	Frequency (f)	Cumulative frequency (cf)	Percentage cumulative frequency ($\%cf$)
55–<57	2	2	$\dfrac{2}{36} \times \dfrac{100}{1}\% \approx 6\%$
57–<59	6	$2 + 6 = 8$	$\dfrac{8}{36} \times \dfrac{100}{1}\% \approx 22\%$
59–<61	12	$8 + 12 = 20$	$\dfrac{20}{36} \times \dfrac{100}{1}\% \approx 56\%$
61–<63	11	$20 + 11 = 31$	$\dfrac{31}{36} \times \dfrac{100}{1}\% \approx 86\%$
63–<65	5	$31 + 5 = 36$	$\dfrac{36}{36} \times \dfrac{100}{1}\% = 100\%$

b. 1. Plot the percentage cumulative frequency curve. For the first interval (55–< 57), plot the minimum value for the interval (55) against 0%.

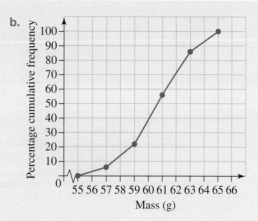

b.

2. Plot the maximum value for each interval against the percentage cumulative frequency for the interval.
- For the first interval, plot 57 against 6%.
- For the second interval, plot 59 against 22%.

c. 1. The 70th percentile is read from the graph by following across from 70% on the vertical axis to the curve and then down to the horizontal axis, as shown in green.

c.

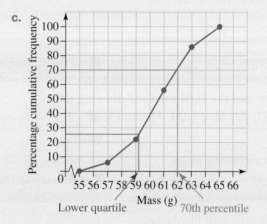

Lower quartile   Mass (g)   70th percentile

2. The lower quartile is the 25th percentile and is read from the graph by following across the graph from 25% on the vertical axis to the curve and then down to the horizontal axis, as shown in pink.

3. Answer the question.

The 70th percentile is 62 grams. The lower quartile is approximately 59.2 grams.

7.3.2 The range and interquartile range

The **range** is the difference between the highest and lowest values of the data set.

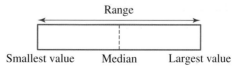

Smallest value   Median   Largest value

Range

Range = largest value − smallest value

The **interquartile range (IQR)** is the range of the middle 50% of the data set. It measures the spread of the middle 50% of data.

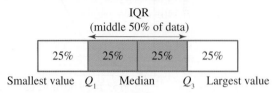

Interquartile range

Interquartile range $(IQR) = Q_3 - Q_1$

The lower quartile (Q_1) is the median of the lower half of the data and the upper quartile (Q_3) is the median of the upper half of the data.

If the median is one of the actual data values, it is not considered to be in either the upper or lower half of the data.

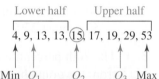

WORKED EXAMPLE 7

Calculate the range and the IQR for the following data.
26, 32, 15, 12, 35, 27, 22, 31, 38, 20, 41, 26, 17, 29

THINK	WRITE
1. Write the data in ascending order.	12, 15, 17, 20, 22, 26, 26, 27, 29, 31, 32, 35, 38, 41
2. The range is found by subtracting the smallest number from the largest number.	Range $= 41 - 12$ $= 29$
3. The median value will split the data into the top and bottom half. There are 14 values, so the median is at the $\dfrac{14 + 1}{2} = 7.5$th value.	12, 15, 17, 20, 22, 26, 26, | 27, 29, 31, 32, 35, 38, 41 Median $= \dfrac{26 + 27}{2} = 26.5$
4. List the lower half of the data. Find the median of the lower half of the data. This is Q_1. There are 7 values, so the median is at the $\dfrac{7 + 1}{2} = 4$th value.	Lower half: 12, 15, 17, 20, 22, 26, 26 $Q_1 = 20$
5. List the upper half of the data. Find the median of the upper half of the data. This is Q_3. There are 7 values, so the median is at the $\dfrac{7 + 1}{2} = 4$th value.	Upper half: 27, 29, 31, 32, 35, 38, 41 $Q_3 = 32$

6. Calculate the IQR.

$$IQR = Q_3 - Q_1$$
$$= 32 - 20$$
$$= 12$$

Since the lower 25% and upper 25% of observations are not included, the interquartile range is generally not affected by extreme values. This makes it a more useful measure of spread than the range.

7.3.3 The standard deviation

The **standard deviation**, like the IQR, is a measure of the spread of the data. The standard deviation measures the spread of data about the mean. The standard deviation of a population and the standard deviation of a sample are calculated differently:

Population standard deviation: $\sigma = \sqrt{\dfrac{\sum (x - \mu)^2}{N}}$	where: σ = population standard deviation x = a data value μ = the population mean N = the total number of population data values.
Sample standard deviation: $\sigma = \sqrt{\dfrac{\sum (x - \bar{x})^2}{n}}$	where: s = population standard deviation x = a data value $\bar{x}$ = the sample mean n = the total number of population data values.

These formulae use the square of the distance between data points and the mean as part of the calculation. Therefore, a small standard deviation shows that the data is clustered closer to the mean, whereas a large standard deviation indicates that the distribution is more spread out about the mean. The sample standard deviation is more commonly used because questions usually involve samples instead of whole populations. When asked to calculate the standard deviation, assume it is referring to the sample standard deviation.

WORKED EXAMPLE 8

The number of lollies in a sample of 8 packets is

11, 12, 13, 14, 16, 17, 18, 19.

Calculate the standard deviation correct to two decimal places.

THINK

1. Calculate the mean.

WRITE

$$\bar{x} = \frac{11 + 12 + 13 + 14 + 16 + 17 + 18 + 19}{8}$$

$$= \frac{120}{8}$$

$$= 15$$

2. To calculate the deviations $(x - \bar{x})$, set up a table as shown and complete by subtracting the mean from each data value.

No. of lollies (x)	($x - \bar{x}$)
11	$11 - 15 = -4$
12	-3
13	-2
14	-1
16	1
17	2
18	3
19	4
Total	

3. Add another column to the table to calculate the square of the deviations, $(x - \bar{x})^2$ by squaring each value in the second column. Then sum the results: $\sum (x - \bar{x})^2$.

No. of lollies (x)	($x - \bar{x}$)	($x - \bar{x}$)^2
11	$11 - 15 = -4$	16
12	-3	9
13	-2	4
14	-1	1
16	1	1
17	2	4
18	3	9
19	4	16
Total		$\sum (x - \bar{x})^2 = 60$

4. To calculate the standard deviation, divide the sum of the squares by one less than the number of data values, then take the square root of the result.

$$s = \sqrt{\dfrac{\sum (x - \bar{x})^2}{n - 1}}$$

$$= \sqrt{\dfrac{60}{7}}$$

$$\approx 2.93 \text{ (correct to two decimal places)}$$

5. Interpret the result.

The standard deviation is 2.93, which means that the number of lollies in each pack differs from the mean by an average of 2.93

Calculations for standard deviation can be completed using different digital technologies, such as calculators, statistics programs or spreadsheets. This is generally the preferred method for calculating the standard deviation.

WORKED EXAMPLE 9

The number of chocolate drops in a sample of different bags was recorded. These values were as follows.

271, 211, 221, 288, 209, 285, 230, 220, 296, 216

Calculate the standard deviation, correct to one decimal place, of the number of chocolate drops in a bag.

THINK

1. Enter the data into the digital technology of your choice. This example will be done in Excel.

	A	B	C
1	**Chocolate drops**		
2	271		
3	211		
4	221		
5	288		
6	209		
7	285		
8	230		
9	220		
10	296		
11	216		
12			

2. To calculate the sample standard deviation, enter the formula '=STDEV()' and select the values to be included in the calculation. The data starts in cell A2 and finishes in cell A11.
Alternatively, in a blank cell type '=STDEV(A2:A11)'.

	A	B
1	**Chocolate drops**	
2	271	
3	211	
4	221	
5	288	
6	209	
7	285	
8	230	
9	220	
10	296	
11	216	
12		
13		
14	=STDEV(A2:A11)	
15		

3. Press Enter to calculate the sample standard deviation.

	A	B
1	**Chocolate drops**	
2	271	
3	211	
4	221	
5	288	
6	209	
7	285	
8	230	
9	220	
10	296	
11	216	
12		
13		
14	35.65903345	
15		

4. Write the answer correct to one decimal place.

WRITE

The standard deviation is 35.7.

Exercise 7.3 Measures of spread

1. Put these expressions in order from the one with the smallest value to the one with the largest value: upper quartile; minimum; median; maximum; lower quartile.

2. **WE5** The number of steps per minute runners took at a local park run event was recorded. The cumulative frequency graph is as shown.

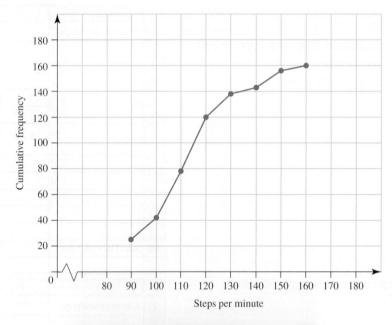

Calculate:

a. the median

b. Q_1

c. Q_3.

d. A local running group invited all people over the 65th percentile to join the group. How many steps per minute would a runner need to take to be invited to join the running group?

3. People who train for triathlons run a great distance every day. The distances (in km) that 60 athletes cover are shown in the frequency table.

Class interval	Frequency
10–15	3
15–20	7
20–25	12
25–30	18
30–35	14
35–40	6

a. Draw a cumulative frequency curve (ogive) for this data.
b. How many of the distances run were less than 19 km?
c. Determine the median for the data.
d. Calculate the 30th percentile and the upper quartile.

4. **WE6** Sometimes when you go through a fast-food outlet drive-through they don't have your order ready and you have to wait in the waiting bay for it to be brought out to you. The waiting times of 50 customers were recorded in minutes and seconds, as shown in the table.

Class intervals	Frequency
2:00–< 2:30	3
2:30–< 3:00	2
3:00–< 3:30	8
3:30–< 4:00	8
4:00–< 4:30	9
4:30–< 5:00	7
5:00–< 5:30	2
5:30–< 6:00	4
6:00–< 6:30	0
6:30–< 7:00	2
7:00–< 7:30	2
7:30–< 8:00	1
8:00–< 8:30	2

a. Draw a percentage cumulative frequency table.
b. Construct a percentage cumulative frequency curve.
c. The fast-food outlet claims that 80% of customers spend less than 4 minutes 30 seconds in the waiting bay. Do you agree or disagree with this statement? Justify your response.

5. Calculate the range of each of the following sets of data.
   a. 4, 6, 8, 11, 15
   b. 1.7, 1.9, 2.5, 0.5, 3.1, 1.9, 1.7, 1.6, 1.2

c.

Score (x)	Frequency (f)
110	12
111	9
112	18
113	27
114	5

6. **WE7** The speed of 20 cars (in km/h) is monitored along a stretch of road that is a designated 80 km/h zone. Calculate the range and IQR of the data.

80, 82, 77, 75, 80, 80, 81, 78, 79, 78, 80, 80, 85, 70, 79, 81, 81, 80, 80, 80

7. 30 pens are randomly selected off the conveyor belt at the factory and are tested to see how long they will last (values given in hours). Find the range and IQR of the data.

20, 32, 38, 22, 25, 34, 47, 31, 26, 29, 30, 36, 28, 40, 31, 26, 37, 38, 32, 36, 35, 25, 29, 30, 40, 35, 38, 39, 37, 30

Use the following data set to answer questions 8–10.
171, 122, 182, 153, 167, 184, 171, 177, 189, 175, 128, 190, 135, 147, 171

8. **MC** What is the range of the data?
   A. 67
   B. 68
   C. 69
   D. 70
9. **MC** What is the IQR of the data?
   A. 32
   B. 33
   C. 34
   D. 35
10. **WE8** **MC** What is the sample standard deviation of the data (correct to one decimal place)?
    A. 19.5
    B. 21.3
    C. 22.0
    D. 24.3
11. a. Calculate the maximum value of a data set if its minimum value is 23 and its range is 134.
    b. Calculate the minimum value of a data set if its range is 32 and its maximum value is 101.

12. Aptitude tests are often used by companies to help them decide who to employ. An employer gave 30 potential employees an aptitude test with a total of 90 marks. The scores achieved are shown below.

67, 67, 68, 68, 68, 69, 69, 72, 72, 73, 73, 74, 74, 75, 75, 77, 78, 78, 78, 79, 79, 79, 81, 81, 81, 82, 83, 83, 83, 86

Only applicants who score above the 80th percentile receive an interview. Use your knowledge of percentiles to work out how many interviews the employer will have to run.

13. **WE9** The ages of patients (in years) that came into a hospital emergency room during an afternoon were recorded. Using a digital technology of your choice, calculate the sample standard deviation of the data.

14, 1, 3, 87, 27, 42, 19, 91, 17, 73, 68, 83, 62, 29, 32, 2

7.4 Outliers

7.4.1 Calculating outliers

Outliers are extreme values on either end of a data set that appear very different from the rest of the data. They may be extremely large or extremely small compared to other data values in the set.

Outliers can be calculated by considering the distance a data point is from the mean or median compared to the rest of the data. If the value of a data point is clearly far away from the mean or median, as well as from other data points, it may considered to be an outlier. For example, in the data set 2, 2, 4, 6, 7, 7, 100, it is reasonable to conclude that 100 is an outlier, as it is much larger than the rest of the data and is far away from the median of 5.

It is not always obvious if a data point is an outlier.

The IQR can be used in a calculation to determine if a data point is far enough away from the median to be considered an outlier.

> **Determining outliers**
> To determine if a data point is an outlier, first calculate the upper and lower fences:
>
> $$\text{Lower fence} = Q_1 - 1.5 \times \text{IQR}$$
> $$\text{Upper fence} = Q_3 + 1.5 \times \text{IQR}$$
>
> where: Q_1 is the lower quartile, Q_3 is the upper quartile and IQR is the interquartile range.
> Any data point below the **lower fence** is an outlier.
> Any data point above the **upper fence** is an outlier.

Consider the following stem plot.

Key: $1|4 = 14$

Stem	Leaf
0	13
1	00134446899
2	2255678889
3	4577
4	01129
5	
6	
7	1

Calculate the IQR and use this to determine if there are any outliers.

THINK

1. To calculate the IQR, you must first calculate Q_1 and Q_3. Start by locating the median. The median is the middle data value, so the median occurs at the $\frac{33+1}{2} = $ 17th value.

WRITE

Stem	Leaf
0	13
1	00134446899
	Median
2	2255678889
3	4577
4	01129
5	
6	
7	1

Median=25

2. Locate the lower quartile. Q_1 is the middle value of the lower half of the data. There are 16 data values in the lower group, so Q_1 occurs at the $\frac{16+1}{2} = $ 8.5th value in the lower group.

Take the average of the 8th and 9th values.
$\frac{14+14}{2} = 14$

Stem	Leaf
0	13
	Q_1
1	001344|46899
2	225567 8889
3	4577
4	01129
5	
6	
7	1

$Q_1 = 14$

3. Locate the upper quartile. Q_3 is the middle value of the upper half of the data. There are 16 data values in the upper group, so Q_3 occurs at the $\frac{16+1}{2} = 8.5$th value in the upper group.

Take the average of the 8th and 9th values.
$$\frac{35 + 37}{2} = 36$$

Stem	Leaf
0	1 3
1	0 0 1 3 4 4 4 6 8 9 9
2	2 2 5 5 6 7 8 8 8 9
	Q_3
3	4 5 | 7 7
4	0 1 1 2 9
5	
6	
7	1

$Q_3 = 36$

4. Calculate the IQR.

$$\begin{aligned} \text{IQR} &= Q_3 - Q_1 \\ &= 36 - 14 \\ &= 22 \end{aligned}$$

5. Calculate the lower fence.

$$\begin{aligned} \text{Lower fence} &= Q_1 - 1.5 \times \text{IQR} \\ &= 14 - 1.5 \times 22 \\ &= -19 \end{aligned}$$

Are there any values below the lower fence?   No values are below this point.

6. Calculate the upper fence.

$$\begin{aligned} \text{Upper fence} &= Q_3 + 1.5 \times \text{IQR} \\ &= 36 + 1.5 \times 22 \\ &= 69 \end{aligned}$$

Are there any values above the upper fence?   There is a value above this point; therefore 71 is an outlier.

7.4.2 The effect of outliers on the mean and median

Outliers can affect the summary statistics for a sample.

The median is calculated using the number of data points, not the value of each data point. Therefore, outliers do not have much of an influence on the median.

The mean involves a summation of the values of the data points, so extremely high or low values can greatly affect the mean.

If outliers are present in a data set, the median is a better measure of central tendency, as it is less affected by outliers.

WORKED EXAMPLE 11

The number of times employees in a small company took sick days per year was recorded. The results were as follows.
5, 2, 4, 3, 26, 1, 1, 6, 4, 29, 0, 2, 1

a. Which, if any, of these values could be outliers?
b. For the data set, calculate:
   i. the mean   **ii. the median.**
c. Exclude the outlier(s) in the data set and calculate:
   i. the mean   **ii. the median.**
d. Comment on the difference in results between parts b and c.

THINK	WRITE
a. 1. Arrange the values in ascending order.	**a.** $0, 1, 1, 1, 2, 2, 3, 4, 4, 5, 6, 26, 29$
2. Outliers are extreme values on either end of the data set.	$0, 1, 1, 1, 2, 2, 3, 4, 4, 5, 6, 26, 29$ 26 and 29 are possible outliers.
b. i. 1. Calculate the sum of all the values.	**b.** $5 + 2 + 4 + 3 + 26 + 1 + 1 + 6 + 4 +$ $29 + 0 + 2 + 1 = 84$
2. Count the number of values.	Number of values $= 13$
3. The mean is the sum of all the values divided by the number of values.	$\bar{x} = \dfrac{84}{13} = 6.5$ The mean number of sick days employees took in a year was 6.5.
ii. 1. Arrange the values in ascending order. The median is the middle value.	$0, 1, 1, 1, 2, 2, 3, 4, 4, 5, 6, 26, 29$
2. There are 13 values, so the median is the $\dfrac{13 + 1}{2} = 7\text{th value.}$	$0, 1, 1, 1, 2, 2, 3, 4, 4, 5, 6, 26, 29$ The median number of sick days employees took in a year was 3.
c. i. 1. Write out the data set excluding the outliers. Calculate the sum of all the values.	**c.** $0, 1, 1, 1, 2, 2, 3, 4, 4, 5, 6$ $5 + 2 + 4 + 3 + 1 + 1 + 6 +$ $4 + 0 + 2 + 1 = 29$
2. Count the number of values.	Number of values $= 11$
3. The mean is the sum of all the values divided by the number of values.	$\bar{x} = \dfrac{29}{11} = 2.6$ The mean number of sick days employees took in a year, when excluding outliers, was 2.6.
ii. 1. Arrange the values in ascending order. The median is the middle value.	$0, 1, 1, 1, 2, 2, 3, 4, 4, 5, 6$
2. There are 11 values, so the median is the $\dfrac{11 + 1}{2} = 6\text{th value.}$	$0, 1, 1, 1, 2, 2, 3, 4, 4, 5, 6$ The median number of sick days employees took in a year, when excluding outliers, was 2.
d. 1. Comment on the difference between the mean in parts **b** and **c**.	**d.** The mean sick days per year taken by employees reduced from 6.5 days to 2.6 days when outliers were excluded. Excluding the large outliers significantly decreased the mean.
2. Comment on the difference between the median in parts **b** and **c**.	The median sick days per year taken by employees changed from 3 days to 2 days when outliers were excluded. Excluding the large outliers did not significantly change the median.
3. Make a conclusion about the effect of outliers on the mean and median.	The outliers had a greater effect on the mean than on the median.

Exercise 7.4 Outliers

1. The following data represents car sales each month at a car yard.

   12, 9, 15, 10, 11, 23, 14, 8, 6, 11, 13, 15
   a. Calculate the IQR
   b. Calculate if there are any outliers.

2. **WE10** Consider the following stem plot.

   Key: 1 | 6 = 16

Stem	Leaf
0	1
1	5
2	
3	
4	2
5	6 7
6	0 1 4 5 6
7	1 1 2 3 5 7 9
8	2 2 4 4 4 8 8
9	3 5

   Calculate the IQR and use this to determine if there are any outliers.

3. From the following data, calculate the lower and upper fence.

   34, 41, 53, 39, 48, 41, 33, 39, 40, 34, 44

4. From the following data: 2.3, 3.1, 3.6, 1.8, 6.7, 4.4, 3.9, 2.8, 3.7, 4.0
   a. calculate the lower and upper fence
   b. determine if there are any outliers in the data set.

5. The number of swimmers at a pool was recorded over a fortnight. The number of swimmers was as follows.

   56, 69, 59, 113, 9, 100, 80, 111, 94, 77, 57, 166, 101, 96
   Determine if any outliers exist and justify your answer with a calculation.

6. **WE11** Below is the data collected during a survey of the number of pets people own.

   2, 1, 0, 3, 1, 1, 2, 1, 2, 3, 3, 2, 0, 2, 1, 2, 3, 4, 1, 2, 1, 1, 3, 3, 1, 3, 2, 1, 2, 2, 2, 3, 21, 1, 2, 3
   a. Are there any possible outliers?
   b. Calculate the mean and median.
   c. Remove any outliers and recalculate the mean and median.
   d. Comment on the difference in means and medians in parts b and c.

7. From the following set of numbers, select possible outliers.
   a. 3, 5, 2, 6, 15, 1, 5, 4
   b. 21, 33, 44, 34, 27, 3, 29, 30, 6, 31, 25, 36

8. From the data in question **7**, state if the mean would increase or decrease with the outliers taken out of the data set.

9. From question **7** calculate the mean before and after taking out the outliers.

10. From the following set of numbers, select possible outliers.
    a. 0.03, 0.05, 0.07, 0.03, 0.9, 0.03, 0.04
    b. $-2.3, -3.6, -1.9, -2.1, 5.2, -3.9, -3.9, -2.7, 6.1, -1.7, -3.3, -2.4$

11. From the data in question **10**, state if the median would increase or decrease with the outliers taken out of the data set.

12. From question **10** calculate the median before and after taking out the outliers.

13. **MC** The mean of a data set was calculated to be 117. The median was calculated to be 107. The data set has two outliers of 178 and 190. If the outliers were replaced with values of 118 and 120:
    A. the median would increase, but the mean would stay the same.
    B. the mean would increase, but the median would remain the same.
    C. the median would decrease, but the mean would stay the same.
    D. the mean would decrease, but the median would stay the same.

14. A statistician investigated the monthly household sales in a suburb over a year. The Q_1, median and Q_3 were calculated to be 18, 27, and 35 respectively.
    a. Prove that 64 houses per month is an outlier.
    b. The statistician realised he made a mistake and the outlier of 64 was recorded incorrectly. The actual value was 46 houses per month. Using your answer from **a** to help you, explain what would happen to the values of Q_1, Q_3 and the IQR if the data point was changed from 64 to 46 houses per month.

7.5 Applications of measures of centre and spread

Besides locating the centre of the data, using the mean or median, any analysis of data must measure the extent of the spread of the data (range, interquartile range and standard deviation). Two data sets may have centres that are very similar but be quite differently distributed.

When analysing data or comparing data sets it is common to discuss or compare measures of central tendency (mean or median) and the spread (range, IQR or standard deviation). We are aware that measures of central tendency give information about the centre of the distribution and that the spread provides information on how spread out (widely spread or tightly packed) the data are. Another feature of the spread is that it can provide important information on how consistent the data are. For example, if the spread of one set of data was tightly packed compared to a set of data that was widely spread (dispersed), we could say that the data set that is tightly packed is more consistent than the set of data that is widely spread. This is because when a distribution's spread is tightly packed the data collected would cluster around a certain value, implying that the data collected were consistent.

Given that there are several measures of centre and spread to select from when analysing data, it is often not necessary to calculate them all.

Decisions need to be made about which measure of centre and which measure of spread to use when analysing and comparing data.

The mean is calculated using every data value in the set. The median is the middle score of an ordered set of data, so it does not include every individual data value in its calculaion. The mode is the most frequently occurring data value, so it also does not include every individual data value in its calculation.

The range is calculated by finding the difference between the maximum and minimum data values, so it includes outliers. It provides only a rough idea about the spread of the data and is inadequate in providing sufficient detail for analysis. It is useful, however, when we are interested in extreme values such as high and low tides or maximum and minimum temperatures.

The interquartile range is the difference between the upper and lower quartiles, so it does not include every data value in its calculation, but it will overcome the problem of outliers skewing data.

The standard deviation is calculated using every data value in the set.

Mean or median?

If the data has no outliers and is symmetric, either the mean or the median can be used as the measure of centre.

If the data is clearly skewed and/or there are outliers, it is more appropriate to use the median as the measure of centre.

Range, IQR or standard deviation?

The most appropriate measure of spread does depend on the type of data; however, the following can be used as a guide.

If the data has no outliers and is symmetric, the standard deviation would be the preferred measure of spread.

If the data is clearly skewed and/or there are outliers, it is more appropriate to use the IQR as the preferred measure of spread.

Although the above information can be applied in most cases, selecting the most appropriate measure of centre and spread may depend on other factors such as wanting to include every data value in a set even if outliers are present.

WORKED EXAMPLE 12

Below are samples of scores achieved by two students in eight Mathematics tests throughout the year.
John: 45, 62, 64, 55, 58, 51, 59, 62
Penny: 84, 37, 45, 80, 74, 44, 46, 50

a. **Determine the most appropriate measure of centre and measure of spread to compare the performance of the students.**
b. **Which student had the better overall performance on the eight tests?**
c. **Which student was more consistent over the eight tests?**

THINK	WRITE
a. In order to include all data values in the calculation of measures of centre and spread, calculate the mean and standard deviation.	**a.** John: $\bar{x} = 57$, $s = 6.41$ Penny: $\bar{x} = 57.5$, $s = 18.62$
b. Compare the mean for each student. The student with the higher mean performed better overall.	**b.** Penny performed slightly better on average as her mean mark was higher than John's.

c. Compare the standard deviation for each student. The student with the lower standard deviation performed more consistently.

c. John was the more consistent student because his standard deviation was much lower than Penny's. This means that his test results were closer to his mean score than Penny's were to hers.

Exercise 7.5 Applications of measures of centre and spread

1. The number of M&M's in 20 packets was recorded by a group of students.
   30, 32, 33, 35, 37, 37, 38,
   38, 39, 39, 40, 40, 40, 41,
   41, 41, 41, 41, 41, 50

   a. Calculate the mean number of M&M's per packet.
   b. Calculate the median number of M&M's per packet.
   c. Calculate the modal number of M&M's per packet.
   d. Based on your calculations, how many M&M's would you expect to find in a packet?
   e. If you were advertising M&M's, which summary statistic (mean, median or mode) would you use? Explain.
2. Explain how the mean, median and mode for both ungrouped and grouped data are similar and different.
3. **a.** Calculate the mean and median of the two data sets below.
      Data set A: 20, 24, 29, 33, 37, 42, 51, 53, 96
      Data set B: 20, 24, 29, 33, 37, 42, 51, 53, 66
   b. Comment on the means and medians of the data sets and explain any similarities and differences you see.
   c. Write a short statement to explain when to use the mean as a measure of central tendency and when to use the median.
4. **WE12** The Mathematics test results for two students over a year were recorded as follows.
   Student A: 49, 52, 51, 50, 54, 49, 100
   Student B: 65, 71, 64, 63, 60, 81, 0
   a. Determine the most appropriate measure of centre and spread to compare the performance of the students.
   b. Which student had the better overall performance over the year?
   c. Which student was more consistent over the year?
5. Create a data set that fits the following descriptions.
   a. 8 data values with a range of 43
   b. 6 data values with a lower quartile of 5 and an upper quartile of 12
   c. 9 data values with a lower quartile of 7 and an upper quartile of 13
6. A Mathematics teacher wanted to design a question for her class. She wanted the question to have 10 data points, a lower quartile of 14 and an IQR of 20. State a data set that could be used in her question.

7. The number of trees in several different parks was recorded, and the mean number of trees was calculated to be 15. The council wanted to determine whether the majority of parks contained close to 15 trees, or whether there was a large difference in the number of trees. Describe how the council could use a measure of spread to investigate this problem.

8. The scores below show the number of points scored by two AFL teams over the first 10 games of the season.

| Sydney Swans: | 110 | 95 | 74 | 136 | 48 | 168 | 120 | 85 | 99 | 65 |
| Brisbane Lions: | 125 | 112 | 89 | 111 | 96 | 113 | 85 | 90 | 87 | 92 |

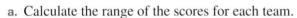

a. Calculate the range of the scores for each team.
b. Based on the results above, which team would you say is the more consistent?

9. Two machines are used to put approximately 100 Smarties into boxes. A check is made on the operation of the two machines. Ten boxes filled by each machine have the number of Smarties in them counted. The results are shown below.

Machine A: 100, 99, 99, 101, 100, 101, 100, 100, 101, 108
Machine B: 98, 104, 96, 97, 103, 96, 102, 100, 97, 104

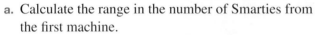

a. Calculate the range in the number of Smarties from the first machine.
b. Calculate the range in the number of Smarties from the second machine.
c. Ralph is the quality control officer and he argues that machine A is more consistent in its distribution of Smarties. Explain why.

10. The following frequency distribution gives the prices paid by a car wrecking yard for a sample of 40 car wrecks.

Price ($)	Frequency
0 to < 500	2
500 to < 1000	4
1000 to < 1500	8
1500 to < 2000	10
2000 to < 2500	7
2500 to < 3000	6
3000 to < 3500	3

Find the mean and standard deviation of the price paid for these wrecks.

11. The following stem plot represents the lifespan of different animals at an animal sanctuary.

Determine which measure of centre is best to represent the data set.

Key: 1|2 = 12

Stem	Leaf
0	3 5 9
1	2 4 6 8
2	0 1 4 5 5 7 9
3	0 2 6
4	
5	
6	0 3

12. The following data set represents the salaries (in $000s) of workers at a small business.

45, 50, 55, 55, 55, 60, 65, 65, 70, 70, 75, 80, 220

a. Calculate the mean of the salaries correct to three decimal places.
b. Calculate the median of the salaries.
c. When it comes to negotiating salaries, the workers want to use the mean to represent the data and the management want to use the median. Explain why this might be the case.

13. A sample of crime statistics over a two-year period are shown in the following table.

Crime	Year 1	Year 2
Theft from motor vehicle	46 700	42 900
Theft from shop	19 800	20 600
Theft of motor vehicle	15 650	14 670
Theft of bicycle	4 200	4 660
Theft (other)	50 965	50 650

a. Calculate the interquartile range and standard deviation (correct to one decimal place) for both years.
b. Recalculate the interquartile range and standard deviation for both years after removing the smallest category.
c. Comment on the effect of removing the smallest category on the interquartile ranges and standard deviations.

14. Data collected on the number of daylight hours in Alice Springs is as shown.

10.3, 9.8, 9.6, 9.5, 8.5, 8.4, 9.1, 9.8, 10.0, 10.0, 10.1, 10.0,
10.1, 10.1, 10.6, 8.7, 8.8, 9.0, 8.0, 8.5, 10.6, 10.8, 10.5,
10.9, 8.5, 9.5, 9.3, 9.0, 9.4, 10.6, 8.3, 9.3, 9.0, 10.3,
8.4, 8.9

a. Calculate the range of the data.
b. Calculate the interquartile range of the data.
c. Comment on the difference between the two measures and what this indicates.

7.6 Review: exam practice

7.6.1 Summarising and interpreting data summary

The mean

- For a sample data set, the mean is calculated using the formula: $\bar{x} = \dfrac{\sum x}{n}$
- When the data are presented in a frequency table the mean can be calculated using the formula $\bar{x} = \dfrac{\sum (f \times x)}{n}$.
- The mean can also be calculated using the statistical function on your calculator.

Median and mode

- The median is the middle score of a data set, or the average of the two middle scores, when the scores are arranged in numerical order (smallest to largest).
- The mode is the score with the highest frequency.

Summary statistics

- The summary statistics are the mean, median and mode.
- Each summary statistic must be examined in the context of the statistical analysis to determine which is the most relevant.

Range and interquartile range

- The range is the difference between the highest score and the lowest score.
  Range = largest value − smallest value.
- The interquartile range is the difference between the scores at the lower quartile and the upper quartile.
  $IQR = Q_3 - Q_1$.
- The range and interquartile range measure the spread of data around the median.

Standard deviation

- The standard deviation is a measure of the spread of data around the mean.
- The smaller the standard deviation, the smaller the spread of the data set.
- The sample standard deviation is found using the statistical function on your calculator or by using the formula $s = \sqrt{\dfrac{\sum (x - \bar{x})^2}{n - 1}}$.
- When the analysis is conducted on the entire population, the population standard deviation is used.
- When the analysis is conducted on a sample of the population, the sample standard deviation is used.

Outliers

- Outliers are extreme values on either end of a data set that appear very different from the rest of the data.
- Outliers can affect the summary statistics for a sample.
- The median is calculated using the number of data points, not the value of each data point. Therefore, outliers do not have much of an influence on the median.
- The mean involves a summation of the values of the data points, so extreme high or low values can greatly affect the mean.
- If outliers are present in a data set, the median is a better measure of central tendency to use, as it is less affected by outliers.
- The IQR can be used in a calculation to determine if a data point is far away enough from the median to be considered an outlier.
- To determine if a data point is an outlier, first calculate the upper and lower fences:

  Lower fence $= Q_1 - 1.5 \times \text{IQR}$
  Upper fence $= Q_3 + 1.5 \times \text{IQR}$

  where: Q_1 is the lower quartile, Q_3 is the upper quartile and IQR is the interquartile range.
- Any data point below the lower fence is an outlier.
- Any data point above the upper fence is an outlier.

Exercise 7.6 Review: exam practice

Simple familiar

1. Calculate the mean, median and mode of each of the following data sets.
   a. 4, 7, 2, 6, 9, 3, 6
   b. 11, 63, 24, 36, 25, 61, 29, 42
   c.

Score	Frequency
6	2
8	4
10	10

2. The following data was collected on the number of days people go away during the holidays.

   2, 10, 5, 7, 9, 14, 2, 0, 6, 7, 0, 7, 14, 7, 8, 10, 12, 5, 2, 1, 16, 12, 10

   a. Calculate the range of this data.
   b. Calculate the mean and median.

3. The screen time per day for a group of 0–14-year-olds is listed in the table. Calculate the mean number of minutes of screen time for 0–14-year-olds.

Age (years)	Screen time/day (minutes)
0–2	45
3–5	73
6–8	98
9–11	124
12–14	142

4. Match the following terms with the descriptions below.
   Mean   Range   Median   IQR   Standard deviation   Mode
   a. A measure of spread of the distribution that uses the highest and lowest value only
   b. A measure of central tendency that uses the middle value of the data set by position
   c. A measure of spread that contains exactly 50% of the distribution
   d. A measure of central tendency that uses the sum of all data points divided by the number of data points
   e. A measure of spread that uses the distance from the mean in the calculation
   f. A summary statistic that uses the most common value

5. Below is a data set showing the number of doors people have in their houses.
   18, 11, 9, 14, 11, 16, 14, 16, 6, 18, 19, 9, 10, 17, 36
   a. State the minimum and maximum number of doors, and hence calculate the range.
   b. Calculate the mean and median.
   c. Are there any outliers? If yes, remove these and recalculate the mean and median.
   d. Compare your answers to parts a and c.

6. The number of books borrowed by 20 students over the period of a month is monitored. Determine the mean, median, mode, range and IQR of the data.
   12, 8, 6, 10, 4, 5, 2, 7, 8, 0, 6, 4, 8, 13, 5, 3, 0, 8, 7, 9

7. Calculate the mean (to two decimal places), median, mode, range and IQR of the following data collected when the temperature of the soil around 25 germinating seedlings was recorded.
   28.9, 27.4, 23.6, 25.6, 21.1, 22.9, 29.6, 25.7, 27.4, 23.6, 22.4, 24.6, 21.8, 26.4, 24.9, 25.0, 23.5, 26.1, 23.6, 25.3, 29.5, 23.5, 22.0, 27.9, 23.6

8. The following data shows the number of ice creams sold nightly over two weeks.
   4, 6, 48, 29, 39, 48, 44, 45, 39, 47, 48, 32, 31, 51
   Determine if there are any outliers. Justify your answer with appropriate calculations.

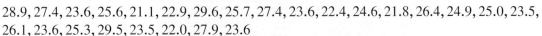

9. For the dot plot at right, calculate:
   a. the mean (to two decimal places)
   b. the median
   c. the IQR
   d. the standard deviation
   e. the range.

10. Consider the following stem plot.

    Key: $11|2 = 112$

Stem	Leaf
5	0 1 2
6	4 5 8 9
7	0 7
8	0 3 3
9	5 5 5 7 8
10	1 1 2 3 4 6 6
11	0 1 3 3 4 5 5 5 7 9
12	1 1 2 2 3 3 4 7
13	2 6 9

    Calculate:
    a. the mean
    b. the median
    c. the IQR
    d. the standard deviation.

11. A survey of Year 9 students was conducted to find out the average travel time to school. The results (in minutes) were as follows.

    12, 18, 17, 15, 15, 12, 14, 14, 14, 17, 10, 14, 11, 16, 14, 14, 11, 15, 10, 9, 13, 20

    a. Calculate the mean, median and mode of this data set.
    b. Determine if any outliers exist in the data.
    c. Which measure of centre would you choose as the average travel time? Why?

12. The price of a return plane ticket from Melbourne to London from nine different airlines is as follows:

    $3400, $2800, $3500, $4100, $2900, $5200, $3900, $4000, $4575

    a. Determine the minimum, lower quartile, median, upper quartile and maximum.
    b. Calculate the interquartile range.

The test marks for a class of students are shown in the histogram and frequency distribution table. Use this information to answer questions **13–16**.

Class interval	Frequency (f)
0–< 5	2
5–< 10	2
10–< 15	5
15–< 20	5
20–< 25	6
25–< 30	3
Total	23

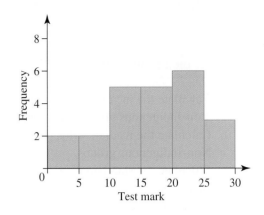

13. **MC** The median test mark for the class falls into which category?
    A. 5–10   **B.** 10–15   **C.** 15–20   **D.** 20–25
14. **MC** What is the mean of this data, to one decimal place?
    A. 14.3   **B.** 16.8   **C.** 22.5   **D.** 19.3
15. **MC** What is the mode of this data?
    A. 5–10   **B.** 10–15   **C.** 15–20   **D.** 20–25
16. **MC** What is the range of this data?
    A. 5   **B.** 15   **C.** 20   **D.** 30

Complex unfamiliar

17. A data set contains the numbers 5, 6, 6 and 10. A fifth number is added to the set. If this fifth number is a whole number, determine the possible values of this fifth number if:
    a. the mean is now equal to the median
    b. the mean is now greater than the mode
    c. the mean is now greater than the range.

18. Kris and Loren were discussing the following histogram showing the number of magazines sold in a convenience store. Kris said that the best way to describe the central tendency for magazine sales would be to look at the median. Loren disagreed and said to use the mean. Who is correct and why?

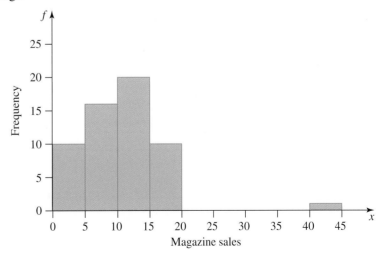

19. The data set used to generate the histogram relating to magazine sales in question **18** was as follows.
    0, 1, 1, 2, 2, 3, 3, 4, 4, 4, 5, 5, 5, 6, 6, 7, 7, 7, 7, 7, 8, 8, 8, 8, 9, 9, 10, 10, 10, 10, 10, 11,
    11, 11, 11, 11, 11, 11, 12, 12, 12, 12, 13, 13, 14, 14, 15, 15, 15, 15, 16, 17, 18, 18, 18, 19, 44
    Is the 44 magazine sales in a day an outlier? Justify your answer using calculations.

20. Use the data shown to answer the following questions.

Women who gave birth and Indigenous status by states and territories, 2009

Status	NSW	Vic	Qld	WA	SA	Tas	ACT	NT	Aust
Indigenous	2904	838	3332	1738	607	284	107	1474	11 284
Non-Indigenous	91 958	70 328	57 665	29 022	18 994	5996	5601	2369	281 933

a. Calculate the mean births per state/territory of Australia in 2009 for both Indigenous and Non-Indigenous groups. Give your answers correct to one decimal place.

b. Calculate the median births per state/territory of Australia in 2009 for both Indigenous and Non-Indigenous groups.

c. Calculate the standard deviation (correct to one decimal place) and IQR for the data on births per state/territory of Australia in 2009 for both Indigenous and Non-Indigenous groups.

d. Comment on the measures of centre and spread you have calculated for this data.

Answers

7 Summarising and interpreting data

Exercise 7.2 Measures of central tendency and mode

1. a. i. 6   ii. 6   iii. 8
   b. i. 10   ii. 12   iii. 12
   c. i. 26.2
      ii. 28
      iii. 11, 21, 28, 29, 42 (no mode)
   d. i. 6   ii. 6.5   iii. 9
   e. i. 5   ii. 5   iii. 4, 5
   f. i. 3.66 ii. 3.7   iii. 3.5, 3.8
2. a. Mean: 22.6; median: 23; mode: 23
   b. Mean: 7.7; median: 8; mode:
      2, 4, 6, 7, 8, 9, 10, 11, 12 (no mode)
   c. Mean: 48; median: 49; mode: 50
   d. Mean: 1.9; median: 1.7; mode: 1.4, 1.7
3. B
4. 47
5. Mean: 104 km/h; median: 102 km/h; mode: 101 km/h
6. 4.8
7. 27
8. a. Mean: 2.4; median: 2; mode: 1
   b. Mean: 26.9; median: 27; mode: 26
   c. Mean: 2.6; median: 2.5; mode: 3.0
9. 73.5
10. 171
11. a. Mean

b. Median. There are two possible outliers of 8 and 12 goals, which were not typical goal numbers for the netballer. The median is less affected than the mean by these two atypically high scores, so it is the best measure to use.

12. Sample responses can be found in the Worked Solutions in the online resources.
13. 435 cm
14. There must have been an even number of students so the median would be the average between the middle two test scores.

Exercise 7.3 Measures of spread

1. Minimum, lower quartile, median, upper quartile, maximum
2. a. 110   b. 99
   c. 120   d. > 117 steps/minute
3. a.

   b. 9
   c. 27.5 km
   d. The 30th percentile is 24 km. The upper quartile is 33.5 km.
4. a. * See the table at the bottom of the page.

---

*4. a.

Class intervals	Frequency	Cumulative frequency (*cf*)	Percentage cumulative frequency (% *cf*)
2:00 − <2:35	3	3	6%
2:30 − <3:00	2	5	10%
3:00 − <3:30	8	13	26%
3:30 − <4:00	8	21	42%
4:00 − <4:30	9	30	60%
4:30 − <5:00	7	37	74%
5:00 − <5:30	2	39	78%
5:30 − <6:00	4	43	86%
6:00 − <6:30	0	43	86%
6:30 − <7:00	2	45	90%
7:00 − <7:30	2	47	94%
7:30 − <8:00	1	48	96%
8:00 − <8:30	2	50	100%

b.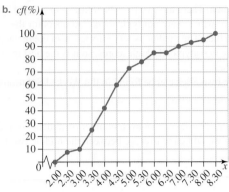

c. This statement is incorrect because the cumulative percentage curve show that 80% of customers must wait over five minutes for their food.

5. a. 11   **b.** 2.6   **c.** 4
6. Range $= 15$ km/h, IQR$= 2$ km/h
7. Range $= 27$ hours, IQR $= 8$ hours
8. B
9. D
10. C
11. a. 157   **b.** 69
12. 5
13. 32.0 years

Exercise 7.4 Outliers

1. a. 5
  b. 23
2. IQR $= 23$; outliers are 1 and 15
3. 19 and 59
4. a. 1 and 5.8
  b. 6.7
5. 166 is an outlier. (It is above the upper fence of 164.)
6. a. 21
  b. Mean: 2.4; median: 2
  c. Mean: 1.9; median: 2
  d. Mean value lowered, median remained unchanged
7. a. 15 is larger than the rest of the data.
  b. 3 and 6 are smaller than the rest of the data.
8. a. Taking 15 out, that is larger than the rest of the data, thus the mean would decrease.
  b. Taking out 3 and 6 that are smaller than the rest of the data, thus the mean would increase.
9. a. With:
    Mean $= 5.125$
    Without:
    Mean $= 3.71$
  b. With:
    Mean $= 26.58$
    Without:
    Mean $= 31$
10. a. 0.9 is larger than the rest of the data.
  b. 5.2 and 6.1 are larger than the rest of the data.
11. a. Taking out 0.9 that is large would therefore decrease the median.
  b. Taking out 5.2 and 6.1 that are large would therefore decrease the median.
12. a. With:
    Median $= 0.04$
    Without:
    Median $= 0.035$

b. With:
  Median $= -2.3$
  Without:
  Median $= -2.4$
13. D
14. a. IQR $= 35 - 18 = 17$
    Upper fence $= Q_3 + 1.5 \times$ IQR $= 35 + 1.5 \times 17 = 60.5$
    Any value above 60.5 will be an outlier, so 64 is an outlier.
  b. Q_1, Q_3, and IQR would remain unchanged.

Exercise 7.5 Applications of measures of centre and spread

1. a. 39
  b. 40
  c. 41
  d. 40
  e. The mode, as it gives the highest number of M&M's in a packet
2. The mean of grouped data is calculated using the midpoint of the class interval, making it an approximation. The mean of ungrouped data will give an exact value that is not an approximation. Using both grouped and ungrouped data, a single value can be obtained for the mean.
    The median for grouped data will be a class interval, rather than an exact value as for ungrouped data.
    The mode is calculated the same way for grouped and ungrouped data: the value with the greatest frequency.
3. a. Data set A: mean: 42.8, median: 37; data set B: mean: 39.4, median: 37
  b. The mean of data set A is larger than the mean of data set B. This is due to the last value being 96 in data set A, compared to 66 in data set B. The medians of the two data sets are the same as each data set has the same number of data values and all data values are the same, except the last value.
  c. The mean or median can be used as good measures of central tendency when there are no outliers or extreme values. The median is best to use when there are outliers or extreme values in the data.
4. a. Student A: median $= 51$, IQR $= 5$; student B: median $= 64$, IQR $= 11$
  b. Student B
  c. Student A
5. Answers will vary. Example answers are shown.
  a. 2, 5, 17, 21, 29, 35, 39, 45. (Several different data sets are possible.)
  b. 1, **5**, 6, 8, **12**, 14. (Several different data sets are possible, but the numbers in bold must be present in the same position.)
  c. 5, 6, 8, 9, 10, 11, 12, 14, 15. (Several different data sets are possible.)
6. 5, 10, **14**, 21, 22, 23, 31, **34**, 35, 40. (Several different data sets are possible, but the numbers in bold must be present in the same position.)
7. Calculate the standard deviation. If the standard deviation is small, then the majority of parks will contain close to 15 trees
8. a. Sydney $= 120$
    Brisbane $= 40$
  b. Brisbane since lower range.

9. a. Machine A range: 9 smarties
   b. Machine B range: 8 smarties
   c. Only 2 boxes out of 10 sampled from machine A contain less than 100 smarties. The box with 108 smarties causes the range to be higher than normal.

10.

Price ($)	Class centre	Frequency
0 to < 500	250	2
500 to < 1000	750	4
1000 to < 1500	1250	8
1500 to < 2000	1750	10
2000 to < 2500	2250	7
2500 to < 3000	2750	6
3000 to < 3500	3250	3

Use the statistical function on a calculator. The sample standard deviation should be used.
Mean = $1825, sample SD = $797

11. The median, as the data set has two clear outliers
12. a. $74 231
    b. $65 000
    c. It would be in the workers' interest to use a higher figure when negotiating salaries, whereas it would be in the management's interest to use a lower figure.
13. a. Year 1: interquartile range = 38 907.5, standard deviation = 20 382.8
    Year 2: interquartile range = 37 110, standard deviation = 19 389.01
    b. Year 1: interquartile range = 31 107.5, standard deviation = 18 123.5
    Year 2: interquartile range = 29 140, standard deviation = 17 289.2
    c. Both values are reduced by a similar amount, but there is a bigger impact on the standard deviation than the interquartile range.
14. a. 2.9
    b. 1.25
    c. The range is less than double the value of the interquartile range. This indicates that the data is quite tightly bunched with no outliers.

Exercise 7.6 Review: exam practice

1. a. Mean: 5.3; median: 6; mode: 6
   b. Mean: 36.4; median: 32.5; no mode
   c. Mean: 9; median: 10; mode: 10
2. a. 16 days
   b. Mean: 7.2; median: 7
3. 96.4 minutes
4. a. Range
   b. Median
   c. IQR
   d. Mean
   e. Standard deviation
   f. Mode
5. a. Minimum: 6; maximum: 36; range: 30
   b. Mean: 14.9; median: 14
   c. Possible outlier 36; mean: 13.4; median: 14
   d. The median was unchanged by removing the outlier, whereas the mean decreased.
6. Mean: 6.25; median: 6.5; mode: 8; range: 13; IQR: 4
7. Mean: 25.0; median: 24.9; mode: 23.6; range: 8.5; IQR: 3.4
8. 4 is an outlier.
9. a. 3.45   b. 4   c. 3
   d. 2.1   e. 7
10. a. 101.0
    b. 106
    c. 37
    d. 23.5
11. a. Mean: 13.9; median: 14; mode: 14
    b. 20 is an outlier
    c. Any measure could be used in this case, as they all give similar results. However, since the data contains an outlier the median is the more appropriate measure of centre.
12. a. 2800, 3150, 3900, 4337.5, 5200
    b. 1187.5
13. C
14. B
15. D
16. D
17. a. 3
    b. Any number greater than 3
    c. Any number from 4 to 12
18. Kris. There is a possible outlier at 40–45, so the median is a better measure of central tendency, as it is less affected by outliers than the mean.
19. 44 is above the upper fence of 23.5, so the data value is an outlier.
20. a. Indigenous mean = 1410.5, Non-Indigenous mean = 35 241.6
    b. Indigenous median = 1156, Non-Indigenous median = 24 008
    c. Indigenous: standard deviation = 1193.8, IQR = 1875.5
    Non-Indigenous: standard deviation = 33 949.03, IQR = 58 198
    d. The median and IQR are probably more appropriate due to the presence of potential extreme values in the data

8 Comparing data sets

8.1 Overview

LEARNING SEQUENCE

8.1 Overview
8.2 Constructing box plots
8.3 Comparing parallel box plots
8.4 Comparing back-to-back stem plots
8.5 Comparing histograms
8.6 Review: exam practice

CONTENT

In this chapter, students will learn to:
- complete a five-number summary for different data sets
- construct box plots using a five-number summary
- compare parallel box plots and back-to-back stem plots for different data sets [complex]
- compare the characteristics of the shape of histograms using symmetry, skewness and bimodality, where applicable [complex].

Fully worked solutions for this chapter are available in the Resources section of your eBookPLUS at www.jacplus.com.au.

8.2 Constructing box plots

8.2.1 Five-number summary

A **five-number summary** is a list consisting of the lowest score, lower quartile, median, upper quartile and highest score of a set of data.

A five-number summary gives information about the centre and spread of a set of data. The convention is not to detail the numbers with labels but to present them in order; so, for example, the five-number summary:

$$4, 15, 21, 23, 28$$

would be interpreted as lowest score 4, lower quartile 15, median 21, upper quartile 23 and highest score 28.

WORKED EXAMPLE 1

From the following five-number summary determine:
a. the median   **b. the interquartile range**   **c. the range.**

$$29, 37, 39, 44, 48$$

THINK	WRITE
The figures are presented in the order of lowest score, lower quartile, median, upper quartile, highest score.	lowest $= 29$, $Q_1 = 37$, median $= 39$, $Q_3 = 44$, highest $= 48$
a. The median is the third number in the list.	a. Median $= 39$
b. The interquartile range is the difference between the upper and lower quartiles.	b. $\begin{aligned} \text{IQR} &= Q_3 - Q_1 \\ &= 44 - 37 \\ &= 7 \end{aligned}$
c. The range is the difference between the highest score and the lowest score.	c. $\begin{aligned} \text{Range} &= \text{highest} - \text{lowest} \\ &= 48 - 29 \\ &= 19 \end{aligned}$

8.2.2 Box plots

A box plot (or box-and-whisker plot) is a graphical representation of the five-number summary. It is a powerful way to show the centre and spread of data. Box plots consist of a central divided box with attached 'whiskers'. The box spans the interquartile range. The median is marked by a vertical line inside the box. The whiskers indicate the range of scores.

Box plots are *always drawn to scale*. They are presented either with the five-number summary figures attached as labels (below left) or with a scale presented alongside the box plot (below right).

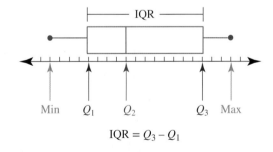

$$\text{IQR} = Q_3 - Q_1$$

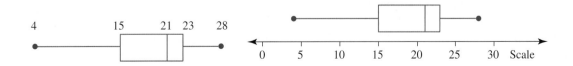

8.2.3 Interpreting a box plot

The box plot neatly divides the data into four sections. One-quarter of the scores lie between the lowest score and the lower quartile, one-quarter between the lower quartile and the median, one-quarter between the median and the upper quartile, and one-quarter between the upper quartile and the highest score.

The reader can easily see where clustering of the data occurs. For example, a small box with relatively long whiskers would indicate that half of the data (from Q_1 to Q_3) would be confined to a small range and the data could be described as clustered. A wide box with relatively short whiskers would indicate that half of the data (from Q_1 to Q_3) would be spread over a wide range and the data could be described as spread out.

Box plots also clearly show how a set of data is distributed. Data that are evenly spaced around a central point can be described as symmetrical. Negatively skewed data has larger amounts of data at the higher end (the median is right of centre, closer to Q_3), and positively skewed data has larger amounts of data at the lower end (the median is left of centre, closer to Q_1). A box plot that is neither positively nor negatively skewed is described as symmetrical (the median is in the middle of the box).

Consider the following box plots with their matching histograms.

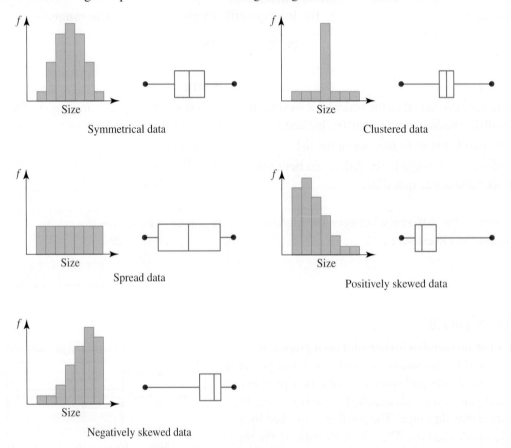

Symmetrical data

Clustered data

Spread data

Positively skewed data

Negatively skewed data

8.2.4 Identification of extreme values

Extreme values (**outliers**) often make the whiskers appear longer than they should and hence give the appearance that the data are spread over a much greater range than they really are.

If an outlier occurs in a set of data, it can be denoted by a small cross on the box plot. The whisker is then shortened to the next largest (or smallest) value.

The box plot below shows that the lowest score was 5. This was an outlier as the rest of the scores were located within the range 15 to 42. The second lowest score was 15. This was not an outlier, so the left-hand whisker extends down to 15.

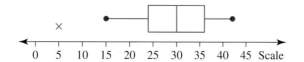

Recall from section 7.4.1 that extreme values (outliers) can be identified by calculating the values of the lower and upper fences. Values that lie below the lower fence or above the upper fence are outliers.

$$\text{Lower fence} = Q_1 - 1.5 \times \text{IQR}$$
$$\text{Upper fence} = Q_3 + 1.5 \times \text{IQR}$$

WORKED EXAMPLE 2

The stem plot (stem-and-leaf plot) below gives the speed of 25 cars caught by a roadside speed camera.
a. Prepare a five-number summary of the data.
b. Draw a box plot of the data. (Identify any extreme values.)
c. Describe the distribution of the data in terms of shape, centre and spread.

Key: $8|2 = 82\,\text{km/h}$
$8^*|6 = 86\,\text{km/h}$

Stem	Leaf
8	2 2 4 4 4 4
8*	5 5 6 6 7 9 9 9
9	0 1 1 2 4
9*	5 6 9
10	0 2
10*	
11	4

THINK

a. 1. First identify the median and upper and lower quartiles.
There are 25 pieces of data. The median is the $\dfrac{25 + 1}{2}$th score; that is, the 13th score. A box has been place around the 13th score.
The Q_1 is halfway between the 6th and 7th scores in the lower half of the data. The Q_3 is halfway between the 6th and 7th scores in the upper half of the data.

2. Identify the lowest and highest data values.

WRITE

a. Key : $8|2 = 82\,\text{km/h}$
$8^*|6 = 86\,\text{km/h}$

Stem	Leaf
8	2 2 4 4 4 4 $|_{Q_1}$
8*	5 5 6 6 7 9 $\boxed{9}$ 9
9	0 1 1 2 4 $|_{Q_3}$
9*	5 6 9
10	0 2
10*	
11	4

$Q_1 = 84.5\,\text{km/h}$
$\text{Median} = 89\ \text{km/h}$
$Q_3 = 94.5\,\text{km/h}$

$\text{Minimum} = 82\,\text{km/h}$
$\text{Maximum} = 114\,\text{km/h}$

3. Write the five-number summary:
   The lowest score is 82. The lower quartile is 84.5.
   The median is 89. The upper quartile is 94.5. The
   highest score is 114.

   Five-number summary:
   82, 84.5, 89, 94.5, 114

b. 1. Identify any outliers by calculating the upper and
   lower fences.

   Lower fence $= Q_1 - 1.5 \times IQR$

   Upper fence $= Q_3 + 1.5 \times IQR$

b. $IQR = 94.5 - 84.5$
   $= 10$
   Lower fence $= 84.5 - 1.5 \times 10$
   $\qquad = 69.5$
   Upper fence $= 94.5 + 1.5 \times 10$
   $\qquad = 109.5$
   114 is higher than the upper fence;
   hence it is an outlier.

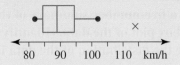

2. To draw the box plot, start by ruling a suitable
   scale. Remember to include the units of
   measurement. The box represents the interquartile
   range so it runs from 84.5 to 94.5. The median is a
   vertical line in the box at 89. The whiskers should
   extend to the lowest score (82) and the highest
   score that is not an outlier (102). The outlier at 114
   should be indicated by a cross.

c. 1. Because an outlier is present in this set of data, the
   most useful measure of centre is the median, and
   best measure of spread is the interquartile range.
      The position of the outlier and the shape of the
   box plot should also be commented on.

c. The distribution of car speeds is
   positively skewed with an outlier at
   114 km/h. The average speed, as
   measured by the median, is 89 km/h.
   The spread of speeds, as measured by
   the interquartile range, is 10 km/h.

 Resources

Interactivities Box plots (int-6245)
   Box-and-whisker plots (int-4623)

Exercise 8.2 Constructing box plots

1. **WE1** From the following five-number summary determine:
   a. the median
   b. the interquartile range
   c. the range.

   6, 11, 12, 16, 32

2. From the following five-number summary determine:
   a. the median
   b. the interquartile range
   c. the range.

   101, 119, 122, 125, 128

3. From the following five-number summary determine:
   a. the median
   b. the interquartile range
   c. the range.

   39.2, 46.5, 49.0, 52.3, 57.8

---

4. The box plot below shows the distribution of final points scored by a football team over a season's roster.

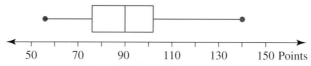

a. What was the team's greatest points score?
b. What was the team's least points score?
c. What was the team's median points score?
d. What was the range of points scored?
e. What was the interquartile range of points scored?

5. The box plot below shows the distribution of data formed by counting the number of honey bears in each of a large sample of packs.

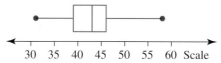

a. What was the largest number of honey bears in any pack?
b. What was the smallest number of honey bears in any pack?
c. What was the median number of honey bears in any pack?
d. What was the range of numbers of honey bears per pack?
e. What was the interquartile range of honey bears per pack?

Questions 6 to 8 refer to the following box plot.

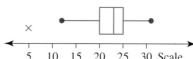

6. **MC** What is the median of the data?
   A. 5 **B.** 20 **C.** 23 **D.** 25
7. **MC** What is the interquartile range of the data?
   A. 23 **B.** 26 **C.** 5 **D.** 20 to 25
8. **MC** Which of the following is *not* true of the data represented by the box plot?
   A. One-quarter of the scores are between 5 and 20.
   B. Half of the scores are between 20 and 25.
   C. The lowest quarter of the data is spread over a wide range.
   D. Most of the data are contained between the scores of 5 and 20.
9. The number of sales made each day by a salesperson is recorded over a 2-week period:

$$25, 31, 28, 43, 37, 43, 22, 45, 48, 33$$

a. Prepare a five-number summary of the data. (There is no need to draw a stem plot of the data.)
b. Draw a box plot of the data.

10. The data below show monthly rainfall in millimetres.

J	F	M	A	M	J	J	A	S	O	N	D
10	12	21	23	39	22	15	11	22	37	45	30

a. Prepare a five-number summary of the data.
b. Draw a box plot of the data.

11. **WE2** The stem plot below details the age of 25 offenders who were caught during random breath testing.
    a. Prepare a five-number summary of the data.
    b. Draw a box plot of the data.
    c. Describe the distribution of the data in terms of shape, centre and spread,

Key: $1|8 = 18$ years

Stem	Leaf
1	8 8 9 9 9
2	0 0 0 1 1 3 4 6 9
3	0 1 2 7
4	2 5
5	3 6 8
6	6
7	4

12. The stem plot details the price at which 30 apartments in a particular suburb sold for.
    a. Prepare a five-number summary of the data.
    b. Draw a box plot of the data.

Key: $32|4 = \$324\,000$

Stem	Leaf
32	4 7 9
33	0 0 2 5 5
34	0 0 2 3 5 5 7 9 9
35	0 0 2 3 7 7 8
36	0 2 2 5 8
37	5

13. The following data detail the number of hamburgers sold by a fast-food outlet every day over a 4-week period.

M	T	W	T	F	S	S
125	144	132	148	187	172	181
134	157	152	126	155	183	188
131	121	165	129	143	182	181
152	163	150	148	152	179	181

a. Prepare a stem plot of the data. (Use a class interval of size 10.)
b. Draw a box plot of the data.

14. The following data show the ages of 30 mothers upon the birth of their first baby.

22, 18, 17, 22, 24, 25,
32, 19, 23, 28, 31, 19,
23, 25, 23, 21, 33, 23,
24, 20, 29, 18, 22, 24,
20, 22, 17, 48, 18, 20

a. Prepare a stem plot of the data. (Use a class interval of size 5.)
b. Draw a box plot of the data. Indicate any extreme values appropriately.
c. Describe the distribution in words. What does the distribution say about the age that mothers have their first baby?

8.3 Comparing parallel box plots

Parallel box plots are drawn one above the other, using the same scale. The set of parallel box plots shown provides a means of comparing the results of four classes in a Maths test.

Reporting on a parallel box plot involves comparing the individual box plot features, including:
- central tendency (median)
- spread (both IQR and range)
- outliers
- shape of the distribution.

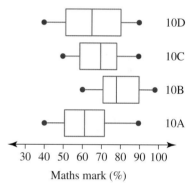

Comparisons made on parallel box plots should always be in context. For example, to compare the above parallel box plots, comment on the median, spread, shape and outliers in terms of Maths marks for the different classes.

Values of the median, IQR, range and/or outliers should always be included in a comparison of parallel box plots.

WORKED EXAMPLE 3

The following parallel box plots show the number of hats produced in a factory per hour over a two-day period in 2007 and 2017.
a. State the five-number summary for the hourly hat production in both 2007 and 2017.
b. i. The top 25% of hats produced in 2017 were above which value?
  ii. The middle 50% of hats produced in 2007 were between which values?
c. Compare the hourly hat production for the two different years.

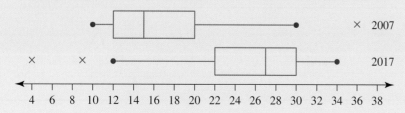

THINK	WRITE
a. 1. Look at the box plot for 2007. Using the scale and the box plot, determine the values for the five-number summary.	**a.** 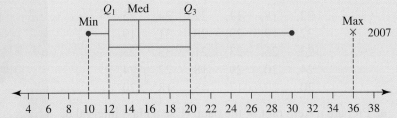

Five-number summary for 2007: 10, 12, 15, 20, 36

2. Look at the box plot for 2017. Using the scale and the box plot, determine the values for the five-number summary.	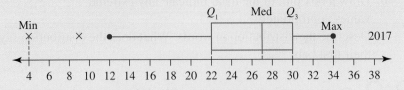

Five-number summary for 2017: 4, 22, 27, 30, 34

b. i. The top 25% on a box plot occurs above the third quartile, Q_3.

b. i The top 25% of hourly hats produced in 2017 occurred above 30 hats per hour.

ii. The middle 50% on a box plot is represented by the IQR, or the length of the box.

ii. The middle 50% of hourly hats produced in 2007 occurred between 12 and 20 hats per hour.

c. 1. Write a sentence, comparing the two box plots for the median, IQR, range, shape and outliers.

c. The distribution of hat production in 2007 was positively skewed with an outlier at 36 hats per hour. The distribution of hat production in 2017 was negatively skewed with outliers at 4 and 9 hats per hour. The parallel box plots show that median hourly hat production in 2007 (15 hats per hour) was lower than in 2017 (27 hats per hour). The IQR for both 2007 and 2017 was identical (8 hats per hour). The consistency of hat production was similar as evidenced by the IQR.

2. Write a sentence summarising the comparison.

Overall, the hourly hat production over the two-day period in 2017 was generally greater than it was in 2007 as evidenced by the median. Hat production was slightly more consistent in 2007 due to the smaller IQR.

 Resources

 Interactivity Parallel box plots (int-6248)

Exercise 8.3 Comparing parallel box plots

1. A gardener recorded how long it took to mow each lawn on the weekly route (in minutes). The data set is shown.

   32, 45, 56, 28, 19, 38, 26, 47, 54, 21, 33, 40, 17, 58, 21

   State the five-number summary for the data set.

2. **MC** What are the five-number summary values for the following data set of ages for a family with 6 members, 45, 42, 12, 9, 7, 4, in order of minimum, Q_1, median, Q_3 and maximum?
   A. 4, 7, 10.5, 42, 45
   B. 4, 9, 12, 44, 45
   C. 4, 7, 12, 42, 45
   D. 4, 9, 10.5, 42, 45

3. The weights of 14 boxes (in kilograms) being moved from one house to another are as follows.

   3, 6, 13, 15, 15, 15, 17, 17, 18, 20, 20, 22, 26, 30

   Draw a box plot to display this data set.

4. The prices of 10 mobile phones are viewed online and are as follows.

   $349, $469, $265, $497, $159, $52, $999, $489, $599, $577

   Draw a box plot to represent this data set.

5. The speeds of 15 cars in a school zone road were as follows.

   39, 51, 60, 42, 44, 38, 75, 45, 40, 52, 41, 42, 46, 41, 39

   Draw a box plot to represent this data set.

6. A restaurant trialled three different menus for a month each to work out which type of cuisine its patrons enjoy the most. The restaurant has asked for some help analysing data that they have collected.
   a. Draw a box plot for each month of data on the same scale.
   b. Write a comparison of the box plots to help the restaurant decide which menu will be the most successful.

	Week 1	Week 2	Week 3	Week 4
Month 1: Greek	75	62	48	50
Month 2: Italian	48	17	9	68
Month 3: Spanish	56	43	37	28

The following parallel box plots, showing the ages at which boys and girls learn to ride a 2-wheel bike, should be used to answer questions 7–9.

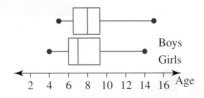

7. **MC** Looking at the parallel box plots, at what age did the youngest girl and the oldest boy, respectively, learn to ride a two-wheel bike?
   A. 15 and 4   **B.** 4 and 15   **C.** 5 and 14   **D.** 14 and 5

8. **MC** Which value is the same for both box plots?
   A. Q_1 **B.** Q_3 **C.** IQR   **D.** Range

9. **MC** What is the difference between the medians of the box plots?
   A. 1 year   **B.** 2 years   **C.** 3 years   **D.** 4 years

10. a. What does it mean if the median on a box plot is not placed exactly halfway between the upper and lower quartiles?
    b. What does it mean if the median on a box plot is closer to the upper quartile than the lower quartile?
    c. Write a statement explaining what the position of the median can tell you about a data set.

11. **WE3** The parallel box plots below show the weekly sales for jars of peanut butter and Nutella in a supermarket.

a. State the five-number summary for the weekly sales of peanut butter and Nutella.
b. i. The lowest 25% of weekly peanut butter jar sales were below which value?
    ii. The middle 50% of Nutella weekly jar sales were between which values?
c. Compare the weekly sales for the two different products.

12. The following data was collected from a company that is comparing two different compounds for sunscreen to see how long (measured in minutes) they remain effective in water before they need to be reapplied.

Compound 1	176	156	123	169	149	133	164	127	139	121	172	168
Compound 2	130	125	128	121	118	120	122	127	130	122	131	126

a. Draw parallel box plots for this data set.
b. Compare the two compounds.

13. A comparison of Year 11 students' achievements (measured as a number out of 100) in History and English was recorded and the following results were obtained.

History	75	78	42	92	59	67	78	82	84	64	77	98
English	78	80	57	96	58	71	74	87	79	62	75	100

a. Draw parallel box plots for this data set.
b. What do the parallel box plots tell you about Year 11 students' achievements in History and English?

14. The heights of Year 11 and Year 12 students (to the nearest centimetre) are being investigated. The results of some sample data are shown.

Year 11	160	154	157	170	167	164	172	158	177	180	175	168	159	155	163	163	169	173	172	170
Year 12	160	172	185	163	177	190	183	181	176	188	168	167	166	177	173	172	179	175	174	180

a. Draw parallel box plots for this data set.
b. Comment on what the plots tell you about the heights of Year 11 and Year 12 students.

8.4 Comparing back-to-back stem plots

We have seen how to construct a stem plot for a set of univariate data. We can also extend a stem plot so that it displays bivariate data. Specifically, we shall create a stem plot that displays the relationship between a numerical variable and a categorical variable. We shall limit ourselves in this section to categorical variables with just two categories; for example, gender. The two categories are used to provide two, back-to-back leaves of a stem plot.

> **Back-to-back stem plots**
> A back-to-back stem plot is used to display bivariate data involving a numerical variable and a categorical variable with 2 categories.

The girls and boys in Grade 4 at Kingston Primary School submitted projects on the Olympic Games. The marks they obtained out of 20 are shown.

Girls' marks	16	17	19	15	12	16	17	19	19	16
Boys' marks	14	15	16	13	12	13	14	13	15	14

Display the data in the form of a back-to-back stem plot.

THINK

1. Identify the highest and lowest scores in order to decide on the stems.

2. Create an unordered stem plot first. Put the boys' scores on the left, and the girls' scores on the right.

3. Now order the stem plot. The scores on the left should increase in value from right to left, while the scores on the right should increase in value from left to right.

WRITE

Highest score = 19
Lowest score = 12
Use a stem of 1, and divide into five class intervals. (Values of 12 and 13 will be included in the first class interval, values of 14 and 15 will be included in the next class interval etc.)

Key: 1|2 = 12

Leaf: Boys	Stem	Leaf: Girls
3 2 3 3	1	2
4 5 4 5 4	1	5
6	1	6 7 6 7 6
	1	9 9 9

Key: 1|2 = 12

Leaf: Boys	Stem	Leaf: Girls
3 3 3 2	1	2
5 5 4 4 4	1	5
6	1	6 6 6 7 7
	1	9 9 9

The back-to-back stem plot allows us to make some visual comparisons of two distributions. In Worked example 4, the centre of the distribution for the girls is higher than the centre of the distribution for the boys. The spread of each of the distributions seems to be about the same. For the boys, the marks are grouped around 12–15 marks; for the girls, they are grouped around 16–19 marks. Overall, we can conclude that the girls obtained better marks than the boys did.

To get a more precise picture of the centre and spread of each of the distributions we can use summary statistics. Specifically, we are interested in:
1. the mean and the median (to measure the centre of the distributions), and
2. the interquartile range and the standard deviation (to measure the spread of the distributions).
   The calculation of these summary statistics is very straightforward and rapid using technology.
   If the distribution of either category is skewed and/or contains outliers, then the median and interquartile range are better measures of centre and spread than the mean and the standard deviation.

The number of 'how-to-vote' cards handed out by various Australian Labor Party and Liberal Party volunteers during the course of a polling day is shown below.

Labor	180	233	246	252	263	270	229	238	226	211
	193	202	210	222	257	247	234	226	214	204
Liberal	204	215	226	253	263	272	285	245	267	275
	287	273	266	233	244	250	261	272	280	279

a. **Display the data using a back-to-back stem plot.**
b. **Use the stem plot, together with summary statistics, to compare the distributions of the number of cards handed out by the Labor and Liberal volunteers.**

THINK

a. 1. Identify the highest and lowest scores in order to decide on the stems.

2. Construct the stem plot.

WRITE

a. The highest score is 287, and the lowest score is 180. So use stems 18 to 28.

Key: $18|0 = 180$

Leaf: Labor	Stem	Leaf: Liberal
0	18	
3	19	
4 2	20	4
4 1 0	21	5
9 6 6 2	22	6
8 4 3	23	3
7 6	24	4 5
7 2	25	0 3
3	26	1 3 6 7
0	27	2 2 3 5 9
	28	0 5 7

b. 1. Use a form of technology to calculate the summary statistics: the mean, the median, the standard deviation and the interquartile range for each party. Enter each set of data as a separate list.

b. For the Labor volunteers:
Mean = 227.9
Median = 227.5
Interquartile range = 36
Standard deviation = 23.9
For the Liberal volunteers:
Mean = 257.5
Median = 264.5
Interquartile range = 29.5
Standard deviation = 23.4

2. Comment on the relationship.

From the stem plot we see that the Labor distribution is symmetric and therefore the mean and the median are very close, whereas the Liberal distribution is negatively skewed.

Since the distribution is skewed, the median is a better indicator of the centre of the distribution than the mean.

Therefore, comparing the medians, we have the median number of cards handed out for Labor at 228 and for Liberal at 265.

The standard deviations were similar, as were the interquartile ranges. There was not a lot of difference in the spread of the data.

Overall, the Liberal Party volunteers handed out a lot more 'how-to-vote' cards than the Labor Party volunteers did.

---

on Resources

🧩 **Interactivities** Back-to-back stem plots (int-6252)

Comparing data sets (int-4625)

Exercise 8.4 Comparing back-to-back stem plots

1. **WE4** The marks (out of 50), obtained for the end-of-term test by the students in German and French classes are given below. Display the data in the form of a back-to-back stem plot.

German	20	38	45	21	30	39	41	22	27	33	30	21	25	32	37	42	26	31	25	37
French	23	25	36	46	44	39	38	24	25	42	38	34	28	31	44	30	35	48	43	34

2. The birth masses of 10 boys and 10 girls (in kilograms, to the nearest 100 grams) are recorded in the table. Display the data using a back-to-back stem plot.

Boys	3.4	5.0	4.2	3.7	4.9	3.4	3.8	4.8	3.6	4.3
Girls	3.0	2.7	3.7	3.3	4.0	3.1	2.6	3.2	3.6	3.1

---

3. **WE5** The number of delivery trucks making deliveries to a supermarket each day over a 2-week period was recorded for two neighbouring supermarkets — supermarket A and supermarket B. The data are as follows.

A	11	15	20	25	12	16	21	27	16	17	17	22	23	24
B	10	15	20	25	30	35	16	31	32	21	23	26	28	29

a. Display the data using a back-to-back stem plot.
b. Use the stem plot, together with summary statistics, to compare the distributions of the number of trucks delivering to supermarkets A and B.

4. The marks out of 20 for males and females on a Science test for a Year 11 class are as follows.

Females	12	13	14	14	15	15	16	17
Males	10	12	13	14	14	15	17	19

a. Display the data using a back-to-back stem plot.
b. Use the stem plot, together with summary statistics, to compare the distributions of the marks of the males and the females.

5. The end-of-year English marks for 10 students in an English class were compared over 2 years. The marks for the first year and for the same students in the second year are shown.

First year	30	31	35	37	39	41	41	42	43	46
Second year	22	26	27	28	30	31	31	33	34	36

a. Display the data using a back-to-back stem plot.
b. Use the stem plot, together with summary statistics, to compare the distributions of the marks obtained by the students over the 2 years.

6. The age and gender of a group of people attending a fitness class were recorded as follows.

Female	23	24	25	26	27	28	30	31
Male	22	25	30	31	36	37	42	46

a. Display the data using a back-to-back stem plot.
b. Use the stem plot, together with summary statistics, to compare the distributions of the ages of the female to male members of the fitness class.

7. The scores on a board game are recorded for a group of kindergarten children and for a group of children in a preparatory school.

Kindergarten	3	13	14	25	28	32	36	41	47	50
Prep. school	5	12	17	25	27	32	35	44	46	52

a. Display the data using a back-to-back stem plot.

b. Use the stem plot, together with summary statistics, to compare the distributions of the scores of the kindergarten children compared to the preparatory school children.

8. **MC** The pair of variables that could be displayed on a back-to-back stem plot is
   A. the height of a student and the number of people in the student's household.
   B. the time put into completing an assignment and a pass or fail score on the assignment.
   C. the weight of a businessman and his age.
   D. the religion of an adult and the person's head circumference.

9. **MC** A back-to-back stem plot is a useful way of displaying the relationship between
   A. the proximity to markets (km) and the cost of fresh foods on average per kilogram.
   B. height and head circumference.
   C. age and attitude to gambling (for or against).
   D. weight and age.

10. The two dot plots shown display the latest Maths test results for two Year 11 classes. The results show the marks out of 20.

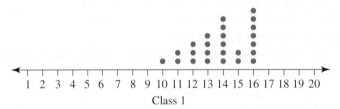

Class 1

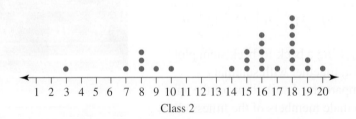

Class 2

a. How many students are in each class?

b. For each class, how many students scored 15 out of 20 for the test?

c. For each class, how many students scored more than 10 for the test?

d. Use the dot plots to describe the performance of each class on the test.

11. The following data show the ages of male and female players at a ten-pin bowling centre. Draw a back-to-back stem plot of the data.

Male: 20, 36, 16, 38, 32, 18, 19, 21, 25, 45, 29, 60, 31, 21, 16, 38, 52, 43, 17, 28, 23, 23, 43, 17, 22, 23, 32, 34

Female: 21, 23, 30, 16, 31, 46, 15, 17, 22, 17, 50, 34, 65, 25, 27, 19, 15, 43, 22, 17, 22, 16, 48, 57, 54, 23, 16, 30, 18, 21, 28, 35

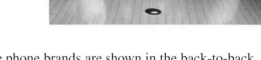

12. The comparisons between the battery lives of two mobile phone brands are shown in the back-to-back stem plot below. Which mobile phone brand has the better battery life? Explain.

Key: 6|1 = 61 hours

Brand A	Stem	Brand B
8 8 7 5	0	7
9 7 4 1 0	1	0 5 5 5 7 9
2 2 2 1	2	0 2 2 6 7
8 6 4 2 0	3	0 2 4 6 8
	4	
	5	6
1	6	
	7	5

13. The winning times in seconds for the women's and men's 100-metre sprint in the Olympics are shown in the table.

   a. Display the winning times for women and men using a stem plot.
   b. Is there a large difference in winning times? Explain your answer.

Year	Women's 100-m sprint	Men's 100-m sprint
1928	12.20	10.80
1932	11.90	10.30
1936	11.50	10.30
1948	11.90	10.30
1952	11.50	10.40
1956	11.50	10.50
1960	11.00	10.20
1964	11.40	10.00
1968	11.00	9.90
1972	11.07	10.14
1976	11.08	10.06
1980	11.60	10.25
1984	10.97	9.99
1988	10.54	9.92
1992	10.82	9.96
1996	10.94	9.84
2000	10.75	9.87
2004	10.93	9.85
2008	10.78	9.69

14. The following data sets show the rental price (in $) of two-bedroom apartments in two different suburbs of Wollongong.

Suburb A
215, 225, 211, 235, 244, 210, 215, 210, 256, 207, 200, 200, 242, 225, 231, 205, 240, 205, 235, 200

Suburb B
235, 245, 231, 232, 240, 280, 280, 270, 255, 275, 275, 285, 245, 265, 270, 255, 260, 258, 251, 285

a. Draw a back-to-back stem plot to compare the data sets.
b. Compare and contrast the rental price in the two suburbs.

8.5 Comparing histograms

8.5.1 Frequency histograms

A histogram is a useful way of displaying large, numerical data sets (say, more than 50 observations). The vertical axis on the histogram displays the frequency and the horizontal axis displays class intervals of the variable (e.g. height or income). The frequency for each class interval determines the height of the column.

When data are given in raw form — that is, just as a list of numbers in no particular order — it is helpful to first construct a frequency table.

WORKED EXAMPLE 6

The data show the distribution of masses (in kilograms) of 60 students in Year 7 at Northwood State High School.

a. Complete a frequency distribution table for these data.
b. Construct a frequency histogram to display these data.

45.7, 34.2, 56.3, 38.7, 52.4, 45.7, 48.2, 52.1, 58.7, 62.3,

45.8, 52.4, 60.2, 48.5, 54.3, 39.8, 36.2, 54.3, 39.7, 46.3,

45.9, 52.3, 44.2, 49.6, 48.6, 42.5, 47.2, 51.3, 43.1, 52.4,

48.2, 51.8, 53.8, 56.9, 53.7, 42.9, 46.7, 51.9, 56.2, 61.2,

48.3, 45.7, 43.5, 43.8, 58.7, 59.2, 58.7, 54.6, 43.0, 48.2,

48.4, 56.8, 57.2, 58.3, 57.6, 53.2, 53.1, 58.7, 56.3, 58.3

THINK

1. First construct a frequency table. The lowest data value is 34.2 and the highest is 62.3. Divide the data into class intervals. If we started the first class interval at, say, 30 kg and ended the last class interval at 65 kg, we would have a range of 35. If each interval was 5 kg, we would then have 7 intervals which is a reasonable number of class intervals. While there are no set rules about how many intervals there should be, somewhere between about 5 and 15 class intervals is usual. So, in this example, we would have class intervals of 30–<35 kg, 35–<40 kg, 40–<45 kg and so on. Count how many observations fall into each of the intervals and record these in a table.

2. Check that the frequency column totals 60.

3. A histogram can be constructed. Since we are dealing with grouped data, each column has the range of the values included in the class interval on either edge of the column. The column heights are determined by the frequency for each class interval.

WRITE

Class interval	Frequency
30–<35	1
35–<40	4
40–<45	7
45–<50	16
50–<55	15
55–<60	14
60–<65	3
Total	**60**

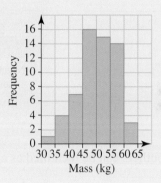

WORKED EXAMPLE 7

The marks out of 20 received by 30 students for a book-review assignment are given in the frequency table.

Mark	12	13	14	15	16	17	18	19	20
Frequency	2	7	6	5	4	2	3	0	1

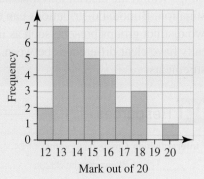

Display these data in the form of a histogram.

THINK

In this case we are dealing with integer values, not grouped data. The integer values are written underneath the centre of each column. The column heights are determined by the frequency for each class interval.

WRITE

8.5.2 Describing the shape of a histogram

Symmetric distributions

The data shown in the histogram to the right can be described as *symmetric*.

There is a single peak and the data trail off on both sides of this peak in roughly the same fashion.

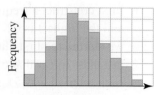

Skewed distributions

Each histogram below shows an example of a skewed distribution.

- The figure below left shows data that are *negatively skewed*. The data in this case peak to the right and trail off to the left.
- The figure below right shows *positively skewed* data. The data in this case peak to the left and trail off to the right.

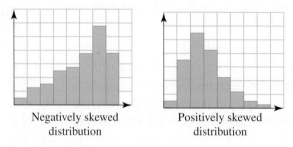

Negatively skewed distribution   Positively skewed distribution

Outliers

Possible outliers can be identified as a column that lies well away from the main body of the histogram.

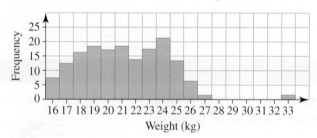

Bimodality

The mode of a distribution is the data value or class interval that has the highest frequency. This will be the column or row on the display that is the longest. When there is more than one mode, the data distribution is multimodal. This can indicate that there may be subgroups within the distribution that may require further investigation. Bimodal distributions can occur when there are two distinct groups present, such as in data values that typically have clear differences between male and female measurements.

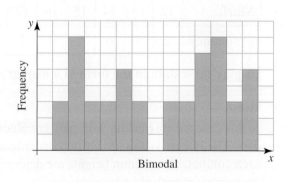

Bimodal

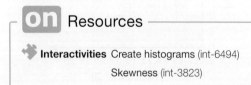

Resources

Interactivities Create histograms (int-6494)

Skewness (int-3823)

Exercise 8.5 Comparing histograms

1. Construct a frequency table for each of the following sets of data.
   a. 3, 4, 4, 5, 5, 6, 7, 7, 7, 8, 8, 9, 9, 10, 10, 12
   b. 4.3, 4.5, 4.7, 4.9, 5.1, 5.3, 5.5, 5.6, 5.2, 3.6, 2.5, 4.3, 2.5, 3.7, 4.5, 6.3, 1.3
   c. 11, 13, 15, 15, 16, 18, 20, 21, 22, 21, 18, 19, 20, 16, 18, 20, 16, 10, 23, 24, 25, 27, 28, 30, 35, 28, 27, 26, 29, 30, 31, 24, 28, 29, 20, 30, 32, 33, 29, 30, 31, 33, 34
   d. 0.4, 0.5, 0.7, 0.8, 0.8, 0.9, 1.0, 1.1, 1.2, 1.0, 1.3, 0.4, 0.3, 0.9, 0.6
2. Using the frequency tables from question 1, construct a histogram for each set of data.
3. Using a form of technology, construct a histogram for each of the sets of data given in question 1. Compare this histogram with the one drawn for question 2.
4. A class of 30 students sat for a Mathematics test. Their results out of 100 are shown below.

   68, 72, 58, 45, 69, 92, 38, 51, 70, 65, 69, 73, 52, 76, 48,
   69, 73, 41, 42, 73, 80, 50, 60, 49, 65, 94, 88, 85, 53, 60

   a. Use these results to copy and complete the frequency table.

   b. Construct a histogram to display these data.

Score	Tally	Frequency
30–39		
40–49		
50–59		
60–69		
70–79		
80–89		
90–99		

5. **WE6** A farmer measures the heights of his tomato plants. The results, in metres, are shown below.

   0.93, 1.21, 2.03, 1.40, 1.17, 1.53, 1.82, 1.77, 1.65, 0.63, 1.24,

   1.99, 0.80, 2.14, 1.53, 2.07, 1.96, 1.05, 0.94, 1.23, 1.72, 1.34,

   0.75, 1.17, 1.50, 1.41, 1.74, 1.86, 1.55, 1.42, 1.52, 1.39, 1.76,

   1.67, 1.28, 1.43, 2.13

   a. Use the class intervals 0.6–<0.8, 0.8–1.0, 1.0–<1.2, ... etc. to complete a frequency distribution table for these data.
   b. Construct a frequency histogram to display these data

6. The following data give the times (in seconds) taken for athletes to complete a 100-m sprint.

12.2,   12.0,   11.9,   12.0,   12.6,   11.7,

11.4,   11.0,   10.9,   11.7,   11.2,   11.8,

12.2,   12.0,   12.7,   12.9,   11.3,   11.2,

12.8,   12.4,   11.7,   10.8,   13.3,   11.7,

11.6,   11.7,   12.2,   12.7,   13.0,   12.2

a. Construct a frequency distribution table for the data. Use class intervals of 0.5 seconds.

b. Construct a histogram to display these data.

7. **WE7** The marks out of 20 received by 26 students for a History assignment are given in the frequency table.

Score	Frequency
13	2
14	9
15	3
16	5
17	6
18	1

Display these data in the form of a histogram.

8. For each of the following histograms, describe the shape of the distribution of the data and comment on the existence of any outliers.

a.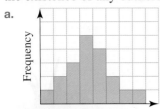

b.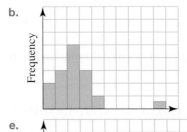

c.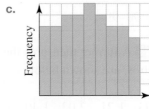

d.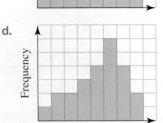

e.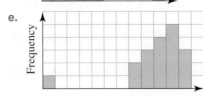

f.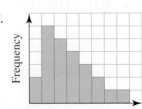

9. **MC** The distribution of the data shown in this histogram could be described as

A. negatively skewed.

B. negatively skewed with outliers.

C. positively skewed.

D. positively skewed with outliers.

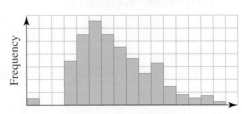

10. **MC** The histogram below is
  A. unimodal and positively skewed.
  B. bimodal and positively skewed.
  C. unimodal and negatively skewed.
  D. bimodal and negatively skewed.

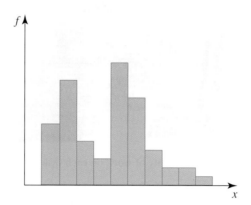

11. The average number of product enquiries per day received by a group of small businesses who advertised in the *Yellow Pages* telephone directory is given below. Describe the shape of the distribution of these data and comment on the existence of any outliers.

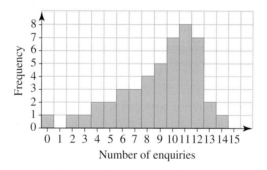

12. The amount of pocket money (to the nearest 50 cents) received each week by students in a Year 6 class is illustrated in this histogram.

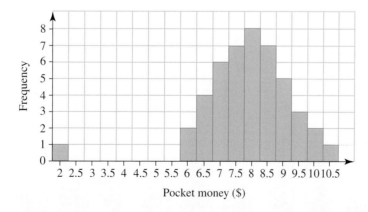

a. Describe the shape of the distribution of these data and comment on the existence of any outliers.
b. What conclusions can you reach about the amount of pocket money received weekly by this group of students?

13. Ten workers were required to complete two tasks. Their supervisor observed the workers and gave them a score for the quality of their work on each task, where higher scores indicated better-quality work. The results are indicated in the following side-by-side bar chart.

Score for quality of work on each task

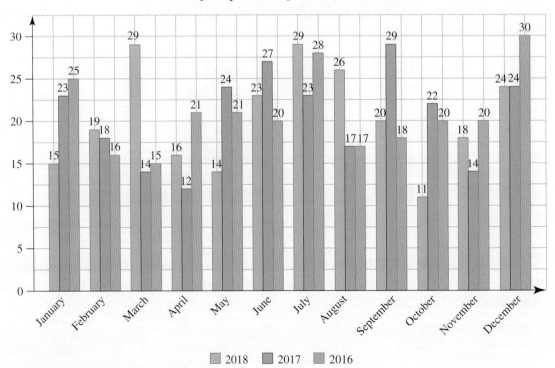

a. Which worker had the largest difference between scores for the two tasks?

b. How many workers received a lower score for task B than task A?

14. The side-by-side bar chart shows a monthly comparison of road fatalities in Queensland.

Monthly comparison of Queensland road fatalities

Source: Australian Government, Department of Infrastructure, Regional Development and Cities, Australian Road Deaths Database

a. Which year had the most fatalities on Queensland roads?
b. Which year had the most fatalities from January to March?
c. Give a possible reason as to why there were a large number of fatalities in the month of December 2016.

15. A coffee bar serves either skim, reduced-fat or whole milk in coffees. The coffees sold on a particular day are shown in the table below, sorted by the type of milk and the gender of the customers.

	Gender	
Type of milk	Male	Female
Skim	87	124
Reduced fat	55	73
Whole	112	49

a. How many coffees were sold on this day?
b. Represent these data in the form of a horizontal side-by-side bar chart.
c. What percentage of males used skim milk for their coffees? Give your answer to the nearest whole number.
d. What percentage of coffees sold contained reduced-fat milk? Give your answer to the nearest whole number.
e. If this was the daily trend of sales for the coffee bar, what percentage of the coffee bar's customers would you expect to be female? Give your answer to the nearest whole number.

8.6 Review: exam practice

8.6.1 Comparing data sets summary

Parallel box plots
- The five-number summary consists of:
  minimum value, lower quartile Q_1, median Q_2, upper quartile Q_3; maximum value

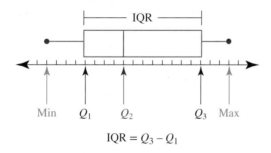

$$IQR = Q_3 - Q_1$$

- Box plots show the distribution of data, and are a graphical representation of the five-number summary.
- The position of the median inside the box indicates whether the box plot is symmetrical, positively skewed or negatively skewed.
- Parallel box plots can display a numerical variable and a categorical variable, where the categorical variable can involve any number of categories.
- Parallel box plots can be used to compare data as long as the data is comparable.

Back-to-back stem plots

- Back-to-back stem plots have a central stem with leaves on either side.
- Back-to-back stem plots display a numerical variable and a categorical variable where the categorical variable involves two categories.
- Back-to-back stem plots are useful when comparing two sets of numerical data.

Comparing histograms

- Modality is a measure of the number of obvious peaks in a data set. Data sets can be described as unimodal (one peak), bimodal (two peaks) or multimodal (more than two peaks).
- Symmetric distributions are when the two halves of a data set are distributed roughly equally about the mean. For a symmetric distribution the median and the mean occur at approximately the same position.
- Positively skewed data tapers off in a long tail to the right-hand side of the peak of the distribution. The median is less than the mean.
- Negatively skewed data tapers off in a long tail to the left-hand side of the peak of the distribution. The median is greater than the mean.
- Measures of central tendency (mean and median) and the mode can be calculated from a histogram.

Exercise 8.6: Review: exam practice

Simple familiar

The test marks for a class of students are shown in the histogram below. Use this information to answer questions 1–4.

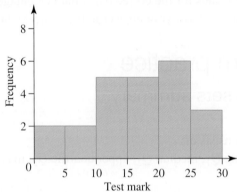

1. **MC** The median test mark for the class falls into which category?
   A. 5–10   **B.** 10–15   **C.** 15–20   **D.** 20–25
2. **MC** What is the mean of this histogram, to one decimal place?
   A. 14.3   **B.** 16.8   **C.** 22.5   **D.** 19.3
3. **MC** What is the mode of this histogram?
   A. 5–10   **B.** 10–15   **C.** 15–20   **D.** 20–25
4. **MC** What is the range of this histogram?
   A. 5   **B.** 15   **C.** 20   **D.** 30
5. Identify the modality and shape of each of the following graphs.
   a. f

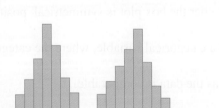

   b. f

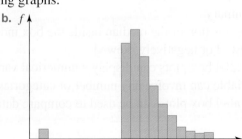

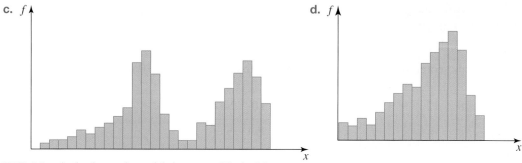

c. *f*

d. *f*

6. **MC** Match the box plot with its most likely histogram.

A. *f*

Size

B. *f*

Size

C. *f*

Size

D. *f*

Size

7. The monthly sales of a particular chocolate bar were recorded over the period of a year. A box plot representing these sales is as shown.

80 90 100 110 120 130 140 150 160 170 180 190 200 210 220 230 240 250 260 270 280 290 300 310 320 330 340 350

Weekly chocolate bar sales

a. State the five-number summary for this data.
b. What is the IQR?
c. Are there any outliers?

8. Use the following two data sets to construct parallel box plots.
   Data set A: 2, 3, 6, 7, 9, 10, 1, 7, 7, 19
   Data set B: 1, 2, 3, 3, 4, 1, 2, 5, 3, 6

9. The number of goals scored by two members of a basketball team during 10 matches is as follows.

Player 1	6	8	12	9	5	15	2	1	10	24
Player 2	6	8	7	8	7	5	8	6	7	7

   a. Draw parallel box plots for this data.
   b. Which player would you rather have on your team and why?

10. For the box plot drawn below:

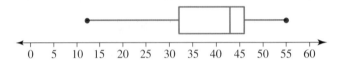

   a. state the median
   b. calculate the range
   c. calculate the interquartile range.

11. The number of babies born each day at a hospital over a year is tabulated and the five-number summary is as follows.
   1. Minimum = 1
   2. Upper quartile = 16
   3. Lower quartile = 8
   4. Maximum = 18
   5. Median = 14
   Show this information in a box plot assuming there are no outliers.

12. The number of hours of counselling received by a group of 9 full-time firefighters and 9 volunteer firefighters after a serious bushfire is given in the table.

Full-time	2	4	3	5	2	4	6	1	3
Volunteer	8	10	11	11	12	13	13	14	15

   a. Construct a back-to-back stem plot to display the data.
   b. Comment on the distributions of the number of hours of counselling of the full-time firefighters and the volunteers.

Complex familiar

13. The IQs of 8 players in 3 different football teams were recorded and are shown below.

Team A	120	105	140	116	98	105	130	102
Team B	110	104	120	109	106	95	102	100
Team C	121	115	145	130	120	114	116	123

   Display the data in the form of parallel box plots.

14. A group of office workers and a group of sports instructors were asked to complete 5 minutes of exercise as part of a study of heart rates. Following the exercise, participants rested for 2 minutes before their pulse rates were measured. The results are set below in the stem plots.

Pulse rates for office workers (beats/min)
Key: $12\,|\,4 = 124$ beats/min

Stem	Leaf
7	6
8	
9	5
10	6 7
11	0 2
12	0 1 2 4 6 7 9
13	0 0 4

Pulse rates for sports instructors (beats/min)
Key: $7\,|\,2 = 72$ beats/min

Stem	Leaf
6	2 4 8 8 9
7	2 2 3 5 7 9
8	2 8
9	6
10	8

a. Describe the shape of each distribution.
b. Calculate the median, the interquartile range, the mode and the range for both.
c. Represent each set of data using a box plot, indicating outliers where present.
d. Calculate the mean and the standard deviation for both sets of data.
e. Using the summary statistics that you have calculated for the office workers, comment on their pulse rates. Do the same for the sports instructors. Comment on any differences between the two groups.

15. To compare two textbooks, a teacher recommends one book to one of his classes and the other book to another class. At the end of the year the classes are each tested. The results are as follows.

Text A (25 students)							Text B (28 students)						
44	52	95	76	13	94	83	65	72	48	63	68	59	68
72	55	81	22	25	64	72	62	75	79	81	72	64	53
35	48	56	59	84	98	84	58	59	64	66	68	42	37
21	35	69	28				39	55	58	52	82	79	55

a. Prepare a back-to-back stem plot of the data.
b. Prepare a five-number summary for each group. (Note that the groups are different sizes.)
c. Prepare parallel box plots of the data.
d. Compare the performance of each of the classes.
e. Which textbook do you think would be best? Why?
f. What other things would you need to take into account before drawing final conclusions?

16. The following data sets show the rental prices (in $) of one-bedroom apartments in two different suburbs of Sydney.

Suburb A

275, 275, 281, 285, 284, 310, 315, 310, 296, 307,

300, 300, 242, 295, 281, 305, 290, 305, 295, 300

Suburb B

235, 225, 231, 232, 240, 280, 300, 310, 295, 275,

275, 285, 245, 305, 270, 255, 270, 228, 241, 285

a. Draw a back-to-back stem plot to compare the data sets.
b. Compare and contrast the rental prices in the two suburbs.
c. The rental prices in a third suburb, suburb C, were also analysed.

Suburb C

335, 325, 351, 335, 340, 360, 300, 390, 395, 285,

357, 385, 354, 305, 375, 345, 270, 358, 340, 365

Compare the rents in the third suburb with the rents in the other two suburbs.

Complex unfamiliar

17. A large company completed a survey of the incomes of their employees. The incomes were recorded for people in their 30s, 40s and 50s. The results are summarised in the parallel box plots below.
Note: The incomes are written in $000s; for example, 110 = $110 000 and 75 = $75 000.

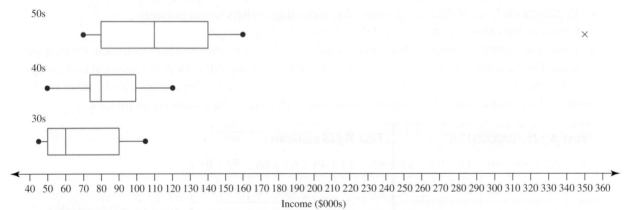

a. **MC** The highest 25% of people in their 30s have an income between
   A. $60 000 and $105 000.
   B. $50 000 and $105 000.
   C. $100 000 and $120 000.
   D. $90 000 and $105 000.
b. The middle 50% of people in their 40s earn between what incomes?
c. How are age and income at this company related?
d. One of the employees surveyed at the company earns $46 000. Which age category do they belong to?
e. The person in their 50s who earns $350 000 was demoted and now earns $190 000. Is that person's income still an outlier? Justify your response.

18. The following table shows the AFL Grand Final statistics for a sample of players who have kicked a total of 5 or more goals from the clubs Carlton and Collingwood.

Player	Team	Kicks	Marks	Handballs	Disposals	Goals	Behinds
Alex Jesaulenko	Carlton	23	11	9	32	11	0
John Nicholls	Carlton	29	3	1	30	13	1
Wayne Johnston	Carlton	78	19	17	95	5	7
Robert Walls	Carlton	19	9	5	24	11	1
Craig Bradley	Carlton	61	11	37	98	6	2
Mark MacLure	Carlton	34	16	14	48	5	4
Stephen Kernahan	Carlton	44	26	8	52	17	5
Ken Sheldon	Carlton	36	5	12	48	5	2
Syd Jackson	Carlton	13	3	1	14	5	1
Rodney Ashman	Carlton	25	4	10	35	5	2
Greg Williams	Carlton	30	6	29	59	6	4
Alan Didak	Collingwood	46	17	24	70	6	2
Peter Moore	Collingwood	42	22	13	55	11	7
Ricky Barham	Collingwood	42	15	16	58	5	5
Travis Cloke	Collingwood	26	16	9	35	5	4
Ross Dunne	Collingwood	17	6	6	23	5	2
Craig Davis	Collingwood	27	8	8	35	6	3

a. Use the data for goals to compare the two clubs using parallel box plots.
b. Comment on what the parallel box plots indicate about the data for goals.
c. Compare the data for kicks and handballs using parallel box plots.
d. Comment on what the parallel box plots indicate about the data for kicks and handballs.

19. The following table shows some key nutritional information about a sample of fruits and vegetables.

Food	Calcium (mg)	Serve weight (g)	Water (%)	Energy (kcal)	Protein (g)	Carbohydrate (g)
Avocado	19	173	73	305	4.0	12.0
Blackberries	46	144	86	74	1.0	18.4
Broccoli	205	180	90	53	5.3	10
Cantaloupe	29	267	90	94	2.4	22.3
Carrots	19	72	88	31	0.7	7.3
Cauliflower	17	62	92	15	1.2	2.9
Celery	14	40	95	6	0.3	1.4
Corn	2	77	70	83	2.6	19.4
Cucumber	4	28	96	4	0.2	0.8
Eggplant	10	160	92	45	1.3	10.6
Lettuce	52	163	96	21	2.1	3.8
Mango	21	207	82	135	1.1	35.2
Mushrooms	2	35	92	9	0.7	1.6
Nectarines	6	136	86	67	1.3	16.0
Peaches	4	87	88	37	0.6	9.6
Pears	19	166	84	98	0.7	25.1
Pineapple	11	155	86	76	0.6	19.2
Plums	10	95	84	55	0.5	14.4
Spinach	55	56	92	12	1.6	2.0
Strawberries	28	255	73	245	1.4	66.1

a. Use a form of technology to convert the water data into its equivalent weight in grams.
b. Compare the data for serve weight with your data for the weight of the water content using parallel box plots.
c. Comment on the parallel box plots from part b.
d. Use a form of technology to compare the data for protein and carbohydrate using parallel box plots.
e. Comment on the parallel box plots from part d.

20. The average maximum temperature (in °C) in Victoria for two 20-year time periods is shown in the following tables.

Time period (1993–2012):

Year	1993	1994	1995	1996	1997	1998	1999	2000	2001	2002
Temp.	22.3	22.6	21.8	22.1	22.4	22.7	22.1	22.7	23.1	23.1

Year	2003	2004	2005	2006	2007	2008	2009	2010	2011	2012
Temp.	22.7	23.4	23.4	23.1	22.7	22.1	22.9	22.6	22.6.	22.7

Time period (1893–1912):

Year	1893	1894	1895	1896	1897	1898	1899	1900	1901	1902
Temp.	20.5	20.9	21.0	20.9	21.5	21.3	21.0	20.9	21.0	21.0

Year	1903	1904	1905	1906	1907	1908	1909	1910	1911	1912
Temp.	20.7	20.9	20.9	21.5	21.4	21.3	21.1	21.4	21.3	21.4

Use a form of technology to display the data:

a. for the period 1993–2012 as a histogram using intervals of 0.4° C commencing with the data value 20° C

b. for the period 1893–1912 as a histogram using intervals of 0.4° C commencing with the data value 20° C.

c. Describe each display and comment on the differences between the two data sets.

Answers

8 Comparing data sets

Exercise 8.2 Constructing box plots

1. **a.** 12   **b.** 5   **c.** 26
2. **a.** 122   **b.** 6   **c.** 27
3. **a.** 49   **b.** 5.8   **c.** 18.6
4. **a.** 140 points   **b.** 56 points   **c.** 90 points
   d. 84 points   **e.** 26 points
5. **a.** 58   **b.** 31   **c.** 43   **d.** 27
   e. 7
6. C
7. C
8. D
9. **a.** 22, 28, 35, 43, 48
   b.
   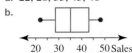
10. **a.** 10, 13.5, 22, 33.5, 45
    b.
    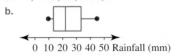
11. **a.** 18, 20, 26, 43.5, 74
    b.
    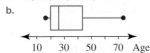
    c. The data is positively skewed with no outliers. The average age of offenders, as measured by the median, is 26. The spread of ages, as measured by the range, is 56.
12. **a.** 324 000, 335 000, 348 000, 357 000, 375 000
    b.
    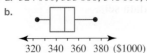

13. **a.** Key: 12|1 = 121

Stem	Leaf
12	1 5 6 9
13	1 2 4
14	3 4 8 8
15	0 2 2 2 5 7
16	3 5
17	2 9
18	1 1 1 2 3 7 8

    b.
    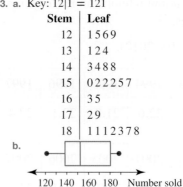

14. **a.** Key: 1*|7 = 17 years

Stem	Leaf
1	7 7 8 8 8 9 9
2	0 0 0 1 2 2 2 2 3 3 3 3 4 4 4
2*	5 5 8 9
3	1 2 3
4	8

    b.
    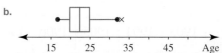
    c. Data is positively skewed with an outlier at 48. The average age to have a first baby, as measured by the median, is 22.5 years of age. Half of all mothers have their first baby between the ages of 20 and 25.

Exercise 8.3 Comparing parallel box plots

1. 17, 21, 33, 47, 58
2. A
3. * See the figure at the bottom of the page.
4. * See the figure at the bottom of the page.
5. * See the figure at the bottom of the page.

---

*3.
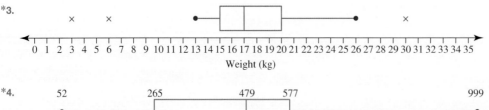

*4.

*5.
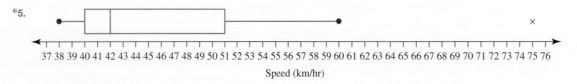

6. a. * See the figure at the bottom of the page.

b. The distribution of Greek food was positively skewed with no outliers. The distributions for Italian and Spanish food were approximately symmetric with no outliers.

    The parallel box plots show that the median number of patrons that preferred the Greek cuisine in month 1 (56) was higher than the median of those that preferred Italian in month 2 (32.5) and Spanish in month 3 (40).

    The *IQR* was smallest in month 1 (14.5) and largest in month 2 (45). The range was also smallest in month 1 (27) compared with the range in month 2 (59) and month 3 (28).

    Overall, patrons preferred Greek food over Italian and Spanish food, as month 1 Greek had a larger median and smaller measures of spread compared to the other two months. The Greek menu will therefore be the most successful.

7. B

8. B

9. A

10. a. The data set is likely to be asymmetric.

b. The data are more clustered between the median and upper quartile (less spread) than between the median and the lower quartile.

c. If the median is halfway between the upper and lower quartiles, the data set has a symmetric distribution; if it is not, then the data set is asymmetric.

11. a. Peanut butter: 2, 16, 19, 21, 26; Nutella: 4, 7, 11, 13, 25

b. i. 16

   ii. 7 and 13

c. The distribution of peanut butter sales was negatively skewed with an outlier at 2 jars. The distribution of Nutella sales is also negatively skewed with an outlier at 25 jars.

    The parallel box plots show that the median weekly jar sales were much higher for peanut butter (19 jars) than for Nutella (11 jars). The IQR was approximately the same for both peanut butter and Nutella (5 and 6 jars, respectively), whereas the range was larger for peanut butter (24 jars) than for Nutella (21 jars). Overall, the weekly jar sales for peanut butter were greater than for Nutella.

12. a. * See the figure at the bottom of the page.

b. The distribution for compound 1 was negatively skewed with no outliers. The distribution for compound 2 was approximately symmetric with no outliers. The parallel box plots show that the median minutes of effectiveness of the compounds in water was much higher for compound 1 (152.5 minutes) than for compound 2 (125.5 minutes). The IQR and ranges were also much higher for compound 1 (38.5 minutes and 55 minutes, respectively), showing more variation in minutes of effectiveness than compound 2 (7.5 minutes and 13 minutes, respectively). Overall, compound 1 was more effective in water before needing to be reapplied than compound 2.

---

*6. a.

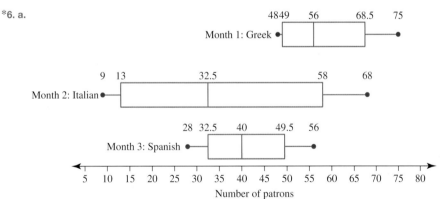

*12. a. Compound 1

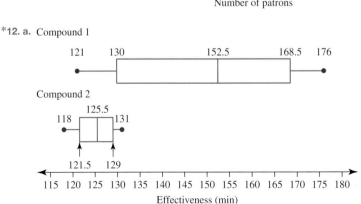

13. a. * See the figure at the bottom of the page.

b. The results for Year 11 students' achievements in History and English are fairly similar, with the range being the only statistic showing an obvious difference. The range for History was 56 but for English it was 43. This means that the average achievement was similar in both subjects; however, student achievement was more consistent in English than History.

14. a. * See the figure at the bottom of the page.

b. The distribution of Year 11 heights was negatively skewed with no outliers. The distribution of Year 12 heights was approximately symmetric with no outliers. The parallel box plots show that the median student height was greater for Year 12 students (175.5 cm) than for Year 11 students (167.5 cm). The IQR and range for both Year 11 (12.5 cm and 26 cm, respectively) and Year 12 (10.5 cm and 30 cm, respectively) students was approximately the same, showing that the spread of heights in a year-level are roughly equal. Overall, height increased with year-level.

Exercise 8.4 Comparing back-to-back stem plots

1. Key: $2|3 = 23$

German		French
2 1 1 0	2	3 4
7 6 5 5	2*	5 5 8
3 2 1 0 0	3	0 1 4 4
9 8 7 7	3*	5 6 8 8 9
2 1	4	2 3 4 4
5	4*	6 8

2. Key: $2|7 = 2.7\,\text{kg}$

Boys		Girls
	2*	6 7
4 4	3	0 1 1 2 3
8 7 6	3*	6 7
3 2	4	0
9 8	4*	
0	5	

3. a. Key: $2|3 = 23$ trucks

A		B
2 1	1	0
7 7 6 6 5	1*	5 6
4 3 2 1 0	2	0 1 3
7 5	2*	5 6 8 9
	3	0 1 2
	3*	5

b. Statistical analysis

Supermarket A	Supermarket B
Mean = 19 $s_x = 4.9$	Mean = 24.4 $s_x = 7.2$
Min = 11 $Q_1 = 16$	Min = 10 $Q_1 = 20$
Median = 18.5 $Q_3 = 23$	Median = 25.5 $Q_3 = 30$
Max = 27 IQR = 7	Max = 35 IQR = 10

For supermarket A the mean is 19, the median is 18.5, the standard deviation is 4.9 and the interquartile range is 7. The distribution is symmetric.

For supermarket B the mean is 24.4, the median is 25.5, the standard deviation is 7.2 and the interquartile range is 10. The distribution is symmetric. The centre and spread of the distribution of supermarket B is higher than that of supermarket A. There is greater variation in the number of trucks arriving at supermarket B.

4. a. Key: $1|7 = 17\,\text{marks}$

Females		Males
	1	0
3 2	1	2 3
5 5 4 4	1	4 4 5
7 6	1	7
	1	9

*13. a.

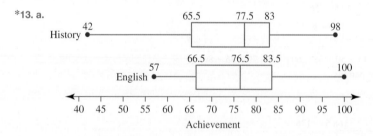

*14. a.

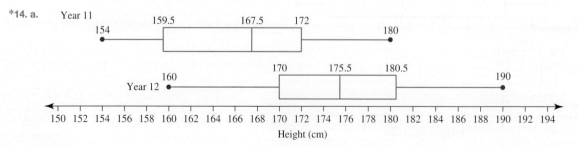

b. **Statistical analysis**

Females	Males
Mean = 14.5 $s_x = 1.6$	Mean = 14.25 $s_x = 2.8$
Min = 12 $Q_1 = 13.5$	Min = 10 $Q_1 = 12.5$
Median = 14.5 $Q_3 = 15.5$	Median = 14 $Q_3 = 16$
Max = 17 IQR = 2	Max = 19 IQR = 3.5

For the marks of the females, the mean is 14.5, the median is 14.5, the standard deviation is 1.6 and the interquartile range is 2. The distribution is symmetric.

For the marks of the males, the mean is 14.25, the median is 14, the standard deviation is 2.8 and the interquartile range is 3.5. The distribution is symmetric.

The centre of each distribution is about the same. The spread of marks for the boys is greater, however. This means that there is a wider variation in the abilities of the boys compared to the abilities of the girls.

5. a. Key: $2|6 = 26$ marks

First year			Second year
	2	2	
	2*	6 7 8	
1 0	3	0 1 1 3 4	
9 7 5	3*	6	
3 2 1 1	4		
6	4*		

b. **Statistical analysis**

First year	Second year
Mean = 38.5 $s_x = 5.2$	Mean = 29.8 $s_x = 4.2$
Min = 30 $Q_1 = 35$	Min = 22 $Q_1 = 27$
Median = 40 $Q_3 = 42$	Median = 30.5 $Q_3 = 33$
Max = 46 IQR = 7	Max = 36 IQR = 6

The distribution of marks for the first year and for the second year are each symmetric.

For the first year marks, the mean is 38.5, the median is 40, the standard deviation is 5.2 and the interquartile range is 7. The distribution is symmetric.

For the second year marks, the mean is 29.8, the median is 30.5, the standard deviation is 4.2 and the interquartile range is 6.

The spread of each of the distributions is much the same but the centre of each distribution is quite different with the centre of the second year distribution quite a lot lower. The work may have become a lot harder!

6. a. Key: $2|5 = 25$ years

Female			Male
4 3	2	2	
8 7 6 5	2*	5	
1 0	3	0 1	
	3*	6 7	
	4	2	
	4*	6	

b. **Statistical analysis**

Female	Male
Mean = 26.75 $s_x = 2.8$	Mean = 33.6 $s_x = 8.2$
Min = 23 $Q_1 = 24.5$	Min = 22 $Q_1 = 27.5$
Median = 26.5 $Q_3 = 29$	Median = 33.5 $Q_3 = 39.5$
Max = 31 IQR = 4.5	Max = 46 IQR = 12

For the distribution of the females, the mean is 26.75, the median is 26.5, the standard deviation is 2.8 and the interquartile range is 4.5.

For the distribution of the males, the mean is 33.6, the median is 33.5, the standard deviation is 8.2 and the interquartile range is 12.

The centre of the distributions is very different: it is much higher for the males. The spread of the ages of the females who attend the fitness class is very small but very large for males. Older males are more likely to the attend fitness classes than females.

7. a. Key: $3|2 = 32$

Kindergarten			Prep.
3	0	5	
4 3	1	2 7	
8 5	2	5 7	
6 2	3	2 5	
7 1	4	4 6	
0	5	2	

b. **Statistical analysis**

Kindergarten	Prep. School
Mean = 28.9 $s_x = 15.4$	Mean = 29.5 $s_x = 15.3$
Min = 3 $Q_1 = 14$	Min = 5 $Q_1 = 17$
Median = 30 $Q_3 = 41$	Median = 29.5 $Q_3 = 44$
Max = 50 IQR = 27	Max = 52 IQR = 27

For the distribution of scores of the kindergarten children, the mean is 28.9, the median is 30, the standard deviation is 15.4 and the interquartile range is 27.

For the distribution of scores for the prep. children, the mean is 29.5, the median is 29.5, the standard deviation is 15.3 and the interquartile range is 27.

The distributions are very similar. There is not a lot of difference between the way the kindergarten children and the prep. children scored.

8. B

9. C

10. a. Class 1:25; Class 2:27
    b. Class 1:2; Class 2:3
    c. Class 1:24; Class 2:20
    d. Eleven students in class 2 scored better than class 1 but 6 students scored worse than class 1. Class 2's results are more widely spread than class 1's results.

11. Key: $1|6 = 16$ years

Male		Female
987766	1	5566677789
9853332110	2	1122233578
8864221	3	00145
533	4	368
2	5	047
0	6	5

12. Brand B seems to have a better battery life, as Brand A has more batteries that have a battery life of less than 10 hours. Brand B has a battery life of up to 75 hours, which is higher than Brand A's battery life maximum of 61 hours.

13. a. Key: $9|69 = 9.69$

Female		Male
	9	69
	9	84 85 87 90 92 96 99
	10	00 06 14 20 25 30 30 30 40
97 94 93 82 78 75 54	10	50 80
40 08 07 00 00	11	
90 90 60 50 50 50	11	
20	12	

b. There is not a large difference within each gender, but there is a large difference in time between the two genders.

14. a. Key: $23|1 = \$231$

Suburb A		Suburb B
7 5 5 0 0 0	20	
5 5 1 0 0	21	
5 5	22	
5 5 1	23	1 2 5
4 2 0	24	0 5 5
6	25	1 5 5 8
	26	0 5
	27	0 0 5 5
	28	0 0 5 5

b. Suburb A has lower rent than Suburb B. Suburb B is a more expensive suburb to rent an apartment.

Exercise 8.5 Comparing histograms

1. a.

Score	Frequency
3	1
4	2
5	2
6	1
7	3
8	2
9	2
10	2
11	0
12	1

b.

Class interval	Frequency
1–1.9	1
2–2.9	2
3–3.9	2
4–4.9	6
5–5.9	5
6–6.9	1

c.

Class interval	Frequency
10–14	3
15–19	9
20–24	10
25–29	10
30–34	10
35–39	1

d.

Score	Frequency
0.3	1
0.4	2
0.5	1
0.6	1
0.7	1
0.8	2
0.9	2
1.0	2
1.1	1
1.2	1
1.3	1

2. a.

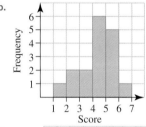

b.

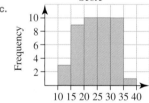

c.

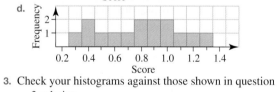

d.

3. Check your histograms against those shown in question 2 solution.

4. a.

Score	Tally	Frequency
30–39	|	1
40–49	ЖΙ	5
50–59	ЖΙ	5
60–69	ЖΙ|||	8
70–79	ЖΙ|	6
80–89	|||	3
90–99	||	2

b.

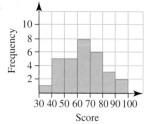

5. a.

Class	Class centre	Tally	Frequency
0.6–<0.8	0.7	||	2
0.8–<1.0	0.9	|||	3
1.0–<1.2	1.1	|||	3
1.2–<1.4	1.3	ЖΙ|	6
1.4–<1.6	1.5	ЖΙ||||	9
1.6–<1.8	1.7	ЖΙ|	6
1.8–<2.0	1.9	||||	4
2.0–<2.2	2.1	|||	3

b.

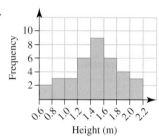

6. a.

Class	Class centre	Frequency
10.5–10.9	10.7	2
11.0–11.4	11.2	5
11.5–11.9	11.7	8
12.0–12.4	12.2	8
12.5–12.9	12.7	5
13.0–13.4	13.2	2

b.

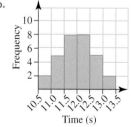

7.

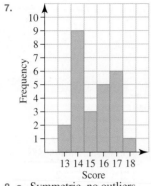

8. a. Symmetric, no outliers
   b. Symmetric, one outlier
   c. Symmetric, no outliers
   d. Negatively skewed, no outliers
   e. Negatively skewed one outlier
   f. Positively skewed, no outliers
9. D
10. B
11.

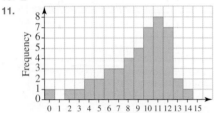

The data on the *Yellow Pages* is negatively skewed with no apparent outliers.
12. a. The data is symmetric, with one outlier at $2.
   b. The majority of students receive about $8.00, or within the range $7 to $9 per week. There is one student, however, who receives $2.
13. a. Worker F
   b. 4
14. a. 2016
   b. 2018
   c. Answers will vary but could include that more people are on the roads during the Christmas break and school holidays.

15. a. 500
   b.

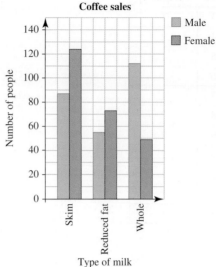

   c. 34%
   d. 26%
   e. 49%

8.6: Review: exam practice

1. C
2. B
3. D
4. D
5. a. Bimodal, symmetric
   b. Unimodal, positively skewed
   c. Bimodal, negatively skewed
   d. Unimodal, negatively skewed
6. C
7. a. 95, 215, 270, 290, 340
   b. 75
   c. Yes, 95
8. * See the figure at the bottom of the page.
9. a. * See the figure at the bottom of the page.
   b. Player 1. Higher median number of goals scored (8.5) than Player 2 (7). Also, over 50% of the number of goals scored by Player 1 across the ten matches lie above the maximum number of goals scored by Player 2 across the 10 matches.

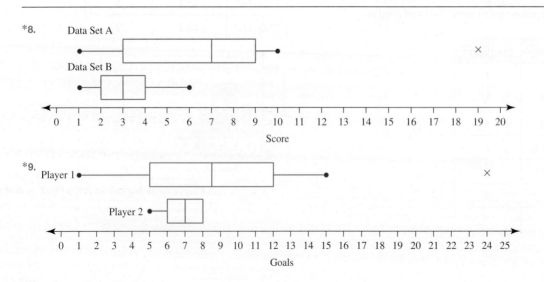

10. a. 43

   b. 43

   c. 14

11. Min $= 1, Q_1 = 8, Q_2 = 14, Q_3 = 16$, Max $= 18$

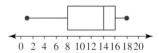

12. a. Key: $1|3 = 13$ hours

Full-time		Volunteer
1	0	
3 3 2 2	0	
5 4 4	0	
6	0	
	0	8
	1	0 1 1
	1	2 3 3
	1	4 5
	1	
	1	

   b. Both distributions are symmetric with the same spread. The centre of the volunteers distribution is much higher than that of the full-time firefighters distribution. Clearly, the volunteers needed more counselling.

13.

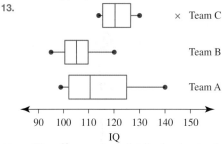

14. a. The office workers distribution is negatively skewed with one outlier (76). The sports instructors distribution is positively skewed positively skewed with one outlier (108).

   b. Office workers:

$$\text{Range} = 58 \text{ beats/min}$$
$$\text{Median} = 121.5 \text{ beats/min}$$
$$\text{IQR} = 19.5 \text{ beats/min}$$
$$\text{Mode} = 130 \text{ beats/min}$$

Sports instructors:

$$\text{Range} = 46 \text{ beats/min}$$
$$\text{Median} = 73 \text{ beats/min}$$
$$\text{IQR} = 14 \text{ beats/min}$$
$$\text{Mode} = 68, 72 \text{ beats/min}$$

   c. Office workers

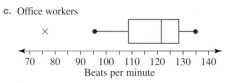

Sports instructors

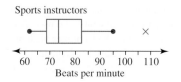

   d. Office workers:

$$\text{Mean} = 116.8125 \text{ beats/min}$$
$$\text{Standard Deviation} = 15.3 \text{ beats/min}$$

Sports instructors:

$$\text{Mean} = 76.87 \text{ beats/min}$$
$$\text{Standard Deviation} = 12.43 \text{ beats/min}$$

   e. Office workers: Pulse rates are generally very high, clustered around 120–130 beats/min. Also, there is one person whose rate was much lower than the rest. This outlier (76) produces a large range and makes the mean slightly lower than the median. As a result the median is a more appropriate measure of the centre of the data rather than the mean.

Sports instructors: Pulse rates are generally low, clustered around 60–70 beats/min, although there are a few people with rates much higher, which makes the mean slightly higher than the median and also produces quite a large range. As a result of the skewed distribution the median is the more appropriate measure of the centre of the data rather than the mean, although there is little difference between these values.

15. a. Key: $1|3 = 13$

Text B		Text A
	1	3
	2	1 2 5 8
9 7	3	5 5
8 2	4	4 8
9 9 8 8 5 5 3 2	5	2 5 6 9
8 8 8 6 5 4 4 3 2	6	4 9
9 9 5 2 2	7	2 2 6
2 1	8	1 3 4 4
	9	4 5 8

   b. Text A $n = 25$

Min	Q_1	Median	Q_3	Max
13	35	59	82	98

Text B $n = 28$

Min	Q_1	Median	Q_3	Max
37	55	63.5	70	82

   c.

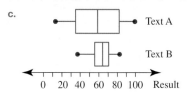

d. Performance on test. Students who used Text A had varied results while the students who used Text B were more consistent and they had a higher median score.
Text A: range = 85, IQR = 47, median = 59
Text B: range = 45, IQR = 15, median = 63.5

e–f. Students' individual differences — that is, studying, doing the work, ability can affect the results. However, you would probably go with the text that produced more consistent results. Therefore, Text B.

16. a. Key: 22 | 5 = 225

Suburb A		Suburb B
	22	5 8
	23	1 2 5
2	24	0 1 5
	25	5
	26	
5 5	27	0 0 5 5
5 4 1 1	28	0 0 5
6 5 5 0	29	5
7 5 5 0 0 0	30	0 5
5 0 0	31	0

b. Answers will vary. Suburb A has higher rent than in suburb B.

c. Answers will vary. Suburb C is higher than suburb A, which is higher than suburb B. The most expensive suburb to rent is suburb C. The least expensive suburb to rent a one-bedroom unit is suburb B.

17. a. D
b. Between $75 000 and $100 000
c. Generally, salary increases with age.

d. 30 s
e. $190 000 is lower than the upper fence, so it is no longer an outlier.

18. a.

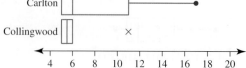

b. Goals scored in Grand Finals for this sample of players is greater but more variable among the Carlton players, as indicated by the larger range and IQR. Collingwood players are concentrated at 5 or 6 with the exception of the one upper outlier of 11.

c. * See the figure at the bottom of the page.

d. Kicks for this sample of players are greater but more variable than the handballs, as indicated by the larger range and IQR. Both are positively skewed with one upper outlier.

19. a. * See the table at the bottom of the page.

b.

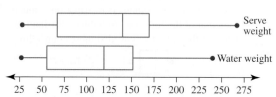

*18. c.

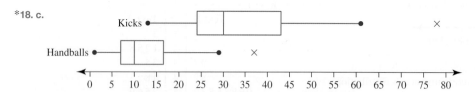

*19. a.

Food	Serve weight (g)	Water (%)	Water weight (g)
Avocado	173	73	126
Blackberries	144	86	124
Broccoli	180	90	162
Cantaloupe	267	90	240
Carrots	72	88	63
Cauliflower	62	92	57
Celery	40	95	38
Corn	77	70	54
Cucumber	28	96	27
Eggplant	160	92	147
Lettuce	163	96	156
Mango	207	82	170
Mushrooms	35	92	32
Nectarines	136	86	117
Peaches	87	88	77
Pears	166	84	139
Pineapple	155	86	133
Plums	95	84	80
Spinach	56	92	52
Strawberries	255	73	186

c. The box plots appear to indicate that there are only slight differences between the serve weights and water weights of the samples. The distributions are very similar in shape, with the water weights being slightly less overall.

d. * See the figure at the bottom of the page.

e. Carbohydrate for this sample of foods is much greater but more variable than protein, as indicated by the larger range and IQR. The protein amounts are all less than the Q_1 for the carbohydrate amounts, with the exception of two upper outliers for protein that lie between the Q_1 and median for carbohydrate. Carbohydrate is positively skewed with one upper outlier.

20. a.

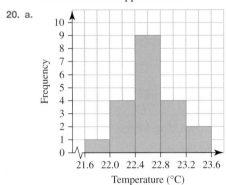

b.

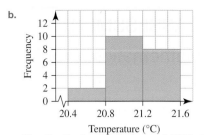

c. The distribution for the interval 1993–2012 is approximately symmetrical, with a slight negative skew. The distribution for the interval 1893–1912 is more symmetrical with a smaller overall range. Both distributions have one mode.

*19. d.

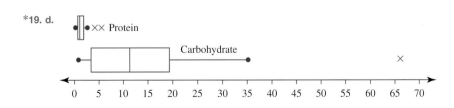

PRACTICE ASSESSMENT 1

Essential Mathematics: Problem solving and modelling task

Unit
Unit 3: Measurement, scales and data

Topic
Fundamental topic: Calculations
Topic 1: Measurement
Topic 2: Scales, plans and models

Conditions

Duration	Mode	Individual/group
5 weeks	Written report	Individual
Resources permitted		**Length**
The use of technology is required, for example: • non-CAS graphics calculator • spreadsheet software • other mathematical software.		• Up to 8 pages (including tables, figures and diagrams) • Maximum of 1000 words • Appendixes can include raw data, repeated calculations, evidence of authentication and student notes (appendixes are not to be marked)

Criterion	Grade
Formulate *Assessment objectives 1, 2, 5	
Solve *Assessment objectives 1, 6	
Evaluate and verify *Assessment objectives 4, 5	
Communicate *Assessment objective 3	
Milestones	
Week 1	
Week 2	
Week 3	
Week 4	
Week 5 (assessment submission)	

* © State of Queensland (Queensland Curriculum & Assessment Authority), *Essential Mathematics Applied Senior Syllabus 2019 v1.1*, Brisbane. For the most up-to-date assessment information, please see www.qcaa.qld.edu.au/senior.

Context

Sustainable living has become more relevant in recent times as more people become aware of the impact we are having on the environment. You have been commissioned to design a house for the Jones family. The house is to accommodate Mr and Mrs Jones and their two children.

Mr and Mrs Jones have given you the following requirements that they need you to meet.
- At least three bedrooms
- A floor plan area of exactly $250\,m^2$
- A minimum of three environmentally sustainable features

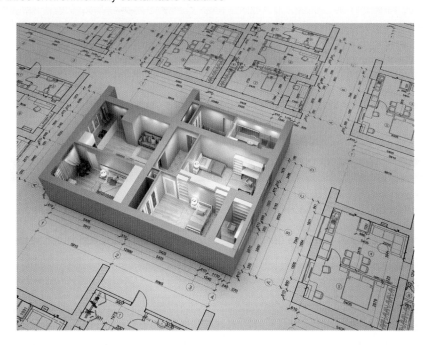

Task

The Jones family have allocated $25\,000 to equip their house with environmentally sustainable features. They are interested to see what options are available to help make their house more environmentally friendly.

Some of the options they are already aware of are:
- solar hot water system
- water tanks to collect rain water and connect to laundry or toilets
- solar energy
- insulation
- double-glazed windows
- recycled materials/furnishing
- aspect (e.g. north–south orientation).

They are also open to other suggestions from your team. They wish to spend all of the budgeted $25\,000 if possible.

The Jones family are very excited about building their new home and are ready to start construction as soon as possible. You will produce a presentation for the Jones family showing the floor plan you have designed and the environmentally friendly features included in the house. Your presentation should explain why you have made the decisions for each of the options and why you think they are the best choices. A budget estimate must be submitted along with the area of each of the rooms in the house.

To complete this task, you must:
- Use the problem-solving and mathematical modelling approach to develop your response.
- Respond with a range of understanding and skills, such as using mathematical language, appropriate calculations, tables of data, graphs and diagrams.
- Provide a response that highlights the real-life application of mathematics.
- Respond using a written report format that can be read and interpreted independently of the instrument task sheet.
- Develop a unique response.
- Use both analytic procedures and technology.

Approach to problem solving and modelling

Formulate

In this task you will investigate different house configurations to satisfy the design brief set out by the Jones family. Your plan should include:

- at least three bedrooms
- a floor plan of exactly $250\,m^2$
- a minimum of three environmentally sustainable features.

Solve

Construct the plan, checking to make sure it is exactly $250\,m^2$.

Review and make any refinements to the plan. Check that it can be analysed and that conclusions can be drawn on any patterns or observations. Link the analysis back to the context. You will make further refinements and comparisons as necessary.

You must use technology efficiently and show detailed calculations demonstrating the procedures used.

Is it solved?

Evaluate and verify

Evaluate the reasonableness of your original solution.

Based on your representation of the plan and sustainable features, consider whether it could be planned better to meet the main brief.

Justify and explain all procedures you have used and decisions you have made. Considering the original task, how valid is your solution?

Is the solution verified?

Communicate

Once you have completed all necessary work, you should consider how you have communicated all aspects of your plan. Communicate using appropriate language that refers to the calculations, tables and graphs included in previous sections. Your response should be coherently and concisely organised.

Ensure you have:

- used mathematical, statistical and everyday language
- considered the strengths and limitations of your solution
- drawn conclusions by discussing your results
- included recommendations.

PRACTICE ASSESSMENT 2

Essential Mathematics: Unit 3 examination

Unit
Unit 3: Measurement, scales and data

Topic
Fundamental topic: Calculations
Topic 1: Measurement
Topic 2: Scales, plans and models
Topic 3: Summarising and comparing data

Conditions

Technique	Response type	Duration	Reading
Paper 1: Simple (27 marks) Paper 2: Complex (13 marks)	Short response	60 minutes	5 minutes

Resources	Instructions
• QCAA formula sheet • Notes not permitted • Scientific calculator permitted	• Show all working. • Write responses using a black or blue pen. • Unless otherwise instructed, give answers to **two decimal places**.

Criterion	Marks allocated	Result
Foundational knowledge and problem solving *Assessment objectives 1, 2, 3, 4, 5 and 6	40	

* © State of Queensland (Queensland Curriculum & Assessment Authority), *Essential Mathematics Applied Senior Syllabus 2019 v1.1*, Brisbane. For the most up-to-date assessment information, please see www.qcaa.qld.edu.au/senior.

A detailed breakdown of the examination marks summary can be found in the PDF version of this assessment instrument in your eBookPLUS.

Part A: Simple — total marks: 27

Question 1 (4 marks)

Evaluate the following:

a. $\dfrac{72}{12} + 48 \div 2^3$

b. $\dfrac{5 \times (50 - 38)}{2^2}$

Question 2 (3 marks)

State the values of the unknown angles

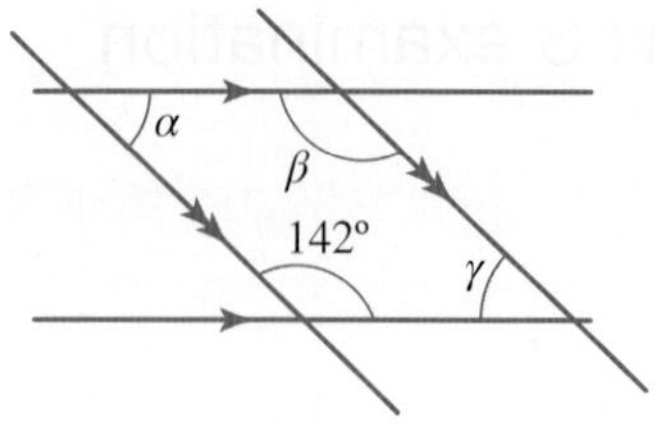

(ruled answer lines)

Question 3 (5 marks)

Calculate the area of the following shape to three decimal places.

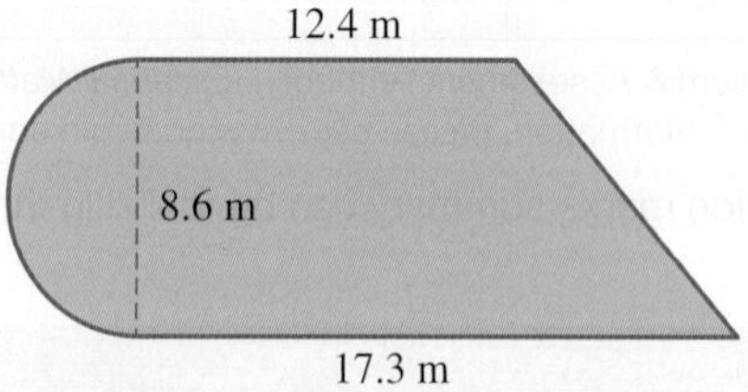

(ruled answer lines)

Question 4 (6 marks)

A fish tank has dimensions $80\,\text{cm} \times 30\,\text{cm} \times 50\,\text{cm}$. How much water would be needed to:

a. one-quarter fill the tank in cm^3

b. one-quarter fill the tank in litres

c. 80% fill the tank in cm^3

d. 80% fill the tank in litres?

Question 5 (4 marks)

A house plan is drawn to a scale of $1 : 400$.

a. What is the actual length of the lounge room if it is represented by 0.8 cm on the plan?

_____

_____

_____

_____

_____

_____

b. The width of the room is 4.8 m. What is the width of the room on the house plan?

_____

_____

_____

_____

_____

_____

Question 6 (3 marks)

A fire is burning in a building and people need to be rescued. The fire brigade ladder must reach a height of 50 m and must be angled at $70°$ to the horizontal. How long must the ladder be to complete the rescue, to one decimal place?

_____

_____

_____

_____

_____

_____

_____

_____

_____

_____

_____

Question 7 (2 marks)

Round the following to three decimal places.

a. 25.793 499

_____

_____

b. 19.011 512

_____

_____

Part B: Complex — total marks: 13

Question 8 (4 marks)

Consider the following stem plot.

Key 11 | 2 = 112

Stem	Leaf
7	0 2
8	3 7 9
9	0 4
10	3 5 8
11	0 0 6 9
12	3 3 6 7
13	0 1

Calculate:

a. the mean

b. the median

c. the IQR.

Question 9 (5 marks)

The number of goals scored by two members of a basketball team over 10 games is shown in the following table.

Player A	8	14	9	7	16	24	12	18	8	14
Player B	14	17	7	14	10	8	6	14	22	15

a. Draw a parallel boxplot of this data.

b. Which player would you rather have on your team and why?

Question 10 (4 marks)

To travel between the two towns of Mathville and Statstown, you need to travel west along a road for 38 km, then north along another road for 68 km. Calculate:

a. the straight-line distance between the two towns

_____

_____

_____

b. the acute angle the straight-line distance makes with the west axis.

_____

_____

9 Cartesian plane

9.1 Overview

LEARNING SEQUENCE

9.1 Overview
9.2 Plotting points on a Cartesian plane
9.3 Generating tables of values for linear functions
9.4 Graphing linear functions
9.5 Review: exam practice

CONTENT

In this chapter, students will learn to:
- demonstrate familiarity with Cartesian coordinates in two dimensions by plotting points on the Cartesian plane
- generate tables of values for linear functions, including for negative values of x
- graph linear functions for all values of x with pencil and paper and with graphing software.

Fully worked solutions for this chapter are available in the Resources section of your eBookPLUS at www.jacplus.com.au.

9.2 Plotting points on a Cartesian plane

The **Cartesian plane** is created by drawing a horizontal number line (the x-axis) and a vertical number line (the y-axis).

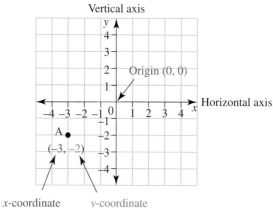

The point where the two axes meet is called the **origin**. The origin is at the point $(0, 0)$.

The position of any point on the Cartesian plane can be located using **coordinates**.

When writing coordinates for a point, the value of the x-coordinate is written first and then the value of the y-coordinate is written. The x- and y-coordinates are separated by a comma and enclosed in round brackets.

The x-coordinate describes how far across from the origin a point is located, and the y-coordinate describes how far up (or down) from the origin a point is located.

Point A on the diagram, positioned at $x = -3$ and $y = -2$, can be described as the point $(-3, -2)$.

WORKED EXAMPLE 1

State the coordinates for the points labelled A, B, C and D on the Cartesian plane below.

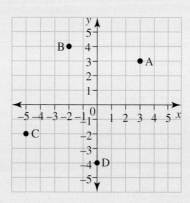

THINK

1. The point marked A is above the 3 on the x-axis and level with the 3 on the y-axis.

WRITE

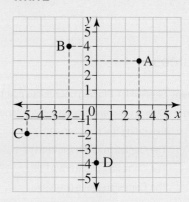

Point A $= (3, 3)$

2. The point marked B is above the -2 on the x-axis and level with the 4 on the y-axis.

Point B $= (-2, 4)$

3. The point marked C is below the -5 on the x-axis and level with the -2 on the y-axis.

Point C $= (-5, -2)$

4. The point marked D is below the 0 on the x-axis and level with the -4 on the y-axis.

Point D $= (0, -4)$

WORKED EXAMPLE 2

Plot the coordinates E $= (3, 5), F = (-3, 5), G = (5, -3)$ and H $= (-5, -3)$ on a single Cartesian plane.

THINK

1. Draw a Cartesian plane that extends from -5 to 5 on both the x- and the y-axis.

2. Point E is at 3 on the x-axis and 5 on the y-axis. From the origin, move across 3 units to the right and up 5 units. Draw a point and label it with the letter E.

3. Point F is at -3 on the x-axis and 5 on the y-axis. From the origin, move across 3 units to the left and up 5 units. Draw a point and label it with the letter F.

4. Point G is at 5 on the x-axis and -3 on the y-axis. From the origin, move across 5 units to the right and down 3 units. Draw a point and label it with the letter G.

5. Point H is at -5 on the x-axis and -3 on the y-axis. From the origin, move across 5 units to the left and down 3 units. Draw a point and label it with the letter H.

WRITE

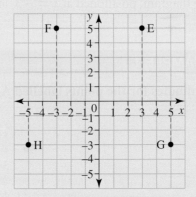

 Resources

Exercise 9.2 Plotting points on a Cartesian plane

1. **WE1** State the coordinates for the points labelled A to H on the Cartesian plane.

2. **WE2** Plot the following points on a single Cartesian plane.
   a. A (1, 3)
   b. B (2, 3)
   c. C (2, 5)
   d. D (5, 0)

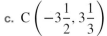

3. Plot the following points on an appropriately scaled Cartesian plane.
   a. A (0, −5)
   b. B (−2, −5)
   c. C (−3, 0)
   d. D (−4, 1)

4. Plot the following points on an appropriately scaled Cartesian plane.
   a. A $\left(\frac{1}{2}, 3\right)$ b. B $\left(2\frac{1}{2}, -4\right)$ c. C $\left(-3\frac{1}{2}, 3\frac{1}{3}\right)$ d. D $\left(0, 2\frac{1}{2}\right)$

5. Plot the following points on an appropriately scaled Cartesian plane.
   a. A (0.8, 2.6)   b. B (2.5, −2.5)   c. C (−1.2, 3)   d. D (−1.6, −2.8)

6. **MC** The point (5, −3) gives a position on the Cartesian plane of
   A. −5 units on the x-axis, −3 units on the y-axis.
   B. 5 units on the x-axis, 3 units on the y-axis.
   C. 5 units on the x-axis, −3 units on the y-axis.
   D. −3 units on the x-axis, 5 units on the y-axis.

7. **MC** What point is described by 2 on the x-axis and −5 on the y-axis?
   A. (−5, 2)   B. (−5, −2)   C. (5, −2)   D. (2, −5)

8. **MC** Which of the following points is the highest on the Cartesian plane?
   A. (0, 6)   B. (10, 2)   C. (6, −7)   D. (8, −11)

9. Draw a Cartesian plane from −10 to 10 on the x-axis and −10 to 15 on the y-axis. Plot and join the following points to uncover the picture.

> START (6, 7) (7.5, 9) (5, 9) (4.5, 12) (2, 11) (0, 13) (−1.5, 10) (−5, 11) (−5, 8) (−8, 6) (−6, 4) (−8, 2) (−6, 1) (−7, −2) (−4, −1.5) (−4, −3.5) (−1.5, −3) (−2, −4) (−4, −7) (−5, −8) STOP
>
> START (−2, −9) (−1, −7) (1, −8) (3, −8) (4, −7.5) (5, −10) STOP
>
> START (4, −7.5) (3.5, −6) (3.5, −4) (4, −3) (5, −2.5) (5, −2) (4, −1.5) (4, −1) (5, 0.5) (7, 1) (8, 2) (8, 2.5) (6.5, 3) STOP
>
> START (4, −2.5) (2, −3) (0.5, −3) (0, −2) (1, −1) (2, −0.5) (3, 0) (7, 1) STOP
>
> START (6, 2.5) (6.5, 3) (6.5, 4) (6, 4) (4, 4) STOP
>
> START (6.5, 4) (7, 5) (7, 6) (6, 7) (5, 7.5) (4, 7) (3, 6) (2, 6) (1, 6) (0, 5) (−1, 4) (0, 2) (1.5, 1.5) (3, 2) (4, 3) (4, 4) (4, 5) (3, 6) STOP
>
> START (1, −1) (5, 0) STOP
>
> DOTS AT (1, 3) AND (5, 5)
>
> LINES (−1, 4) TO (−2, 4.5), (0, 5) TO (−0.5, 6), (1, 6) TO (0.5, 7), (2, 6) TO (2, 7), (4, 7) TO (3.5, 8), (5, 7.5) TO (5, 8.5), (6, 7) TO (6.5, 8), (6.5, 6.5) TO (7, 7)

10. Given the points plotted, correct the mistakes and explain what has been done incorrectly.

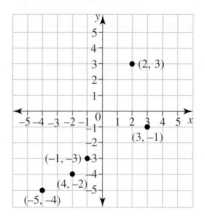

11. Your house is located 6 km north and 3 km west of the city centre.
    a. Represent this on a Cartesian plane, assuming that the city centre is located at the origin.

    Your friend's house is 10 km south and 1 km west of your house.
    b. Represent your friend's house on the Cartesian plane.

    Your school is 2 km north and 5 km east of your friend's house.
    c. Represent your school on the Cartesian plane.
    d. How far north/south and east/west is your school from your house?

12. You are given the following coordinates:
    $(0, 0)$, $(5, 4)$, $(8, 4)$, $(14, 9)$, $(22, 0)$.

    a. Draw these points on a Cartesian plane.
    b. These points are used to describe a journey, which begins at your house. The y-axis represents distance in kilometres from your house and the x-axis represents the time taken for the journey in minutes. Make up a story that describes this journey in detail.

13. a. On an A4 sheet of graph paper, rule up a Cartesian plane to cover the page.
    b. Draw a simple picture on your graph paper.
    c. List the coordinates and write instructions showing where to start and stop joining the dots so that someone could replicate your picture.

9.3 Generating tables of values for linear functions

9.3.1 Algebra review

Algebra is a type of language used in mathematics that consists of **pronumerals** (letters or groups of letters) that are used to represent unknown numbers.

Pronumerals can also be used to describe **variables** (varying values).

Some common words used in linear functions are shown at right.

- **Term:** A group of letters and/or numbers. Terms are separated by an addition or subtraction sign.
  e.g. $5x$ is a term
- **Pronumeral:** The letter part of a term.
  e.g. x is the pronumeral of the term $5x$
- **Coefficient:** The number multiplying (out the front) the pronumeral.
  e.g. 5 is the coefficient of x
- **Constant term:** The term that does not have a pronumeral.
  e.g. 6 is the constant term as there is no letter attached to it
- **Equation:** A mathematical statement containing a left- and right-hand side separated by an equals sign.
  e.g. $y = 5x + 6$ is an equation

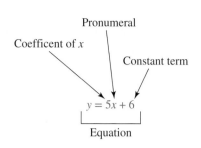

9.3.2 Patterns and rules

A table of values can be used to investigate the relationship between two variables.

WORKED EXAMPLE 3

For the set of shapes below:

a. construct a table to show the relationship between the number of triangles in each figure and the number of matchsticks used to construct it

b. devise a rule in words that describes the pattern relating the number of triangles in each figure and the number of matchsticks used to construct it

c. use your rule to determine the number of matchsticks required to construct a figure made up of 20 such triangles.

THINK

a. 1. Identify the two variables that are being investigated.

2. Three matches were needed for 1 triangle, 6 matches for 2 triangles and 9 matches for 3 triangles.
Construct a table of values displaying the relationship between the number of triangles and the number of matchsticks required.

b. 1. Look for a pattern between the number of triangles and number of matchsticks. Each triangle requires 3 matchsticks.

WRITE

a. The variables are:
- the number of triangles in the shape
- the number of matchsticks used to construct the shape.

Number of triangles	1	2	3
Number of matchsticks	3	6	9

b. The number of matchsticks equals 3 times the number of triangles.

c. 1. Apply the rule 'the number of matchsticks equals 3 times the number of triangles' to determine the number of matchsticks required for 20 triangles.

c. $20 \times 3 = 60$
To construct 20 triangles, 60 matchsticks would be required.

9.3.3 Input and output tables

A rule can be used to produce a set of output numbers from a set of input numbers. We can think of this as a machine that takes any input number and changes it according to a given rule to produce an output number. The rule can involve more than one step; for example, 'multiply each input number by 11 and then add 55'.

WORKED EXAMPLE 4

Complete the table below using the given rules to calculate the correct output numbers.

Input	4	10	38	144
Output				

a. Rule: Subtract 3 from each input number.
b. Rule: Divide each input number by 2.

THINK	WRITE
a. 1. Take the first input number (4) and apply the rule; that is, subtract 3.	**a.** $4 - 3 = 1$
2. Apply the rule to each of the other input numbers.	$10 - 3 = 7$ $38 - 3 = 35$ $144 - 3 = 141$
3. Enter these output values in the table under their corresponding input value.	

Input	4	10	38	144
Output	1	7	35	141

THINK	WRITE
b. 1. Take the first input number (4) and apply the rule; that is, divide by 2.	**b.** $4 \div 2 = 2$
2. Apply the rule to each of the other input numbers.	$10 \div 2 = 5$ $38 \div 2 = 19$ $144 \div 2 = 72$
3. Enter these output values in the table under their corresponding input value.	

Input	4	10	38	144
Output	2	5	19	72

9.3.4 Substitution

A **formula** or **rule** that describes the relationship between the input and output values is often expressed as an equation. In Worked example 4a, the rule 'subtract 3 from each input number' could be written as the following equation.

$$\text{Output} = \text{input} - 3$$

Pronumerals are usually used to represent the variables 'input' and 'output'. If x is used to represent the input values and y is used to represent the output values, then the equation

$$\text{Output} = \text{input} - 3$$

can be rewritten as follows.

$$y = x - 3$$

When a variable in a formula is replaced by a number, we say that the number is substituted into the formula.

For example, if we wanted to determine the output number (y) for the rule $y = x - 3$ when the input number (x) is 2, we would substitute 2 into the formula for x as follows.

$$y = x - 3$$
$$y = 2 - 3$$
$$y = -1$$

When x is equal to 2, y is equal to -1.

WORKED EXAMPLE 5

If $y = 5x - 4$, substitute the given value of x into the formula to find the value of y in each case.
a. $x = 2$
b. $x = 6$

THINK	WRITE
a. 1. Write the formula.	a. $y = 5x - 4$
2. Substitute 2 for x in the formula.	$y = 5 \times 2 - 4$
3. Evaluate the expression for y and write the answer.	$y = 10 - 4$
	$y = 6$
b. 1. Write the formula.	b. $y = 5x - 4$
2. Substitute 6 for x in the formula.	$y = 5 \times 6 - 4$
3. Evaluate the expression for y and write the answer.	$y = 30 - 4$
	$y = 26$

The equation or expression will not always be given to you. You may need to create the equation from the worded problem.

If y represents any number, write expressions for:

a. **3 times the number**   b. **7 less than the number**   c. **12 more than the number.**

THINK	WRITE
a. 1. Think about which operation is used.	a. 'Times' means multiplication.
2. When multiplying we don't show the multiplication sign. Write the number first, then the pronumeral.	$3 \times y = 3y$
b. 1. Think about which operation is used.	b. 'Less than' means subtraction.
2. Since less than, take 7 away from y.	$y - 7$
c. 1. Think about which operation is used.	c. 'More than' means addition.
2. Since more than, add 12 to y.	$y + 12$

The pronumeral may not be specified. In this case you will need to choose an appropriate pronumeral that relates well to the question. For example, you may use C to represent cost.

9.3.5 Creating a table of values

A table of values can be created by substituting multiple input values into an equation and determining the output value that corresponds to each input value.

Generate a table of values for the rule $y = 7 - 4x$ for x values of $-1, 0, 1, 2, 3$.

THINK	WRITE
1. Draw a table with a row for the x values and a row for the y values.	<table><tr><td>x</td><td></td><td></td><td></td><td></td><td></td></tr><tr><td>y</td><td></td><td></td><td></td><td></td><td></td></tr></table>
2. Write the given x values in the first row of the table.	<table><tr><td>x</td><td>-1</td><td>0</td><td>1</td><td>2</td><td>3</td></tr><tr><td>y</td><td></td><td></td><td></td><td></td><td></td></tr></table>
3. Substitute the first x value, -1, into the rule for x and solve for y.	$y = 7 - 4x$ $= 7 - 4 \times -1$ $= 7 + 4$ $= 11$
4. Write the value $y = 11$ in the table underneath $x = -1$.	<table><tr><td>x</td><td>-1</td><td>0</td><td>1</td><td>2</td><td>3</td></tr><tr><td>y</td><td>11</td><td></td><td></td><td></td><td></td></tr></table>

5. Substitute the remaining x values into the rule and determine their corresponding y values.

When $x = 0$:
$y = 7 - 4x$
$= 7 - 4 \times 0$
$= 7$
When $x = 1$:
$y = 7 - 4x$
$= 7 - 4 \times 1$
$= 7 - 4$
$= 3$
When $x = 2$:
$y = 7 - 4x$
$= 7 - 4 \times 2$
$= 7 - 8$
$= -1$
When $x = 3$:
$y = 7 - 4x$
$= 7 - 4 \times 3$
$= 7 - 12$
$= -5$

6. Write the y values in the table underneath their corresponding x values.

x	-1	0	1	2	3
y	11	7	3	-1	-5

---

on Resources

Exercise 9.3 Generating tables of values for linear functions

1. **WE3** For each of the sets of shapes:
   i. construct a table to show the relationship between the number of shapes in each figure and the number of matchsticks used to construct it
   ii. devise a rule in words that describes the pattern relating the number of shapes in each figure and the number of matchsticks used to construct it
   iii. use your rule to determine the number of matchsticks required to construct a figure made up of 20 such shapes.

   a.    b.

---

2. For each of the sets of shapes:
   i. construct a table to show the relationship between the number of shapes in each figure and the number of matchsticks used to construct it
   ii. devise a rule in words that describes the pattern relating the number of shapes in each figure and the number of matchsticks used to construct it.

   a.
   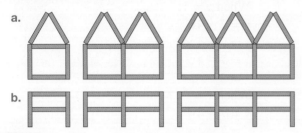
   b.

3. **WE4** Copy and complete the following tables using the rule given in each case to calculate the correct output numbers.

   a. Subtract 4 from each input number.

Input	4	5	6	10	14
Output					

   b. Add 12 to each input number.

Input	1	2	6	13	107
Output					

   c. Multiply each input number by itself.

Input	0	1	5	8	10
Output					

   d. Multiply each input number by itself, then add 4.

Input	0	1	4	7	12
Output					

4. Copy and complete the following tables. Use the rule given in each case to calculate the correct output numbers.

   a. Multiply each input number by 2.

Input	1	10	51	60	144
Output					

   b. Divide each input number by 3.

Input	3	12	21	66	141
Output					

   c. Add 2 to each input number.

Input	1	3	27			
Output				4	55	193

   d. Add 20 to each input number.

Input	3	25	56			
Output				20	94	1773

5. **WE5** If $y = 2x + 7$, substitute the given value of x into the formula to find the value of y in each case.
   a. $x = 3$
   b. $x = 6$
6. If $y = -3x - 2$, substitute the given value of x into the formula to find the value of y in each case.
   a. $x = 1$
   b. $x = -3$
7. **WE6** If Y represents any number, write expressions for:
   a. 7 times the number
   b. 13 less than the number
   c. 5 more than the number.
8. If M represents any number, write expressions for:
   a. 3 times the number
   b. 7 less than the number
   c. 12 more than the number.

9. If a is the number of lollies in a jar, then write the algebraic expression that describes the number of lollies left in the jar in the following situations.
   a. 8 lollies were eaten.
   b. Half of the lollies were eaten.
   c. 23 lollies were taken, 18 eaten and the rest put back in the jar.
   d. 33 lollies were eaten.
   e. What assumption have we made with our answer in part **d**?

In questions **10** and **11**, D represents an unknown number.

10. MC An expression for one less than the number is
    A. $D - 3$.　　**B.** $D + 1$.　　**C.** $D - 1$.　　**D.** D.

11. MC An expression for three more than double the number is
    A. $2D$.　　**B.** $3D + 2$.　　**C.** $D + 3$.　　**D.** $2D + 3$.

12. Complete the tables, given the following rules.
    a. $y = 5x - 4$

x	0	1	2	3
y				

    b. $y = -2x + 16$

x	3	6	9	12
y				

13. Complete the tables, given the following rules.
    a. $y = 4x + 10$

x	-4	-2	0	2
y				

    b. $y = -15x + 35$

x	-8	-4	0	4
y				

14. WE7 Generate a table of values for each of the following rules given the specified x values.
    a. $y = -x + 3$ for $x = -4, -2, 0, 2$
    b. $y = -4x + 18$ for $x = -8, -4, 0, 4$

15. Complete the tables, given the following rules:
    a. $y = -\dfrac{x}{2} + 4$

x	-4	-2	0	2
y				

    b. $y = \dfrac{x}{2} - 2$

x	-8	-4	0	4
y				

16. Generate a table of values for each of the following rules given the specified x values.
    a. $y = -\dfrac{2x}{5} - 1$ for $x = -10, -5, 0, 5$
    b. $y = -\dfrac{x}{3} + 3$ for $x = -6, -3, 0, 9$

9.4 Graphing linear functions

9.4.1 Plotting linear graphs

A set of points may be such that the x- and y-coordinates of each point are connected by the same rule. For example, each of the points $(1, 2)$ $(2, 3)$ $(3, 4)$ $(4, 5)$ have the same relationship (rule) between their x- and y-values; namely, the y-coordinate of every point is one more than its x-coordinate.

If the set of points where x- and y-coordinates are connected by a certain rule is plotted on the Cartesian plane, the points will form a pattern.

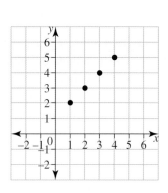

If the pattern formed by the set of points is a straight line, we refer to it as a linear pattern. For example, the points $(1, 2)$ $(2, 3)$ $(3, 4)$ $(4, 5)$, when plotted, will form a linear pattern as shown.

A diagram formed by plotting a set of points on the Cartesian plane is referred to as a graph. If the points form a straight line, then the graph is linear (a straight-line graph).

A set of points with their coordinates can be presented in the form of a table. For example, the points $(-2, 4)$ $(-1, 2)$ $(0, 0)$ $(1, -2)$ $(2, -4)$ can be presented as shown.

x	−2	−1	0	1	2
y	4	2	0	−2	−4

WORKED EXAMPLE 8

a. **Plot the points in the following table on a Cartesian plane.**

x	−4	−3	−2	−1	0	1
y	−2	−1	0	1	2	3

b. **Do the points form a linear graph? If so, what would the next point in the pattern be?**

THINK

a. 1. Look at the x- and y-values of the points and draw a Cartesian plane.

   2. Plot each point.

WRITE

a.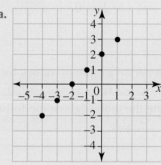

b. 1. Look at the position of the points and answer the question.
   Note: The points form a straight line, so we have a linear graph.

   2. Study the pattern and answer the question.
   Note: The pattern shows that the x-values increase by 1, and the y-values increase by 1. The next x-value will be 2 and the next y-value will be 4.

b. Yes, the points do form a linear graph.

The next point in the pattern is $(2, 4)$.

To plot a linear graph whose equation is given, follow these steps.
1. Create a table of values first.
   - Draw a table.
   - Select some x-values.
   - Substitute the selected x-values into the rule to find corresponding y-values.
2. Draw a Cartesian plane.
3. Plot the points from the table and join them with the straight line.
4. Label the graph.

WORKED EXAMPLE 9

Draw a table of values and plot the graph of $y = 2x + 1$ and label the line.

THINK	WRITE

1. Write the rule.

$y = 2x + 1$

2. Draw a table and choose simple x-values.

x	−2	−1	0	1	2
y					

3. Use the rule to find each y-value and enter them in the table.
   When $x = -2, y = 2 \times -2 + 1 = -3$.
   When $x = -1, y = 2 \times -1 + 1 = -1$.
   When $x = 0, y = 2 \times 0 + 1 = 1$.
   When $x = 1, y = 2 \times 1 + 1 = 3$.
   When $x = 2, y = 2 \times 2 + 1 = 5$.

x	−2	−1	0	1	2
y	−3	−1	1	3	5

4. Draw a Cartesian plane and plot the points.

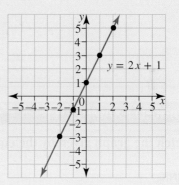

5. Join the points to form a straight line and lable the graph.

It is not necessary to know the coordinates of many points that lie on a straight line in order to sketch a linear graph. In fact, we only need to know the coordinates of two points that lie on the line to sketch a linear graph. The two points can be obtained by substituting any two x-values into the rule to obtain the corresponding y-values. Alternatively, there are two other convenient techniques used to sketch straight lines: (a) the intercept method and (b) the gradient and y-intercept method. These are discussed below.

9.4.2 Sketching linear graphs from intercepts

Recall that the x-intercept is the point where a graph intercepts the x-axis, and the y-intercept is the point where a graph intercepts the y-axis.

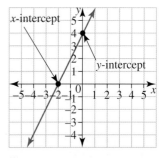

The intercept method involves finding the coordinates of both the x- and the y-intercepts, and plotting them.

At the x-intercept, $y = 0$. Therefore, to obtain the value of the x-intercept, substitute $y = 0$ into the equation and solve the equation for x.

At the y-intercept, $x = 0$. Therefore, to obtain the value of the y-intercept, substitute $x = 0$ into the equation and evaluate.

To sketch the graph, plot both intercepts on the Cartesian axes and join them with the straight line. Label the line.

WORKED EXAMPLE 10

For the linear function $y = 2x + 4$:
a. determine the x- and y-intercepts   **b. sketch the graph of the function.**   ▶

THINK	WRITE
a. 1. The y-intercept is found by substituting $x = 0$ into the equation and solving for y.	**a.** $y = 2 \times 0 + 4$ $\quad = 0 + 4$ $\quad = 4$ y-intercept: $(0, 4)$
2. The x-intercept is found by substituting $y = 0$ into the equation and solving for x. Subtract 4 from both sides of the equation. Divide both sides by 2.	$0 = 2x + 4$ $2x = -4$ $x = -2$ x-intercept: $(-2, 0)$
b. 1. Plot the x- and y-intercepts on a suitably scaled Cartesian plane. Join the two intercepts with a straight line, extending the line in both directions. Write the equation of the line.	**b.** 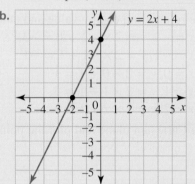

WORKED EXAMPLE 11

Sketch the graph given by the equation $y = 4x$.

THINK	WRITE/DRAW
1. The y-intercept is found by substituting $x = 0$ into the equation and solving for y.	$y = 4x$ $y = 4 \times 0$ $y = 0$ The y-intercept is the point $(0, 0)$.
2. The x-intercept is found by substituting $y = 0$ into the equation and solving for x.	$y = 4x$ $0 = 4x$ $0 = x$ The x-intercept is the point $(0, 0)$.
3. As both intercepts are at the origin, we have only one unique coordinate so far. Choose any other x-value, say $x = 1$, and substitute this into the original equation to find a second point.	$y = 4x$ $\quad = 4 \times 1$ $\quad = 4$ Another point is $(1, 4)$.
4. Plot these two points on a suitably scaled Cartesian plane. Join the two points with a straight line, extending the line in both directions. Write the equation of the line.	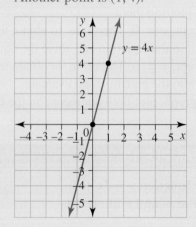

9.4.3 The gradient

Gradient gives a measure of steepness. For example, the gradient of a road going up the hill would inform us of the steepness of the slope of the hill.

The gradient of a straight line shows the change in vertical height as the horizontal distance increases by one unit.

The gradient of a straight line is denoted by the letter m and is given by the formula: $m = \dfrac{\text{rise}}{\text{run}}$, where *rise* is the vertical distance between any two points on the line and *run* is the horizontal distance between the same two points.

To find the gradient of a straight line on a Cartesian plane, draw a right-angled triangle any where along the line as shown at right and use it to measure *rise* and *run*.

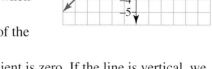

The value of the gradient is given by the ratio $\dfrac{\text{rise}}{\text{run}}$. For example, in the diagram, rise $= 2$ and run $= 2,$ so the gradient $m = \dfrac{2}{2} = 1.$

The gradient of a straight line is constant; that is, it will be the same when measured anywhere along the line.

Since the gradient is the measure of steepness, the greater the value of the gradient, the steeper the line.

If the line is horizontal, there is no slope; hence the value of the gradient is zero. If the line is vertical, we say that its gradient is **infinite** or **undefined**.

m is zero m is undefined

If the line slopes upwards from left to right (that is, it *rises*), the gradient is positive. If it slopes downwards from left to right (that is, it *falls*), the gradient is negative.

m is positive m is negative

rise   rise

run   run

WORKED EXAMPLE 12

Determine:
a. **the gradient, m**
b. **the y-intercept, c, of the linear graph at right.**

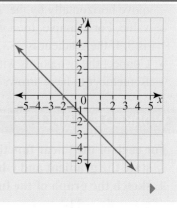

▶

THINK	WRITE

a. 1. Choose two convenient points on the line and draw a triangle to find the rise and the run.

a.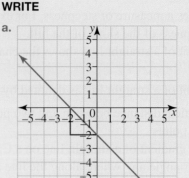

rise $= -2$, run $= 2$

2. Calculate the gradient. In this case the rise is negative since the line slopes down from left to right.

$$\text{Gradient}: m = \frac{\text{rise}}{\text{run}} = \frac{-2}{2}$$
$$= -1$$

b. The y-intercept is where the graph crosses the y-axis, and c is the y-value at this point.

b. $c = -2$

9.4.4 Sketching linear graphs using the gradient and y-intercept

The graph of the linear equation $y = 2x - 4$ is shown at right.
The gradient of the line is 2.

$$\text{Gradient} = \frac{\text{rise}}{\text{run}}$$
$$= \frac{6}{3}$$
$$= 2$$

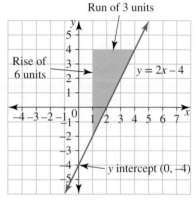

The y-coordinate of the y-intercept is -4.

The equation of a straight line contains the values of both the gradient and the y-coordinate of the y-intercept.

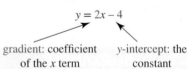

When constructing linear graphs using the gradient and y-intercept:
- Mark the y-intercept on the y-axis.
- Another point is found by using the gradient. From the y-intercept, move horizontally by the 'run' value, then vertically by the 'rise' value. For example, for a gradient of 3, the 'run' is 1 and the 'rise' is 3. If the gradient is -3, the 'rise' will be down by 3.
- Connect the two points with a straight line.
- Remember to have arrows on the ends of the line and to name the graph.

WORKED EXAMPLE 13

For each of the linear functions below:
i. state the gradient and y-intercept
ii. sketch the graph of the function.
   a. $y = 4x - 11$ **b.** $y = -4x$

THINK	**WRITE**
a. i. Determine the value of the gradient and the coordinates of the y-intercept. The gradient is given by the coefficient of x and the y-coordinate of the y-intercept is given by the constant in the equation.	**a.** For $y = 4x - 11$, gradient $= 4$ and y-intercept $= -11$.

ii. • Construct a set of axes and mark the position of the y-intercept.
- The y-intercept is -11, as shown in black.
- The gradient is 4, so $\dfrac{\text{rise}}{\text{run}} = \dfrac{4}{1}$.
- From the y-intercept, move 4 units up and 1 unit to the right, then mark a second point, as shown in pink.
- The two points can now be connected with a straight line.
- Write the equation next to the line.

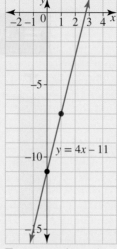

| **b. i.** Determine the value of the gradient and the coordinates of the y-intercept.
The gradient is given by the coefficient of x and the y-coordinate of the y-intercept is given by the constant in the equation. | **b.** For $y = -4x$ or $(y = -4x + 0)$, gradient $= -4$ and y-intercept $= 0$. |

ii. • Construct a set of axes.
- Mark in the position of the y-intercept at 0, as shown in black. The gradient is -4, so $\dfrac{\text{rise}}{\text{run}} = \dfrac{-4}{1}$.
- From the y-intercept, move 4 units down and 1 unit to the right, then mark a second point, as shown in pink.
- The two points can now be connected with a straight line to form the graph.
- Write the equation next to the line.

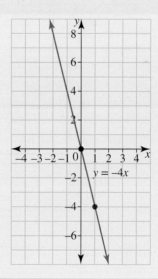

9.4.5 Sketching linear graphs using technology

While it is important to understand the processes behind sketching a straight line using a pencil and paper method, it is also vital to engage in the use of graphing software packages to produce graphs. It is common for many people in the business, finance, statistical and science industries to write reports and draw graphs of information. Since they are in the professional world such graphs would be created using a graphing software package rather than using pencil and paper. There are many freely available graphing software packages on the internet. Some common ones are Desmos, GeoGebra, Excel and of course a graphics calculator.

In this section we will demonstrate how linear functions can be drawn with graphing software.

In Worked example 9 we are asked to plot the graph of $y = 2x + 1$ by constructing a table of values. Using a graphing software package such as Desmos we could answer this question in two ways: (a) by plotting all the points from the table or (b) by graphing the line without the points.

For option (a), after we have constructed the table of values for the rule, we can input the table of values directly into the software using the 'table' feature.

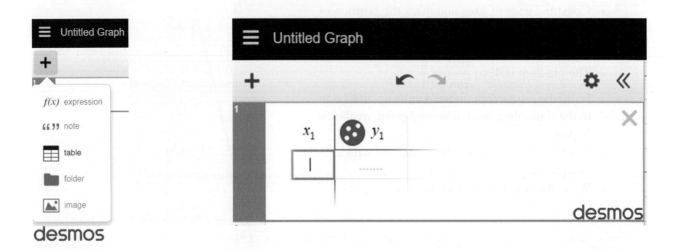

Input the calculated values into the table and take note that these tables are set out slightly differently to those you have seen in this section. The software will automatically plot the points.

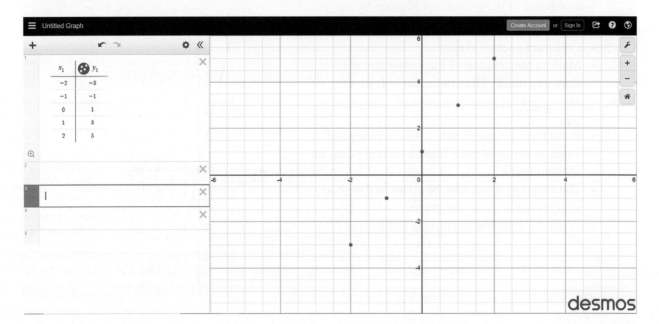

For option (b), we could simply type $y = 2x + 1$ into the program and it will produce the graph.

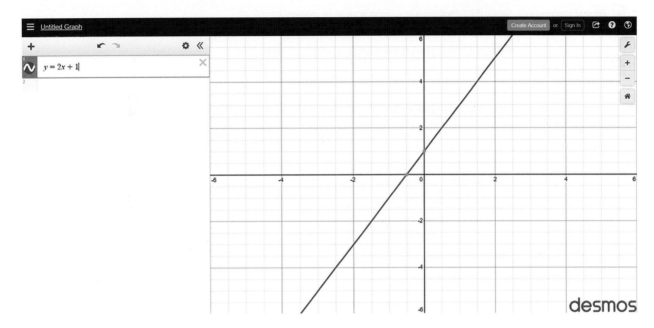

One very handy feature of this program is that once the line is produced, a dot is placed where the graph crosses the x and y axes. You can hover the cursor over these dots to display the coordinates of the x and y intercepts.

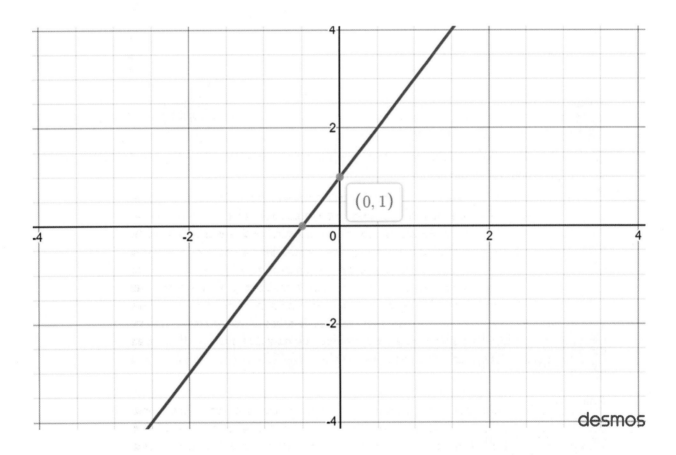

Exercise 9.4 Graphing linear functions

1. **WE8** Plot the following points on a Cartesian plane. Do the points form a linear graph? If so, state the next point in the pattern.
   a. $(-3, -3)$ $(-2, -1)$ $(-1, 1)$ $(0, 3)$ $(1, 5)$ $(2, 7)$ $(3, 9)$
   b. $(-3, -5)$ $(-2, -3)$ $(-1, 0)$ $(0, 1)$ $(1, 4)$ $(2, 5)$ $(3, 7)$
   c.

x	−2	−1	0	1	2
y	3	−1	−2	−3	−4

   d.

x	−2	−1	0	1	2
y	−6	−3	0	3	6

2. **WE9** Complete the following tables of values, plot the points on a Cartesian plane, and join them to make a linear graph. Label the graphs with the rules.
   a. Rule: $y = x + 3$

x	−2	−1	0	1	2
y	1	2	3		

   b. Rule: $y = x - 5$

x	−2	−1	0	1	2
y		−6			−3

   c. Rule: $y = 5x$

x	−2	−1	0	1	2
y	−10		0		

   d. Rule: $y = 2x + 4$

x	−2	−1	0	1	2
y	0			6	

   e. Rule: $y = -3x + 2$

x	−2	−1	0	1	2
y		5			−4

3. **WE10** For each of the linear functions below:
   i. determine the x- and y-intercepts
   ii. sketch the graph of the function.
   a. $y = 3x + 6$ b. $y = 3x - 9$ c. $y = 3x + 12$

4. For each of the straight-line graphs below:
   i. determine the x- and y-intercepts
   ii. sketch the graph of the function.
   a. $y = 4x + 2$ b. $y = 2x - 5$ c. $y = -2x + 1$

5. **WE11** Sketch the graphs given by the following linear equations on a single Cartesian plane.
   a. $y = x$ b. $y = 3x$ c. $y = -3x$

6. State whether each of the following lines has a positive, negative, zero or undefined gradient.

a. 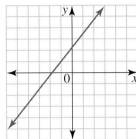   b. 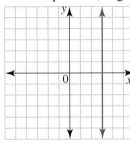   c.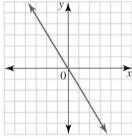

d. 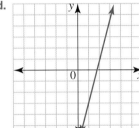   e. 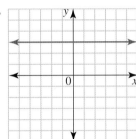   f.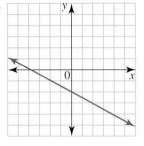

7. **WE12** For each of the following linear graphs, determine
   i. the gradient, m
   ii. the y-intercept, c.

a. 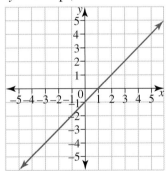   b.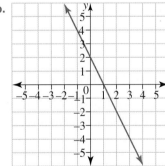

c. 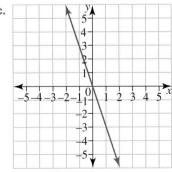   d.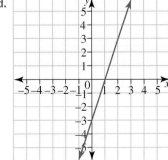

e.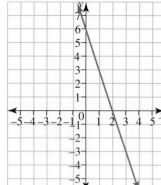

f.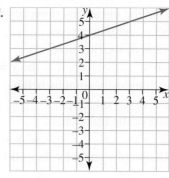

8. **WE13** For each of the linear functions below:
   i. state the gradient and the y-intercept
   ii. sketch the graph of the function.

   a. $y = 2x + 4$ b. $y = 2x + 6$ c. $y = 2x - 6$

   d. $y = x - 6$ e. $y = x + 4$ f. $y = 4x - 8$

9. For each of the linear functions below:
   i. state the gradient and the y-intercept
   ii. sketch the straight-line graph.

   a. $y = 2x + 5$ b. $y = -4x - 3$ c. $y = -3x + 5$

   d. $y = \dfrac{1}{2}x - 3$ e. $y = \dfrac{-3}{2}x + 3$ f. $y = \dfrac{4}{5}x$

10. Using the gradient and y-intercept, sketch the following linear graphs on the same set of axes.

Gradient	y-intercept
3	2
4	−3
−2	0

11. a. Sketch the graph of the linear function with a gradient of $\dfrac{1}{2}$ and y-intercept of 4.

    b. State the equation of this straight-line graph.

12. a. Sketch the graph of the linear function with a gradient of $\dfrac{-2}{3}$ and a y-intercept of -5.

    b. State the equation of this straight-line graph.

13. Sketch the following linear equations on the same set of axes.

    a. $y = 2x + 12$ b. $y = -\dfrac{1}{2}x + 6$ c. $y = x - 1.5$

14. a. Sketch the linear equation $y = -\dfrac{5}{7}x - \dfrac{3}{4}$:

       i. using the y-intercept and the gradient   ii. using the x- and y-intercepts.

    b. Compare and contrast the methods and generate a list of advantages and disadvantages for each method. Which method do you think is best? Why?

15. a. Sketch each of the following straight lines on the same set of axes.

       i. $y = \dfrac{5}{2}x + 5$ ii. $y = -\dfrac{5}{2}x + 5$

    b. Determine the x-intercepts for these straight lines.

    c. What shape is formed by the three axis intercepts?

    d. Find the area enclosed by the axis intercepts.

16. a. Using an appropriate method, sketch each of the following straight lines on the same set of axes.

    i. $y = 4x + 4$ **ii.** $y = 4x - 6$ **iii.** $x + y = 4$ **iv.** $x + y + 1 = 0$

    b. What shape is formed by these four lines?

    c. Write down the coordinates of the vertices (corners) of this shape.

9.5 Review: exam practice

9.5.1 Cartesian plane: summary

Plotting points on a Cartesian plane

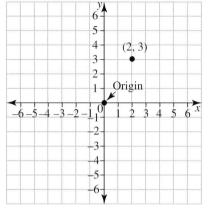

- Cartesian coordinates can be used to locate any point on a plane.
- The Cartesian plane is formed by two perpendicular lines called *axes*. The horizontal axis is called the *x*-axis and the vertical axis is called the *y*-axis. The axes intersect at a point called the *origin*.
- Both axes must be marked (with marks being evenly spaced) and numbered. The axis can extend infinitely in both directions.
- The location of any point on the Cartesian plane is given by its Cartesian coordinates. The Cartesian coordinates are a pair of numbers that are separated by a comma and are shown within brackets. The first number is called the *x*-coordinate of the point; this shows how far across (that is, to the left or to the right) from the origin the point is located. The second number is called the *y*-coordinate; this shows how far up or down from the origin the point is. For example, the point $(2, 3)$ is located 2 units to the right and 3 units up from the origin.
- Review of linear functions terminology
  - *Term*: A group of letters and/or numbers. Terms are separated by an addition or subtraction sign.
  - *Pronumeral*: The letter part of a term.
  - *Coefficient*: The number multiplying (out the front) the pronumeral.
  - *Constant term*: The term that does not have a pronumeral.
  - *Equation*: A mathematical statement containing a left- and right-hand side separated by an equals sign.

Generating tables of values for linear functions

- A table of values can be constructed by repeatedly substituting values into a rule, creating pairs of coordinates.

Graphing linear functions

- The *y*-intercept is the point where the graph crosses the *y*-axis. At the *y*-intercept, $x = 0$.
- The *x*-intercept is the point where the graph crosses the *x*-axis. At the *x*-intercept, $y = 0$.
- Only two distinct points are needed to sketch a straight line.
- To sketch a line using the *x*- and *y*-intercept method:
  1. Calculate the *x*-intercept by substituting $y = 0$ into the rule and solving for *x*.
  2. Calculate the *y*-intercept by substituting $x = 0$ into the rule and solving for *y*.
  3. Plot and label the *x*- and *y*-intercepts and rule a straight line through the two points.
- If the rule is in the form $y = mx + c$, the gradient–intercept method may be used to sketch the graph.
  1. Plot and label the *y*-intercept, *c*.
  2. Write the gradient (*m*) as a fraction and identify the values of rise and run.
  3. Starting from the *y*-intercept, move up or down and to the right or left the number of units suggested by the rise and run, then plot the second point.
  4. Rule a straight line through the two points.

- Lines with a rule of the form $y = mx$ pass through the origin. These graphs can be sketched by determining another point that lies on the line by using one of two methods:
    1. Substitute a value for x into the equation to find y.
       For example, substitute $x = 1$ into the equation to find the corresponding value for y.
    or
    2. Use the gradient and y-intercept method.

Exercise 9.5 Review: exam practice

Simple familiar

1. **MC** The point $(7, -1)$ gives a position on the Cartesian plane of
   A. -7 units on the x-axis, -1 unit on the y-axis.   **B.** -1 unit on the x-axis, 7 units on the y-axis.
   C. -7 units on the x-axis, -1 unit on the y-axis.   **D.** 7 units on the x-axis, -1 unit on the y-axis.

2. **MC** Which point is described by 4 on the y-axis and -3 on the x-axis?
   A. $(4, -3)$ **B.** $(4, 3)$ **C.** $(-3, 4)$ **D.** $(3, -4)$

3. **MC** Which of the following points is the lowest on the Cartesian plane?
   A. $(0, 6)$ **B.** $(12, 2)$ **C.** $(8, 0)$ **D.** $(8, -1)$

4. Complete the following table, given the rule $y = 5x - 4$.

x	-2	0	2	5
y				

5. Complete the following table, given the rule $y = -3x + 15$.

x	-4	-1	2	6
y				

6. **MC** For the rule $y = 3x - 1$, what is y when $x = 2$?
   A. -1 **B.** 1 **C.** 2 **D.** 5

7. **MC** What is the gradient of the linear rule $y = 4 - 6x$?
   A. 6 **B.** -6 **C.** 4 **D.** -4

8. **MC** What is the x-intercept of the graph with the rule $2y - x + 6 = 0$?
   A. -2 **B.** 0 **C.** 2 **D.** 6

9. **MC** What is the y-intercept of the graph with the rule $2y - x + 6 = 0$?
   A. -6 **B.** -3 **C.** 0 **D.** 3

10. **MC** What is the gradient of the line shown?
    A. 5 **B.** -5 **C.** 1 **D.** -1

11. **MC** What is the rule for a line with gradient $= -4$ and y-intercept $= 8$?
    A. $y = -4x + 32$
    B. $y = -4x + 8$
    C. $y = 4x - 32$
    D. $y = 4x - 8$

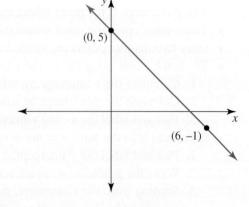

12. **MC** Which point *does not* lie on the line $2y - x = 15$?
    A. $(10, 12.5)$ **B.** $(-9, -3)$ **C.** $(-5, 5)$ **D.** $(3, 9)$

Complex familiar

13. For each of the following linear graphs, what is:
    i. the gradient
    ii. the y-intercept?
    a. $y = 8x - 3$
    b. $y = 5 - 9x$
    c. $2x + y - 6 = 0$
    d. $4x - 2y = 0$
    e. $y = \dfrac{2x - 1}{3}$

14. **MC** The table of values shown is modelled on which linear equation?

x	0	2	4	10
y	4	10	16	34

    A. $y = x + 4$
    B. $y = 4x + 3$
    C. $y = 3x + 4$
    D. $y = 5x + 4$

15. For the following equations, use the gradient–intercept method to sketch linear graphs.
    a. $y = -x + 5$
    b. $y = 4x - 2.5$
    c. $y = \dfrac{2}{3}x - 1$
    d. $y = 3 - \dfrac{5}{4}x$

16. For the following equations, use the x- and y-intercept method to sketch linear graphs.
    a. $y = -6x + 25$
    b. $y = 20x + 45$
    c. $2y + x = -5$
    d. $4y + x - 2.5 = 0$

Complex unfamiliar

17. **MC** If $y = 2x + 1$, then a point that could not be on the line is
    A. $(3, 7)$.
    B. $(-3, -5)$.
    C. $(0, 1)$.
    D. $(-3, 0)$.

18. Complete the table, given the following rule.
    $3x - 2y - 2 = 0$

x	0	2	4	8
y				

19. **MC** The graph shows a straight line that passes through the points $(-7, 12)$ and $(-5, 8)$.
    The coordinates of the point where the line crosses the x-axis are
    A. $(-2, 0)$.
    B. $(-1.5, 0)$.
    C. $(-4, 0)$.
    D. $(-1, 0)$.

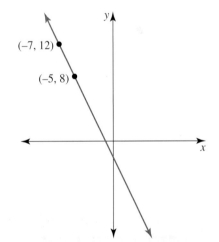

20. Louise owes her friend Sula $400 and agrees to pay her back $15 per week.
    a. State a linear rule that demonstrates this reducing debt schedule and sketch the graph.
    b. How many weeks does it take her to repay the debt?
    c. How much does she owe after 15 weeks?
    d. After how many repayments does she owe $85?

Answers

9 Cartesian plane

Exercise 9.2 Plotting points on a Cartesian plane

1. a. $(2, 3)$ b. $(6, 8)$ c. $(7, 0)$
   d. $(3, -6)$ e. $(0, -2)$ f. $(-7, -3)$
   g. $(-4, -7)$ h. $(-5, 0)$

2.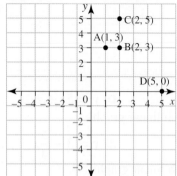

3.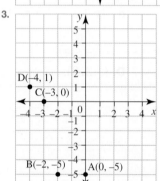

4.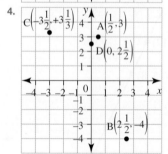

5.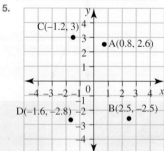

6. C

7. D

8. A

9.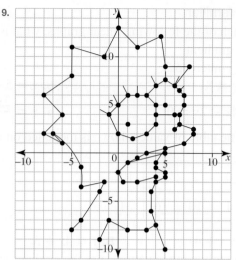

10. The point $(4, -2)$ is in the wrong position; it is in the $(-2, -4)$ position.
    Also, the point $(-5, -4)$ is in the wrong position; it is in the $(-4, -5)$ position.
    The correct plot is:

    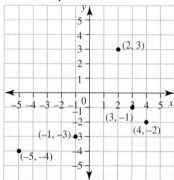

11. a, b, c.

    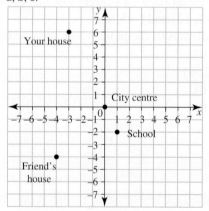

    d. Your school is 4 km east and 8 km south of your house.

12. a.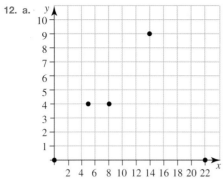

b. Sample responses can be found in the Worked Solutions in the online resources.

13. Sample responses can be found in the Worked Solutions in the online resources.

Exercise 9.3 Generating tables of values for linear functions

1. a. i.

Number of squares	1	2	3
Number of matchsticks	4	7	10

ii. The number of matchsticks equals three times the number of squares and then add one.

iii. 61

b. i.

Number of triangles	1	2	3
Number of matchsticks	3	5	7

ii. The number of matchsticks equals two times the number of triangles and then add one.

iii. 41

2. a. i.

Number of shapes	1	2	3
Number of matchsticks	6	11	16

ii. The number of matchsticks is equal to five times the number of shapes and then add one.

b. i.

Number of shapes	1	2	3
Number of matchsticks	4	7	10

ii. The number of matchsticks equals three times the number of shapes and then add one.

3. a.

Input	4	5	6	10	14
Output	0	1	2	6	10

b.

Input	1	2	6	13	107
Output	13	14	18	25	119

c.

Input	0	1	5	8	10
Output	0	1	25	64	100

d.

Input	0	1	4	7	12
Output	4	5	20	53	148

4. a.

Input	1	10	51	60	144
Output	2	20	102	120	288

b.

Input	3	12	21	66	141
Output	1	4	7	22	47

c.

Input	1	3	27	2	53	191
Output	3	5	29	4	55	193

d.

Input	3	25	56	0	74	1753
Output	23	45	76	20	94	1773

5. a. 13   **b.** 19

6. a. −5   **b.** 7

7. a. $7Y$ **b.** $Y − 13$ **c.** $Y + 5$

8. a. $3M$ **b.** $M − 7$ **c.** $M + 12$

9. a. $a − 8$

   b. $\dfrac{1}{2} \times a = \dfrac{a}{2}$

   c. $a − 18$

   d. $a − 33$

   e. That there was at least 33 lollies in the jar and they have not eaten lollies from anywhere but the jar we are referring to.

10. C

11. D

12. a.

x	0	1	2	3
y	−4	1	6	11

b.

x	3	6	9	12
y	10	4	−2	−8

13. a.

x	−4	−2	0	2
y	−6	2	10	18

b.

x	−8	−4	0	4
y	155	95	35	−25

14. a.

x	−4	−2	0	2
y	7	5	3	1

b.

x	−8	−4	0	4
y	50	34	18	2

15. a.

x	−4	−2	0	2
y	6	5	4	3

b.

x	−8	−4	0	4
y	−6	−4	−2	0

16. a.

x	−10	−5	0	5
y	3	1	−1	−3

b.

x	−6	−3	0	9
y	5	4	3	0

Exercise 9.4 Graphing linear functions

1. a.

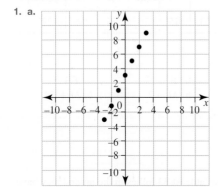

Linear, (4, 11)

b.

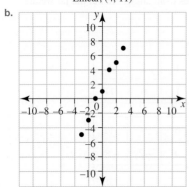

Not linear

c.

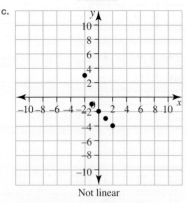

Not linear

d.

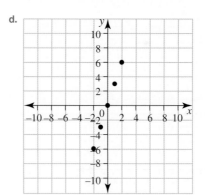

Linear, (3, 9)

2. a.

x	−2	−1	0	1	2
y	1	2	3	4	5

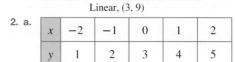

$y = x + 3$

b.

x	−2	−1	0	1	2
y	−7	−6	−5	−4	−3

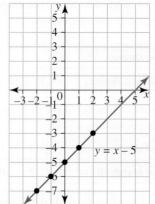

$y = x - 5$

c.

x	−2	−1	0	1	2
y	−10	−5	0	5	10

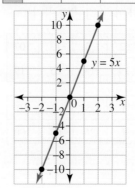

$y = 5x$

d.

x	−2	−1	0	1	2
y	0	2	4	6	8

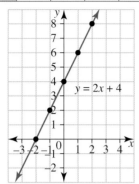

e.

x	−2	−1	0	1	2
y	8	5	2	−1	−4

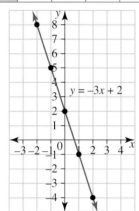

3. a. x-intercept: $(−2, 0)$
   y-intercept: $(0, 6)$

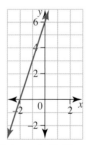

b. x-intercept: $(3, 0)$
   y-intercept: $(0, −9)$

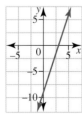

c. x-intercept: $(−4, 0)$
   y-intercept: $(0, 12)$

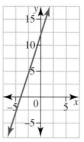

4. a. x-intercept: $(−0.5, 0)$
   y-intercept: $(0, 2)$

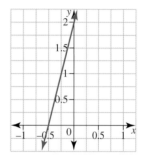

b. x-intercept: $(2.5, 0)$
   y-intercept: $(0, −5)$

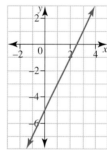

c. x-intercept: $(0.5, 0)$
   y-intercept: $(0, 1)$

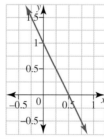

5.

6. a. Positive   b. Undefined   c. Negative
   d. Positive   e. Zero   f. Negative

7. a. i. $m = 1$ ii. $c = -1$
   b. i. $m = -2$ ii. $c = 2$
   c. i. $m = -3$ ii. $c = 0$
   d. i. $m = 3$ ii. $c = -3$
   e. i. $m = -3$ ii. $c = 6$
   f. i. $m = \dfrac{1}{3}$ ii. $c = 4$

8. a. i. Gradient $= 2$; y-intercept $= (0, 4)$

    ii.

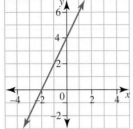

   b. i. Gradient $= 2$; y-intercept $= (0, 6)$

    ii.

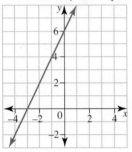

   c. i. Gradient $= 2$; y-intercept $= (0, -6)$

    ii.

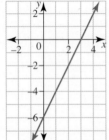

   d. i. Gradient $= 1$; y-intercept $= (0, -6)$

    ii.

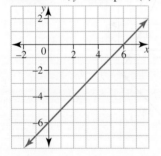

   e. i. Gradient $= 1$; y-intercept $= (0, 4)$

ii.

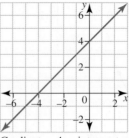

   f. i. Gradient $= 4$; y-intercept $= (0, -8)$

    ii.

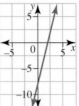

9. a. i. Gradient $= 2$; y-intercept $= (0, 5)$

    ii.

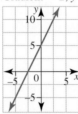

   b. i. Gradient $= -4$; y-intercept $= (0, -3)$

    ii.

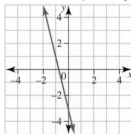

   c. i. Gradient $= -3$; y-intercept $= (0, 5)$

    ii.

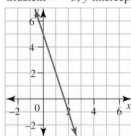

   d. i. Gradient $= \dfrac{1}{2}$; y-intercept $= (0, -3)$

    ii.

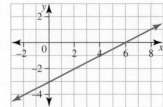

e. i. Gradient $= \dfrac{-3}{2}$; y-intercept $= (0, 3)$

   ii.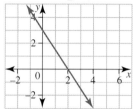

f. i. Gradient $= \dfrac{4}{5}$; y-intercept $= (0, 0)$

   ii.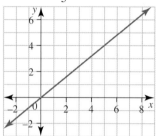

10.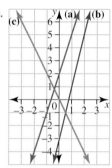

11. a.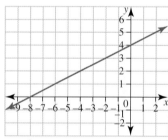

    b. $y = \dfrac{1}{2}x + 4$

12. a.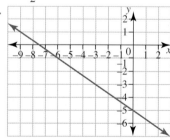

    b. $y = \dfrac{-2}{3}x - 5$

13.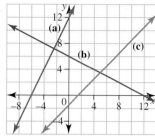

14. a.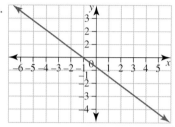

    b. Sample responses can be found in the Worked Solutions in the online resources.

15. a.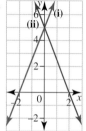

    b. i. $(-2, 0)$
       ii. $(2, 0)$

    c. Isosceles triangle

    d. 10 units^2

16. a.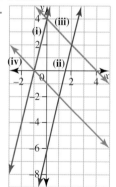

    b. Parallelogram

    c. $(-1, 0), (0, 4), (2, 2)$ and $(1, -2)$

9.5 Review: exam practice

1. D
2. C
3. D
4.

x	-2	0	2	5
y	-14	-4	6	21

5.

x	−4	−1	2	6
y	27	18	9	−3

6. D

7. B

8. D

9. B

10. D

11. B

12. B

13. a. $m = 8$ and y-intercept $= -3$

  b. $m = -9$ and y-intercept $= 5$

  c. $m = -2$ and y-intercept $= 6$

  d. $m = 2$ and y-intercept $= 0$

  e. $m = \dfrac{2}{3}$ and y-intercept $= -\dfrac{1}{3}$

14. C

15. a.

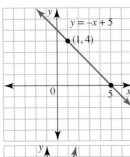

b.

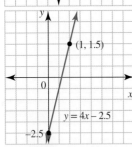

c.

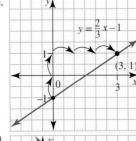

d.

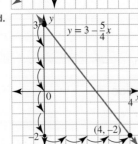

16. a.

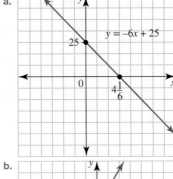

b.

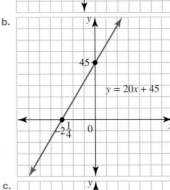

c.

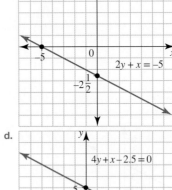

d.

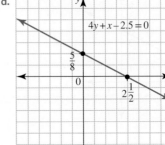

17. D

18.

x	0	2	4	8
y	−1	2	5	11

19. D

20. a. y: the amount owed x: the number of weeks

  $y = 400 - 15x$

  b. $x = 26.67$ weeks

  $x \approx 27$ weeks

  c. $y = \$175$

  d. $x = 21$ weeks

10 Scatterplots and lines of best fit

10.1 Overview

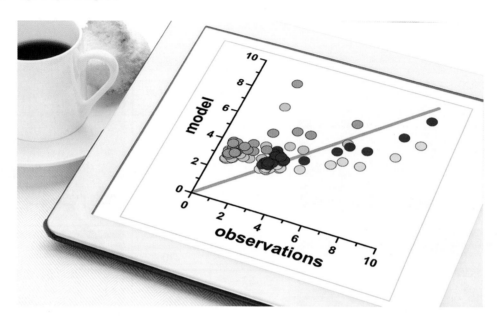

CONTENT

In this chapter, students will learn to:
- describe the patterns and features of bivariate data
- describe the association between two numerical variables in terms of direction (positive/negative), form (linear/non-linear) and strength (strong/moderate/weak)
- identify the dependent and independent variable
- find the line of best fit by eye
- use technology to find the line of best fit [complex]
- interpret relationships in terms of the variables [complex]
- use technology to find the correlation coefficient (an indicator of the strength of linear association) [complex]
- use the line of best fit to make predictions, both by interpolation and extrapolation [complex]
- recognise the dangers of extrapolation [complex]
- distinguish between causality and correlation through examples [complex].

Fully worked solutions for this chapter are available in the Resources section of your eBookPLUS at www.jacplus.com.au.

10.2 Interpreting scatterplots

10.2.1 Bivariate data

The manager of a small ski resort has a problem. He wants to be able to predict the number of skiers using his resort each weekend in advance, so that he can organise additional resort staffing and catering if needed. He knows that good, deep snow will attract skiers in big numbers but limited covering is unlikely to attract a crowd. To investigate the situation further, he collects the following data over 12 consecutive weekends at his resort.

Depth of snow (m)	Number of skiers
0.5	120
0.8	250
2.1	500
3.6	780
1.4	300
1.5	280
1.8	410
2.7	320
3.2	640
2.4	540
2.6	530
1.7	200

As there are two lists of data in this example, they are called **bivariate** data. For each item (weekend), two variables are considered (depth of snow and number of skiers). When analysing bivariate data, we are interested in examining the association (relationship) between the two variables. In the case of the ski resort data, the manager might be interested in answering the following questions.

1. Are visitor numbers related to depth of snow?
2. If there is an association between visitor numbers and depth of snow, is it always true or is it just a guide? In other words, how strong is the association?
3. How much confidence could be placed in the prediction?

In this chapter we shall examine how questions such as these can be answered by the appropriate presentation and analysis of data.

Independent and dependent variables

When examining the association between a pair of variables, we commonly have an **independent variable** and a **dependent variable**. The dependent variable, as the name suggests, is the one whose value depends on the other variable. The independent variable takes on values that do *not* depend on the value of the other variable.

WORKED EXAMPLE 1

i. **For the following pairs of variables, state which variable would be classified as the dependent variable.**

ii. **For each of the pairs of variables, state what is expected to happen to the value of the dependent variable when there is an increase in the value of the independent variable.**

   a. **Height and age**

   b. **Temperature and elevation**

   c. **Blood alcohol level and reaction time**

   d. **IQ and results on an academic test**

   e. **Money earned and hours worked**

   f. **Value of a car and its age**

   g. **Time taken to travel a given distance and speed of travel**

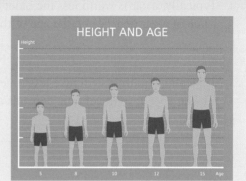

HEIGHT AND AGE

Height

5   8   10   12   15 Age

THINK	WRITE
a. i. Typically, children grow taller as they get older, so a child's height depends on their age.	**a.** The dependent variable is height.
ii.	As age increases, height is expected to increase.
b. i. Typically, the temperature grows colder the higher above sea level the elevation, so temperature depends on elevation.	**b.** The dependent variable is temperature.
ii.	As the elevation increases, the temperature is expected to decrease.
c. i. Typically, a person's reaction time becomes slower as they increase their alcohol consumption, so reaction time depends on blood alcohol level.	**c.** The dependent variable is reaction time.
ii.	As blood alcohol level increases, reaction time is expected to increase.
d. i. Typically, a person with a higher IQ will score a higher result on an academic test, so their result depends on their IQ.	**d.** The dependent variable is the result on an academic test.
ii.	As IQ increases, results on an academic test are expected to increase.

▶

e. i. Typically, a person earns more money when they work more hours, so money earned depends on hours worked.	e. The dependent variable is money earned.
ii.	As the number of hours worked increases, the amount of money earned is expected to increase.
f. i. Typically, a car is worth less the older it gets, so the value of a car depends on the age of the car.	f. The dependent variable is the value of a car.
ii.	As the age of a car increases, its value is expected to decrease.
g. i. Typically, a journey will take less time travelling at a faster speed, so travel time depends on speed.	g. The dependent variable is travel time.
ii.	As the speed increases, the travel time is expected to decrease.

10.2.2 Scatterplots

We shall now look more closely at the data collected by the ski resort manager, and at the three questions he posed. To help answer these questions, the data can be arranged on a **scatterplot**. Each of the data points is represented by a single visible point on the graph.

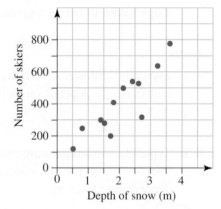
Depth of snow (m)

When drawing a scatterplot, it is important to choose the correct variable to assign to each of the axes. The convention is to place the independent variable on the x-axis and the dependent variable on the y-axis. The independent variable in an experiment or investigation is the variable that is deliberately controlled or adjusted by the investigator. The dependent variable is the variable that responds to changes in the independent variable.

Neither of the variables involved in the ski resort data was controlled directly by the investigator but 'Number of skiers' would be considered the dependent variable because it is likely to change depending on depth of snow. (The snow depth does not depend on numbers of skiers.) As 'Number of skiers' is the dependent variable, we graph it on the y-axis and the 'Depth of snow' on the x-axis.

Notice how the scatterplot for the ski resort data shows a general upward trend. It is not a perfectly straight line, but it is still clear that a general trend or association has formed: as the depth of snow increases, so too does the number of skiers.

> ### Constructing a scatterplot
> When constructing a scatterplot, it is important to place the independent variable along the x-axis and the dependent variable along the y-axis.

The table below shows the height and mass of ten year 12 students. Display this information in the form of a scatterplot.

Height (cm)	120	124	130	135	142	148	160	164	170	175
Mass (kg)	45	50	54	59	60	65	70	78	75	80

THINK

1. Determine the independent and dependent variables. Taller people tend to be heavier, so mass depends on height.

2. Show the height on the *x*-axis and the mass on the *y*-axis.

3. Plot the point given by each pair of data.

WRITE

The independent variable is height and the dependent variable is mass.

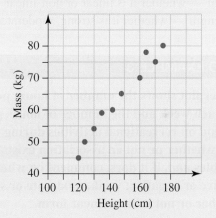

In analysing the scatterplot we look for a pattern in the way the points lie. Certain patterns tell us that certain associations exist between the two variables. This is referred to as **correlation**. We look at what type of correlation exists and how strong it is.

Here is a gallery of scatterplots showing the various patterns we look for

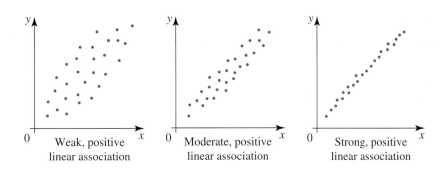

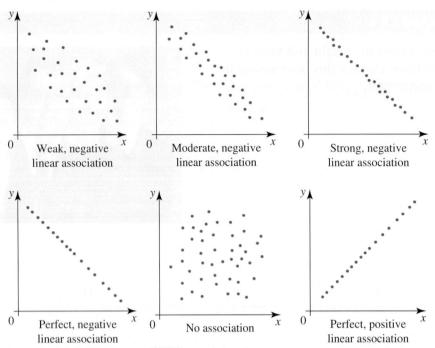

Weak, negative linear association

Moderate, negative linear association

Strong, negative linear association

Perfect, negative linear association

No association

Perfect, positive linear association

When describing the association between two variables displayed on a scatterplot, we need to comment on:

a. the direction — whether it is positive or negative
b. the form — whether it is linear or non-linear
c. the strength — whether it is strong, moderate or weak.

WORKED EXAMPLE 3

The scatterplot shows the number of hours people spend at work each week and the number of hours people get to spend on recreational activities during the week.

Decide whether or not an association exists between the variables and, if it does, comment on whether it is positive or negative; weak, moderate or strong; and whether or not it has a linear form.

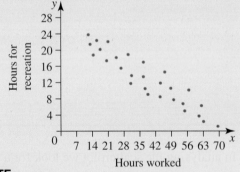

THINK

1. The points on the scatterplot are spread in a certain pattern, namely in a rough corridor from the top left to the bottom right corner. Interpret this pattern.

2. The corridor is straight (i.e. it would be reasonable to fit a straight line into it).

3. The points are neither too tight nor too dispersed.

4. The pattern resembles the central diagram in the gallery of scatterplots shown previously. Describe the relationship between the variables.

WRITE

As the work hours increase, the recreation hours decrease.

There is a moderate, negative linear association between the two variables.

Data showing the average weekly number of hours studied by each student in 12B at Northbank Secondary College and the corresponding height of each student (correct to the nearest tenth of a metre) are given in the table.

Average hours of study	18	16	22	27	15	28	18	20	10	28	25	18	19	17
Height (m)	1.5	1.9	1.7	2.0	1.9	1.8	2.1	1.9	1.9	1.5	1.7	1.8	1.8	2.1

Average hours of study	19	22	30	14	17	14	19	16	14	29	30	30	23	22
Height (m)	2.0	1.9	1.6	1.5	1.7	1.8	1.7	1.6	1.9	1.7	1.8	1.5	1.5	2.1

Construct a scatterplot for the data and use it to comment on the direction, form and strength of any association between the number of hours studied and the height of the students.

Take the independent variable to be height (m).

THINK

1. Plot the point given by each pair of data. There should be a total of 28 points.

2. Comment on the direction of any association.

3. Comment on the form of the association.

WRITE

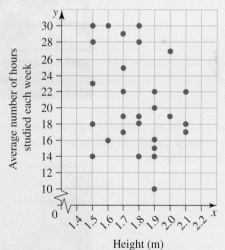

Height (m)

There is no association; the points appear to be randomly placed.

There is no form, just a random placement of points.

| 4. Comment on the strength of any association. | Since there is no association, strength is not relevant. |
| 5. Draw a conclusion. | Clearly, from the graph, the number of hours spent studying has no relation to how tall you might be. |

---

on Resources

📄 **Digital documents** SkillSHEET Determining independent and dependent variables (doc-5411)

SkillSHEET Determining the type of relationship (doc-5413)

Exercise 10.2 Interpreting scatterplots

1. **WE1** For each of the following pairs, decide which of the variables is independent and which is dependent.
   a. Number of hours spent studying for a Mathematics test and the score on that test
   b. Daily amount of rainfall (in mm) and daily attendance at the Botanical Gardens
   c. Number of hours per week spent in a gym and the annual number of visits to the doctor

   d. The amount of computer memory taken by an essay and the length of the essay (in words)
   e. The cost of care in a childcare centre and attendance at the childcare centre

   f. The cost of the property (real estate) and the age of the property
   g. The entry requirements for a certain tertiary course and the number of applications for that course

---

h. The heart rate of the runner and the running speed

2. **WE2** The table below shows the marks obtained by a group of 10 students in History and Geography. Display this information in the form of a scatterplot.

History	36	65	82	72	58	39	58	74	82	66
Geography	45	78	66	72	50	51	61	70	60	88

3. The table below shows the maximum temperature each day, together with the number of people who attend the cinema that day. Display the information in the form of a scatterplot.

Temperature (°C)	25	33	30	22	15	18	27	22	28	20
Number at cinema	256	184	190	312	458	401	200	357	312	423

4. The table below shows the wages of 20 people and the amount of money they spend each week on entertainment. Display this information in the form of a scatterplot.

Wages ($)	370	380	500	510	395	430	535	490	495	550
Amount spent on entertainment ($)	55	85	150	75	145	100	130	115	70	150
Wages ($)	810	460	475	520	530	475	610	780	350	460
Amount spent on entertainment ($)	220	50	100	150	140	160	90	130	40	50

5. **WE3** The scatterplot shown represents the number of hours of basketball practice each week and a player's shooting percentage. Decide whether or not an association exists between the variables and, if it does, comment on whether it is positive or negative; weak, moderate or strong; and whether or not it is linear form.

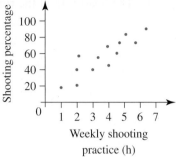

6. The scatterplot shows the hours after 5 pm and the average speed of cars on a freeway. Explain the direction, form and strength of the association of the two variables.

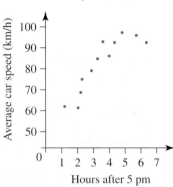

7. For each of the following pairs of variables, write down whether or not you would reasonably expect an association to exist between the pair and, if so, comment on whether it would be a positive or negative association.
   a. Time spent in a supermarket and total money spent
   b. Income and value of car driven
   c. Number of children living in a house and time spent cleaning the house
   d. Age and number of hours of competitive sport played per week
   e. Amount spent on petrol each week and distance travelled by car each week
   f. Number of hours spent in front of a computer each week and time spent playing the piano each week
   g. Amount spent on weekly groceries and time spent gardening each week

8. For each of the scatterplots, describe whether or not an association exists between the variables and, if it does, comment on whether it is positive or negative, whether it is weak, moderate or strong and whether or not it has a linear form.

a.
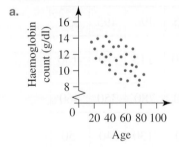

b.
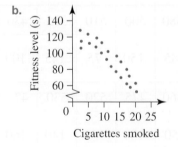

c.
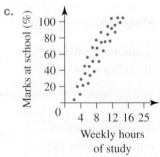

d.

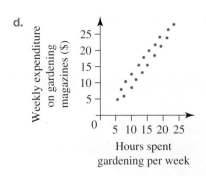

e.

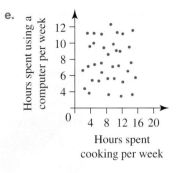

f.

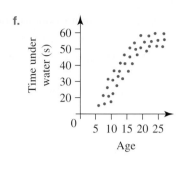

9. **MC** From the scatterplot shown, it would be reasonable to observe that

   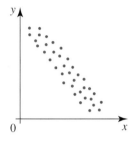

   A. as the value of *x* increases, the value of *y* increases.
   B. as the value of *x* increases, the value of *y* decreases.
   C. as the value of *x* increases, the value of *y* remains the same.
   D. as the value of *x* remains the same, the value of *y* increases.

10. **WE4** Data on the height of a person and the length of their hair is shown. Construct a scatterplot for the data and use it to comment on the direction, form and strength of any association between the height of a person and the length of their hair.

Height (cm)	158	164	184	173	194	160	198	186	166
Hair length (cm)	18	12	5	10	7	3	10	6	14

11. The following table shows data on hours spent watching television per week and age. Use the data to construct a scatterplot and use it to comment on the direction, form and strength of any association between the two variables.

Age (years)	12	25	61	42	18	21	33	15	29
TV per week (hours)	23	30	26	18	12	30	20	19	26

12. The population of a municipality (to the nearest ten thousand) together with the number of primary schools in that particular municipality is given below for 11 municipalities.

Population (×1000)	110	130	130	140	150	160	170	170	180	180	190
Number of primary schools	4	4	6	5	6	8	6	7	8	9	8

Construct a scatterplot for the data and use it to comment on the direction, form and strength of any association between the population and the number of primary schools.

13. The table contains data for the time taken to do a paving job and the cost of the job.
Construct a scatterplot for the data. Comment on whether an association exists between the time taken and the cost. If there is an association describe it.

Time taken (hours)	Cost of job ($)
5	1000
7	1000
5	1500
8	1200
10	2000
13	2500
15	2800
20	3200
18	2800
25	4000
23	3000

10.3 The line of best fit

The process of 'fitting' straight lines to bivariate data (known as linear regression) enables us to analyse associations between the data and possibly make predictions based on the given data set.

10.3.1 Fitting a line of best fit by eye

Consider the set of bivariate data points shown at right. In this case the *x*-values could be the hand lengths of a group of people, while *y*-values could be the lengths of their feet. We wish to determine a linear association between these two random variables.

Of course, there is no single straight line which would go through all the points, so we can only *estimate* such a line.

Furthermore, the more closely the points appear to be on or near a straight line, the more confident we are that such a linear association may exist and the more accurate our fitted line should be.

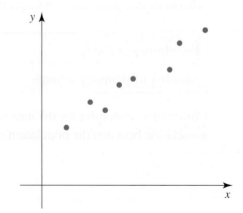

Consider the estimate, drawn 'by eye' in the figure at right. The line extends through all data points, and most of the points are very close to this straight line, indicating a strong linear association between the two variables. This line was easily drawn since the points are very much part of an apparent linear association.

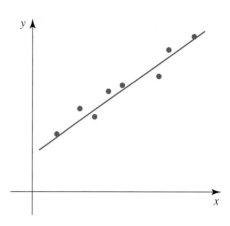

Due to the strong linear association between foot length and hand length, we can conclude from this graph that there is evidence to suggest that a person's foot length is related to the length of their hand.

Regression analysis is concerned with finding these lines of best fit using various methods so that the number of points above and below the lines are 'balanced'.

There are many different methods of fitting a straight line by eye. They may appear logical or even obvious but fitting by eye involves a considerable margin of error. We are going to consider only one method: fitting the line by balancing the number of points.

The technique of balancing the number of points involves fitting a line so that there is an *equal number of points* above and below the line. For example, if there are 12 points in the data set, 6 should be above the line and 6 below it.

WORKED EXAMPLE 5

Draw the line of best fit for the data in the diagram using the equal-number-of-points method.

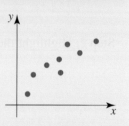

THINK

1. Note that the number of points (*n*) is 8.
2. Fit a line where 4 points (i.e. half *n*) are above the line. Using a clear plastic ruler, try to fit the best line. Make sure the line extends through all data points.
3. The first attempt shown has only 3 points below the line, where there should be 4. Make refinements.
4. The second attempt is an improvement, but the line is too close to the points above it.
5. Improve the position of the line until a better 'balance' between upper and lower points is achieved.

WRITE

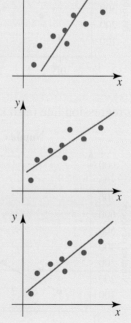

10.3.2 Determining the equation of a line of best fit

Using technology

Using a spreadsheet, the equation of the line of best fit, commonly referred to as a **regression line**, can readily be produced from two lists of data.

Consider the example used at the beginning of the chapter involving skiers and snow depth.

Step 1: Enter the two lists of data into two columns in the spreadsheet.

	A	B	C
1	**Depth of snow (m)**	**Number of skiers**	
2	0.5	120	
3	0.8	250	
4	2.1	500	
5	3.6	780	
6	1.4	300	
7	1.5	280	
8	1.8	410	
9	2.7	320	
10	3.2	640	
11	2.4	540	
12	2.6	530	
13	1.7	200	
14			
15			
16			

Step 2: Highlight the data and insert a scatterplot from the spreadsheet software's chart options.

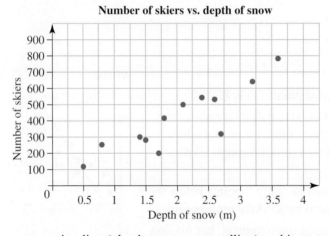

Step 3: Display the linear regression line (also known as a trendline) and its equation on the scatterplot.

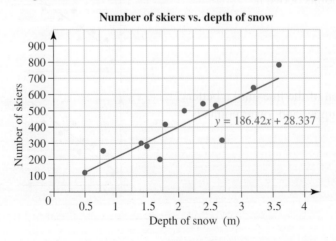

$y = 186.42x + 28.337$

Step 4: Replace the variables x and y in the regression equation with the variable names:

$$\text{Number of skiers} = 186.42 \times \text{Depth of snow} + 28.337$$

Manually

To determine the equation of a line of best fit by hand, we must determine the gradient and y-intercept of the drawn line. First, consider drawing the regression line as shown in Worked example 6.

WORKED EXAMPLE 6

The table below shows the marks of 10 students in a Physics and Chemistry quiz. Using the variable Physics as the independent variable, draw a scatterplot and on the graph show the regression line.

Physics	2	5	6	7	7	8	8	9	9	10
Chemistry	4	7	5	8	6	6	9	7	10	9

THINK

1. Plot the point corresponding to each pair of marks.

2. Add a regression line, making sure that there are an equal number of points above and below the line.

WRITE

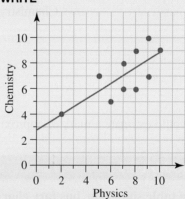

Once the regression line has been drawn, we can determine its equation.

The equation of a straight line has the form $y = mx + c$, where m is the gradient and c is the y-intercept. The gradient of a regression line is best found by calculating the rise and the run from two points on the line of regression. (Note that these two points are not necessarily given data points.) The gradient is then found using:

$$\text{gradient}\,(m) = \frac{\text{rise}}{\text{run}}$$

The y-intercept can then be seen by noting the point where the line crosses the y-axis. The y-intercept can only be determined from the graph if the scale on the horizontal axis starts at 0.

Consider Worked example 6. We can see that the points $(2, 4)$ and $(10, 9)$ lie on the line. The rise between these points is 5 and the run is 8, hence the gradient of the line is:

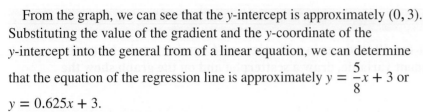

$$\text{gradient}\,(m) = \frac{\text{rise}}{\text{run}}$$

$$= \frac{5}{8}$$

From the graph, we can see that the y-intercept is approximately $(0, 3)$. Substituting the value of the gradient and the y-coordinate of the y-intercept into the general from of a linear equation, we can determine that the equation of the regression line is approximately $y = \frac{5}{8}x + 3$ or $y = 0.625x + 3$.

That is, $\text{Chemistry} = \frac{5}{8} \times \text{Physics} + 3$.

A teacher may then be able to estimate a student's Chemistry mark by using their result in Physics. For example, a student who achieved a mark of 5 in Physics may get:

$$\text{Chemistry} = \frac{5}{8} \times \text{Physics} + 3$$

$$= \frac{5}{8} \times 5 + 3$$

$$= 6\frac{1}{8} \text{ or } 6.125$$

The teacher would estimate a mark of approximately 6 for this student in Chemistry.

WORKED EXAMPLE 7

The following table shows the bus fare charged by a bus company for different distances.

Distance (km)	Fare ($)
1.5	2.10
0.5	2.00
7.5	4.50
6	4.00
6	4.50
2.5	2.60
0.5	2.10
8	4.50
4	3.50
3	3.00

a. **Represent the data using a scatterplot and add the regression line.**
b. **Determine the gradient and y-intercept of the regression line; hence, determine its equation, manually.**
c. **Use technology to determine the equation of the regression line.**

THINK

a. 1. The fare would generally depend on the distance travelled. Show the distance on the x-axis and the fare on the y-axis.

   2. Plot the points given by each pair.

   3. Add the regression line to the graph.

WRITE

a.

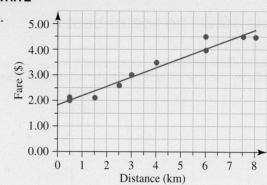

b. 1. Take two points on the regression line and determine the gradient between them by substituting in the formula $m = \dfrac{\text{rise}}{\text{run}}$ and simplifying. Use the points $(0.5, 2)$ and $(6, 4)$.

b.

rise $= 2$, run $= 5.5$

$$m = \frac{2}{5.5}$$

$$\approx 0.36$$

   2. Give an estimate for the y-intercept.

y-intercept ≈ 1.8

   3. Substitute the gradient and y-intercept into the formula $y = mx + c$ and state the equation.

$y = 0.36x + 1.8$

c. 1. Enter distances into a list and fares into a second list in a spreadsheet.

   2. Follow the procedures outlined previously to determine the equation of the regression line.

c. The equation of the regression line is
$y = 0.36x + 1.8$

   3. Write the equation in terms of the variables.

Fare ($\$$) $= 0.36 \times$ Distance $+ 1.8$

Note that there may be some slight discrepancies in the calculated values when manually determining the regression line compared to using technology.

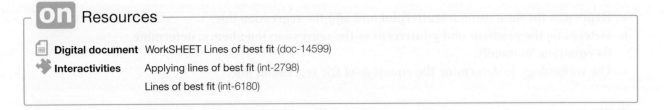
Exercise 10.3 The line of best fit

Technology may be used.

1. **WE5** Draw the line of best fit for the scatterplot shown using the equal-number-of-points method.

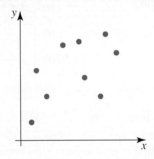

2. Draw the line of best fit for the scatterplot shown using the equal-number-of-points method.

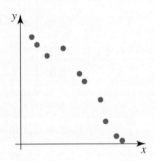

3. Fit a straight line to the data in the scatterplots using the equal-number-of-points method.

a.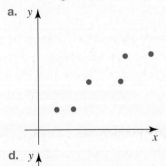

b.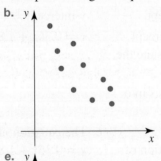

c.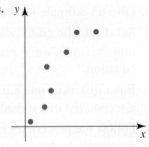

d.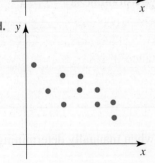

e.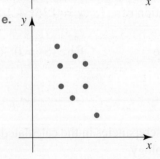

4. **WE6** The table below shows the marks achieved by a class of students in English and Maths.

English	64	75	81	63	32	56	47	59	73	64
Maths	76	62	89	56	49	57	53	72	80	50

Using English as the independent variable, draw a scatterplot and on the graph show the regression line.

5. Position the line of best fit through each of the following graphs and determine the equation of each.

a.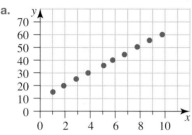

b.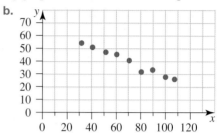

c.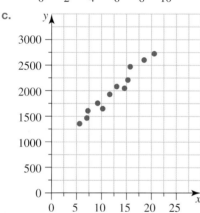

6. **MC** Which of the following equations represents the line of best fit that passes through the points $(3, 8)$ and $(12, 35)$?

A. $y = 3x + 1$ **B.** $y = -3x + 1$ **C.** $y = 3x - 1$ **D.** $y = \dfrac{1}{3}x - 1$

7. **WE7** In an experiment, a student measures the length of a spring (L) when different masses (M) are attached to it. Her results are shown below.

Mass (g)	Length of spring (mm)
0	220
100	225
200	231
300	235
400	242
500	246
600	250
700	254
800	259
900	264

a. Draw a scatterplot of the data and on it insert the regression line.
b. Determine the gradient and y-intercept of the regression line and, hence, state the equation of the regression line. Write your equation in terms of the variables L and M.

8. A scientist who measures the volume of a gas at different temperatures provides the following table of values.

Temperature (°C)	Volume (L)
−40	1.2
−30	1.9
−20	2.4
0	3.1
10	3.6
20	4.1
30	4.8
40	5.3
50	6.1
60	6.7

a. Draw a scatterplot of the data and on it insert the regression line.
b. Determine the equation of the regression line. Write your equation in terms of the variables: volume of gas, V, and its temperature, T.

9. A sports scientist is interested in the importance of muscle bulk to strength. He measures the biceps circumference of ten people and tests their strength by asking them to complete a lift test. His results are given in the following table.

Circumference of biceps (cm)	Lift test (kg)
25	50
25	52
27	58
28	51
30	60
30	62
31	53
33	62
34	61
36	66

a. Draw a scatterplot of the data and draw the regression line.
b. Determine a rule for predicting the ability of a person to complete a lift test, S, from the circumference of their biceps, B.

10. A sports scientist is looking at data comparing the heights of athletes and their performance in the high jump. The following table and scatterplot represent the data they have collected. A line of best fit by eye has been drawn on the scatterplot. Use technology to determine the equation of the line.

Height (cm)	168	173	155	182	170	193	177	185	163	190
High jump (cm)	172	180	163	193	184	208	188	199	174	186

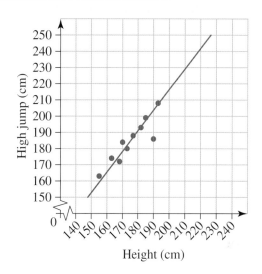

11. Nidya is analysing the data from question 10, but a clerical error means that she only has access to two points of data: (170, 184) and (177, 188).
    a. Determine Nidya's equation for the line of best fit, rounding all decimal numbers to two places.
    b. Add Nidya's line of best fit to the scatterplot of the data.
    c. Comment on the similarities and differences between the two lines of best fit.

12. The following table and scatterplot shows the age and height of a field of sunflowers planted at different times throughout summer.

Age of sunflower (days)	63	71	15	33	80	22	55	47	26	39
Height of sunflower (cm)	237	253	41	101	264	65	218	182	82	140

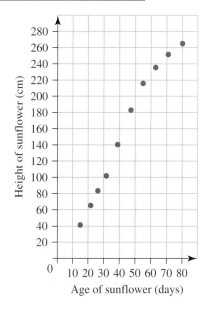

a. Xavier draws a line of best fit by eye that goes through the points $(10, 16)$ and $(70, 280)$. Draw his line of best fit on the scatterplot and comment on his choice of line.
b. Calculate the equation of the line of best fit using the two points that Xavier selected.
c. Patricia draws a line of best fit by eye that goes through the points $(10, 18)$ and $(70, 258)$. Draw her line of best fit on the scatterplot and comment on her choice of line.
d. Calculate the equation of the line of best fit using the two points that Patricia selected.
e. Why is the value of the y-intercept not 0 in either equation?

13. Steve is looking at data comparing the size of different music venues across the country and the average ticket price at these venues. After plotting his data in a scatterplot, he calculates a line of best fit for his data as $y = 0.04x + 15$, where y is the average ticket price in dollars and x is the capacity of the venue.
   a. What does the value of the gradient (m) represent in Steve's equation?
   b. What does the value of the y-intercept represent in Steve's equation?
   c. Is the y-intercept a realistic value for this data?

14. A government department is analysing the population density and crime rate of different suburbs to see if there is a connection. The following table and scatterplot display the data that has been collected so far.

Population density (persons per km²)	3525	2767	4931	3910	1572	2330	2894	4146	1968	5337
Crime rate (per 1000 people)	185	144	279	227	65	112	150	273	87	335

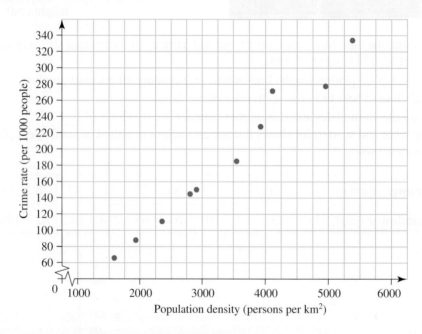

a. Draw a line of best fit on the scatterplot of the data.
b. Using technology or otherwise, determine the equation of the line of best fit.
c. What does the value of the x-intercept mean in terms of this problem?
d. Is the x-intercept value realistic? Explain your answer.

10.4 Making predictions using a line of best fit

If a linear association exists between a pair of variables, it is useful to be able to make predictions of one variable from the other. The line of best fit can be extended and then used to make predictions. When this is done, the line is called a regression line.

There are two methods available:
1. Algebraically
   We can find an equation for the regression line (or line of best fit), substitute the value of the known variable and calculate the value of the unknown variable. The equation of the regression line can be generated by hand or using technology from data entered in lists.
2. Graphically
   We can draw lines horizontally or vertically from the known value to the regression line, then read off the corresponding value of the unknown variable.

10.4.1 Algebraic predictions

Once the equation of the regression line has been determined, it is possible to use this equation to make predictions by substituting the value of the given variable into it.

WORKED EXAMPLE 8

A casino records the number of people, N, playing a jackpot game and the prize money, p, for that game, and plots the results on a scatterplot. The regression line is found to have the equation $N = 0.07p + 220$.

a. Determine the number of people playing when the prize money is $2500.

b. Calculate the likely prize on offer when there are 500 people playing.

THINK

a. 1. Write the equation of the regression line.

2. Substitute 2500 for p in the regression equation.

3. Calculate N.

4. Give a written answer.

b. 1. Write the equation of the regression line.

2. Substitute 500 for N in the regression equation.

3. Solve the equation for p by subtracting 220 from both sides, then dividing the result by 0.07.

4. Give a written answer.

WRITE

a. $N = 0.07p + 220$

$N = 0.07 \times 2500 + 220$

$= 395$

When the prize money is $2500, it is estimated that 395 people will be playing.

b. $N = 0.07p + 220$

$500 = 0.07p + 220$

$280 = 0.07p$

$p = \dfrac{280}{0.07}$

$p = 4000$

When 500 people are playing, the prize is estimated to be $4000.

10.4.2 Graphical predictions

Once the regression line has been drawn, horizontal or vertical lines can be added from the axes to the regression line to read off the value of the unknown variable. Remember that the values determined are only estimates, and not exact values.

Consider the graph drawn in Worked example 7 representing bus fares for travelling various distances. Use the graph to estimate:
a. the fare for a journey of 5 km
b. how far you could expect to travel for a cost of $3.25.

THINK	WRITE
a. 1. Locate the value of 5 km on the *x*-axis.	a. 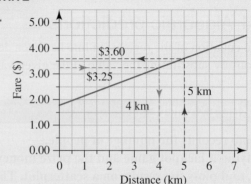
2. Draw a vertical line from this point to meet the regression line. See pink line.	
3. Draw a horizontal line from this point to the *y*-axis. See pink line.	
4. Read this value on the *y*-axis.	A distance of 5 km of bus travel is predicted to cost about $3.60.
b. 1. Locate $3.25 on the *y*-axis.	b.
2. Draw a horizontal line from this point to the regression line. See green line.	
3. From this point, draw a vertical line to the *x*-axis. See green line.	
4. Read the *x*-value at this point.	For $3.25 you could expect to travel about 4 km.

10.4.3 Reliability of predictions

Interpolation

When we use **interpolation**, we are making a prediction from a line of best fit that appears within the parameters of the original data set.

Extrapolation

When we use **extrapolation**, we are making a prediction from a line of best fit that appears outside the parameters of the original data set.

If we plot our line of best fit on the scatterplot of the given data, then extrapolation will occur before the first point or after the last point of the scatterplot.

The more pieces of data there are in a set, the better the line of best fit you will be able to draw. More data points allow more reliable predictions.

Using extrapolation to make predictions is usually unreliable due to the lack of data in that region.

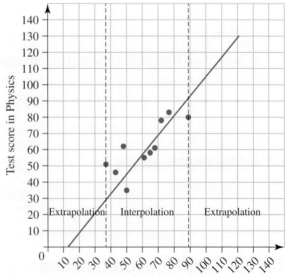

In general, interpolation is a far more reliable method of making predictions than extrapolation. However, there are other factors that should also be considered. Interpolation closer to the centre of the data set will be more reliable than interpolation closer to the edge of the data set. Extrapolation that appears closer to the data set will be much more reliable than extrapolation that appears further away from the data set.

A strong correlation between the points of data will give a more reliable line of best fit to be used. This is shown when all of the points appear close to the line of best fit. The more points there are that appear further away from the line of best fit, the less reliable other predictions will be.

When making predictions, always be careful to think about the data that you are making predictions about. Be sure to think about whether the prediction you are making is realistic or even possible!

WORKED EXAMPLE 10

The following data represent the air temperature (°C) and depth of snow (cm) at a popular ski resort.

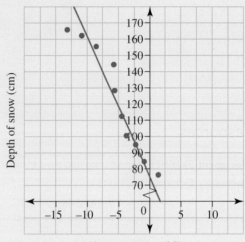

Air temperature (°C)	−4.5	−2.3	−8.9	−11.0	−13.3	−6.2	−0.4	1.5	−3.7	−5.4
Depth of snow (cm)	111.3	95.8	155.6	162.3	166.0	144.7	84.0	77.2	100.5	129.3

The equation of the line of best fit for this data set has been calculated as $y = -7.2x + 84$.
a. Use the line of best fit to estimate the depth of snow if the air temperature is −6.5 °C.
b. Use the line of best fit to estimate the depth of snow if the air temperature is 25.2 °C.
c. Comment on the reliability of your estimations in parts a and b.

THINK

a. 1. Rewrite the regression equation in terms of the variables.

2. Substitute −6.5 for the air temperature into the equation for the line of best fit.

3. Evaluate the expression for the depth of snow.

4. Write the answer.

WRITE

a. $$y = -7.2x + 84$$

Depth of snow $= -7.2 \times$ Air temperature $+ 84$
Air temperature $= -6.5$

Depth of snow $= -7.2 \times$ Air temperature $+ 84$
$= -7.2 \times -6.5 + 84$
$= 130.8$

When the air temperature is −6.5 °C the depth of snow will be approximately 130.8 cm.

b. 1. Substitute 25.2 for the air temperature into the equation for the line of best fit.	**b.** Air temperature $= 25.2$
2. Evaluate the expression for the depth of snow.	$\text{Depth of snow} = -7.2 \times \text{Air temperature} + 84$ $= -7.2 \times 25.2 + 84$ $= -97.4 \text{ (one decimal place)}$
3. Write the answer.	When the air temperature is 25.2 °C the depth of snow will be approximately −97.4 cm.
c. Relate the answers back to the original data to check their reliability.	**c.** The estimate in part **a** was made using interpolation, with the point being comfortably located within the parameters of the original data. The estimate appears to be consistent with the given data and as such is reliable. The estimate in part **b** was made using extrapolation, with the point being located well outside the parameters of the original data. This estimate is clearly unreliable, as we cannot have a negative depth of snow.

 Resources

 Digital document SkillSHEET Substitution into a linear rule (doc-5405)

Interactivities Extrapolation (int-1154)

Interpolation and extrapolation (int-6181)

Exercise 10.4 Making predictions using the line of best fit

1. **WE8** A taxi company adjusts its meters so that the fare is charged according to the following equation: $F = 1.2d + 3$ where F is the fare, in dollars, and d is the distance travelled, in km.

   a. Determine the fare charged for a distance of 12 km.
   b. Determine the fare charged for a distance of 4.5 km.
   c. Determine the distance that could be covered on a fare of $27.
   d. Determine the distance that could be covered on a fare of $13.20.

2. Detectives can use the equation $H = 6.1f - 5$ to estimate the height of a burglar who leaves footprints behind. (H is the height of the burglar, in cm, and f is the length of the footprint.)
   a. Determine the height of a burglar whose footprint is 27 cm in length.
   b. Determine the height of a burglar whose footprint is 30 cm in length.
   c. Determine the footprint length of a burglar of height 185 cm. (Give your answer correct to two decimal places.)
   d. Determine the footprint length of a burglar of height 152 cm. (Give your answer correct to two decimal places.)

3. A football match pie seller finds that the number of pies sold is related to the temperature of the day. The situation could be modelled by the equation $N = 870 - 23t$, where N is the number of pies sold and t is the temperature in degrees Celsius of the day.

   a. Determine the number of pies sold if the temperature was 5 degrees.
   b. Determine the number of pies sold if the temperature was 25 degrees.
   c. Determine the likely temperature if 400 pies were sold.
   d. Determine how hot the day would have to be before the pie seller sold no pies at all.

4. **WE9** The following graph shows the average annual costs of running a car. It includes all fixed costs (registration, insurance etc.) as well as running costs (petrol, repairs etc.).

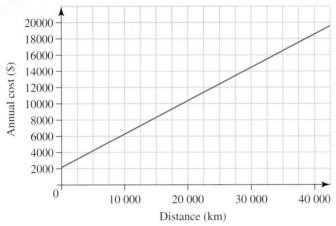

   Use the graph to estimate:
   a. the annual cost of running a car if it is driven 15 000 km
   b. the annual cost of running a car if it is driven 1000 km
   c. the likely number of kilometres driven if the annual costs were $8000
   d. the likely number of kilometres driven if the annual costs were $16 000.

5. A market researcher finds that the number of people who would purchase 'Wise-up' (the thinking man's deodorant) is related to its price. He provides the following table of values.

Price ($)	Weekly sales (× 1000)
1.40	105
1.60	101
1.80	97
2.00	93
2.20	89
2.40	85
2.60	81
2.80	77
3.00	73
3.20	69
3.40	65

   a. Draw a scatterplot of the data.
   b. Draw in the line of best fit.

c. Determine an equation that represents the association between the number of cans of 'Wise-up' sold, N (in thousands), and its price, p.

d. Use the equation to predict the number of cans sold each week if:
   i. the price was $3.10
   ii. the price was $4.60.

e. At what price should 'Wise-up' be sold if the manufacturers wish to sell 80 000 cans?

f. Given that the manufacturers of 'Wise-up' can produce only 100 000 cans each week, at what price should it be sold to maximise production?

6. The following table gives the adult return air fares between some Australian cities.

City	Distance (km)	Price ($)
Melbourne–Sydney	713	580
Perth–Melbourne	2728	1490
Adelaide–Sydney	1172	790
Brisbane–Melbourne	1370	890
Hobart–Melbourne	559	520
Hobart–Adelaide	1144	820
Adelaide–Melbourne	669	570

a. Draw a scatterplot of the data and on it draw the regression line.

b. Determine an equation that represents the association between the air fare, A, and the distance travelled, d.

c. Use the equation to predict the likely air fare (to the nearest dollar) from:
   i. Sydney to the Gold Coast (671 km)
   ii. Perth to Adelaide (2125 km)
   iii. Hobart to Sydney (1024 km)
   iv. Perth to Sydney (3295 km).

7. **WE10** The owner of an ice-cream parlour has collected data relating the outside temperature to ice-cream sales.

Outside temperature (°C)	23.4	27.5	26.0	31.1	33.8	22.0	19.7	24.6	25.5	29.3
Ice-cream sales	135	170	165	212	204	124	86	144	151	188

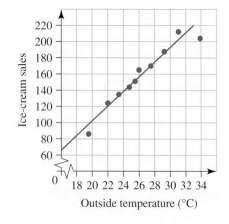

A line of best fit for this data has been calculated as ice-cream sales $= 9 \times$ outside temperature $- 77$.

    a. Use the line of best fit to estimate ice-cream sales if the outside temperature is 27.9 °C.

    b. Use the line of best fit to estimate ice-cream sales if the air temperature is 15.2 °C.

    c. Comment on the reliability of your answers to parts **a** and **b**.

8. Georgio is comparing the cost and distance of various long-distance flights, and after drawing a scatterplot he creates an equation for a line of best fit to represent his data. Georgio's line of best fit is $y = 0.08x + 55$, where y is the cost of the flight and x is the distance of the flight in kilometres.

    a. Estimate the cost of a flight between Melbourne and Sydney (713 km) using Georgio's equation.

    b. Estimate the cost of a flight between Melbourne and Broome (3121 km) using Georgio's equation.

    c. All of Georgio's data came from flights of distances between 400 km and 2000 km. Comment on the suitability of using Georgio's equation for shorter and longer flights than those he analysed. What other factors might affect the cost of these flights?

9. Mariana is a scientist and is collecting data measuring lung capacity (in L) and time taken to swim 25 metres (in seconds). Unfortunately a spillage in her lab causes all of her data to be erased apart from the records of a person with a lung capacity of 3.5 L completing the 25 meters in 55.8 seconds and a person with a lung capacity of 4.8 L completing the 25 meters in 33.3 seconds.

    a. Use the remaining data to construct an equation for the line of best fit relating lung capacity (x) to the time taken to swim 25 metres (y). Give any numerical values correct to two decimal places.

    b. What does the value of the gradient (m) represent in the equation?

    c. Use the equation to estimate the time it takes people with the following lung capacities to swim 25 metres.

      i. 3.2 litres   **ii.** 4.4 litres   **iii.** 5.3 litres

    d. Comment on the reliability of creating the equation from Mariana's two remaining data points.

10. The regression line shown on the scatterplot has the equation $c = 13.33 + 2.097m$, where c is the number of new customers each hour and m is the number of market stalls.

    a. Using the line of best fit, interpolate the data to find the number of new customers expected if there are 30 market stalls.

    b. Use the formula to extrapolate the number of market stalls required in order to expect 150 new customers.

    c. Explain why part **a** is an example of interpolating data, while part **b** demonstrates extrapolation.

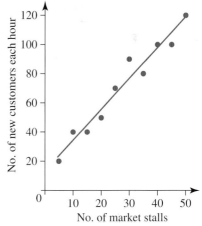

11. Use the data given below to complete the following questions.

x	10	11	12	13	14	15	16	17	18	19
y	22	18	20	15	17	11	11	7	9	8

   a. Draw a scatterplot and a line of best fit by eye.
   b. Determine the equation of the line of best fit. Give values correct to three decimal places.
   c. Extrapolate the data to predict the value of y when $x = 23$.
   d. What assumptions are made when extrapolating data?

12. While camping a mathematician estimated that:
   number of mosquitoes around fire $= 10.2 + 0.5 \times$ temperature of the fire (°C).
   a. Determine the number of mosquitoes that would be expected if the temperature of the fire was 240 °C. Give your answer correct to the nearest whole number.
   b. What would be the temperature of the fire if there were only 12 mosquitoes in the area?
   c. Identify some factors that could affect the reliability of this equation.

13. Data on people's average monthly income and the amount of money they spend at restaurants was collected.

Average monthly income ($000s)	Money spent at restaurants per month ($)
2.8	150
2.5	130
3.0	220
3.1	245
2.2	100
4.0	400
3.7	380
3.8	200
4.1	600
3.5	360
2.9	175
3.6	350
2.7	185
4.2	620
3.6	395

a. Draw a scatterplot of this data.
b. Find the equation of the line of best fit in terms of average monthly income in thousands of dollars (I) and money spent at restaurants in dollars (R). Give values correct to one decimal place.
c. Extrapolate the data to predict how much a person who earns $5000 a month might spend at restaurants each month.
d. Explain why part c is an example of extrapolation.
e. A person spent $265 eating out last month. Estimate their monthly income, giving your answer to the nearest $10. Is this an example of interpolation or extrapolation?

10.5 Causality and correlation

10.5.1 Correlation

Correlation is a measure of the strength of the linear association between two variables.

Pearson's correlation coefficient, r, (also known as Pearson's product–moment correlation coefficient) is a numerical measure of strength. It can be calculated if the scatterplot indicates that a linear association is present between the variables and there are no outliers.

Pearson's correlation coefficient, r, can have values between -1 and $+1$. The figure at right is a guide when using r to describe the strength of a linear association between two variables.

Pearson's correlation coefficient can be calculated using technology.

Graphics calculators can calculate r when computing a linear regression. In Excel, the formula $=$PEARSON(array1,array2) can be used to calculate r.

For example, the formula $=$PEARSON(A1:A15,B1:B5) will calculate the r value of the data set where the x values are stored in cells A1 to A5 and the corresponding y values are stored in cells B1 to B5.

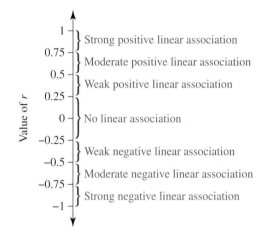

WORKED EXAMPLE 11

Data was collected on the time it takes students to get to school and their ATAR score.

Time (mins)	12	35	19	42	33	31	25	46	45	40	14	44	39	31	22
ATAR score	53	75	97	59	87	70	71	66	37	48	94	68	33	59	42

Using technology:
a. **draw a scatterplot of this data**
b. **determine the equation of the line of best fit, writing the coefficients to two decimal places**
c. **determine Pearson's correlation coefficient, r, correct to two decimal places**
d. **interpret this value.**

▶

THINK

a. 1. In a spreadsheet, enter the time data into cells A2 to A16, and the ATAR data into cells B2 to B16. You could also input the data into lists on a graphics calculator if one is available.

WRITE

a.

	A	B
1	**Time (min)**	**ATAR score**
2	12	53
3	35	75
4	19	97
5	42	59
6	33	87
7	31	70
8	25	71
9	46	66
10	45	37
11	40	48
12	14	94
13	44	68
14	39	33
15	31	59
16	22	42
17		

2. Highlight the two lists of data, then insert a scatterplot from the software's chart options.

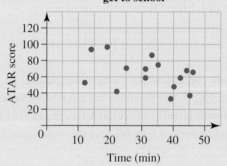

ATAR score vs. time to get to school

b. 1. Display the line of best fit (trendline) with its equation on the scatterplot.

b.

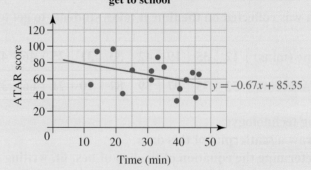

ATAR score vs. time to get to school

$y = -0.67x + 85.35$

2. Rewrite the equation of the line of best fit in terms of the variables ATAR score and time travelled.

ATAR score $= -0.67 \times$ time $+ 85.35$

c. 1. To determine the value of Pearson's correlation coefficient, r, enter the formula
=PEARSON(A2:A16,B2:B16)
into a blank cell and then press ENTER.

c.

	A	B	C
1	**Time (min)**	**ATAR score**	
2	12	53	
3	35	75	
4	19	97	
5	42	59	
6	33	87	
7	31	70	
8	25	71	
9	46	66	
10	45	37	
11	40	48	
12	14	94	
13	44	68	
14	39	33	
15	31	59	
16	22	42	
17			
18	Pearson's r	=PEARSON(A2:A16,B2:B16)	
19			

2. Write the value of r.

$r = -0.39$

d. Interpret the r value.

d. An r value of -0.39 implies that there is a weak negative linear association between the time taken to get to school and ATAR score. Students who live further from school tend to have lower ATAR scores.

10.5.2 Causation

A correlation gives only information about the strength of the association between the two variables. It tells you nothing about the cause of the association. **Causation** is a measure of how much change in one variable is caused by the change in the other variable. If two variables display an association, that is they are correlated, it does not necessarily mean that a change in one variable will *cause* a change in the other.

Consider the following scenario. The heights (in centimetres) of 21 football players were recorded against the number of marks they took in a game of football. The correlation coefficient, r, was equal to 0.86. While we are entitled to say that there is a strong association between the height of a footballer and the number of marks they take, we cannot assert that the height of a footballer causes them to take a lot of marks. Being tall might assist in taking marks, but there will be many other factors which come into play; for example, skill level, accuracy of passes from teammates, abilities of the opposing team, and so on. So, while establishing a high degree of correlation between two variables may be interesting and can often flag the need for further, more detailed investigation, it in no way gives us any basis to comment on whether or not one variable *causes* particular values in another variable.

Data showing the number of tourists visiting a small country in a month and the corresponding average monthly exchange rate for the country's currency against the American dollar are as given.

Exchange rate	1.2	1.1	0.9	0.9	0.8	0.8	0.7	0.6
Number of tourists (×1000)	2	3	4	5	7	8	8	10

a. Construct a scatterplot for the data
b. Comment on the correlation between the number of tourists and the exchange rate
c. Calculate r.
d. Based on the pattern shown in the scatterplot and the value of r, would it be appropriate to conclude that the decrease in tourist numbers is caused by the increasing exchange rate?

THINK

a. 1. Determine the independent and dependent variables

   2. Sketch a scatterplot

b. A negative association can be observed.

c. Use technology to find r.
d. Correlation does not imply causation.

WRITE

a. Exchange rate (independent)
Tourist numbers (dependent)

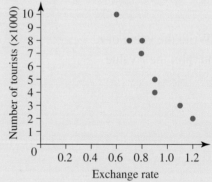

b. From the scatterplot there appears to be a strong negative correlation between the number of tourists visiting the country and the exchange rate.

c. $r = -0.96$
d. Just because $r = -0.96$, it cannot be stated that the decreasing tourist numbers is caused by the increasing exchange rate. There may be other factors that need to be considered. Further investigation is needed before any conclusions can be drawn.

 Resources

 Interactivity Pearson's product–moment correlation coefficient and the coefficient of determination (int-6251)

Exercise 10.5 Causality and correlation

1. Explain the meanings of 'correlation' and 'causation' in your own words.
2. **MC** If Pearson's correlation coefficient, r, is found to be -0.7564, the strength and direction of the association could be described as
   A. weak negative.
   B. strong negative.
   C. moderate negative.
   D. no association.
3. **WE11** Data on people's average monthly income and the amount of money they spend at restaurants was collected.

Average monthly income ($000s)	2.8	2.5	3.0	3.1	2.2	4.0	3.7	3.8	4.1	3.5	2.9	3.6	2.7	4.2	3.6
Money spent at restaurants per month ($)	150	130	220	245	100	400	380	200	600	360	175	350	185	620	395

a. Draw a scatterplot of this data.
b. Determine the equation of the line of best fit.
c. Determine Pearson's correlation coefficient, r.
d. Interpret this value.

4. Data on a student's marks in Geography and their marks in Music were collected.

Geography	65	80	72	61	99	54	39	66	78	89	84	73	68	57	60
Music	91	57	77	89	51	76	62	87	88	64	90	45	60	79	96

a. Draw a scatterplot of this data, using Geography as the independent variable.
b. Determine the equation of the line of best fit.
c. Determine Pearson's correlation coefficient, r.
d. Interpret this value.

5. Data on the daily sales of gumboots and the maximum temperature was collected.

Temperature (°C)	17	16	12	10	14	17	18	22	23	19	17	15	12	15	20
Daily sales (no. of pairs)	2	3	8	16	7	3	2	1	1	2	3	3	12	9	1

a. Draw a scatterplot of this data.
b. Determine the equation of the line of best fit.
c. Determine Pearson's correlation coefficient, r.
d. Interpret this values.

6. The following table outlines the cost of an annual magazine subscription along with the number of magazine issues per year.

No. of magazine issues per year	7	9	10	6	8	4	4	5	11	9	10	5	11	3	7	12	7	6	12
Subscription cost ($)	34	40	52	38	50	25	28	40	55	55	45	28	65	24	38	55	50	33	59

Use technology to determine Pearson's correlation coefficient for this data correct to four decimal places. What does this tell you about the strength of the linear association between the variables?

7. Use your understanding of Pearson's correlation coefficient to explain what the following results indicate.
   a. $r = 0.68$ b. $r = -0.97$ c. $r = -0.1$ d. $r = 0.30$
8. A survey asked random people for their house number and the combined age of the household members. The following data was collected:

House number	Total age of household
14	157
65	23
73	77
58	165
130	135
95	110
54	94
122	25
36	68
101	53
57	64
34	120
120	180
159	32
148	48
22	84
9	69

a. Using the house number as the independent variable, display this data in a scatterplot.
b. Comment on the resulting scatterplot.
c. Determine Pearson's correlation coefficient. Give your answer correct to four decimal places.
d. What conclusions can you draw from this value?

9. **WE12** Data showing the number of people on 8 fundraising committees and the annual funds raised are given in the table.

Number of people oncommittee	3	6	4	8	5	7	3	6
Annual funds raised ($)	4500	8500	6100	12 500	7200	10 000	4700	8800

   a. Construct a scatterplot for the data
   b. Comment on the correlation between the number of people on the committee and the annual funds raised.
   c. Calculate r.
   d. Based on the value of r obtained in part c, would it be appropriate to conclude that the increase in funds raised is caused by an increase in the number of people on the committee?

10. There is a strong positive correlation between rates of diabetes and the number of hip replacements. Does diabetes cause deterioration of the hip joint? What could be the common cause(s) that links these two variables?

10.6 Review: exam practice

10.6.1 Summary: Scatterplots and lines of best fit

Scatterplots and basic correlation

- The independent variable is the variable that explains the change in the dependent variable.
- When constructing a scatterplot, it is important to place the independent variable along the x-axis and the dependent variable along the y-axis.
- Here is a gallery of scatterplots showing the various patterns we look for when analysing the association between two variables.

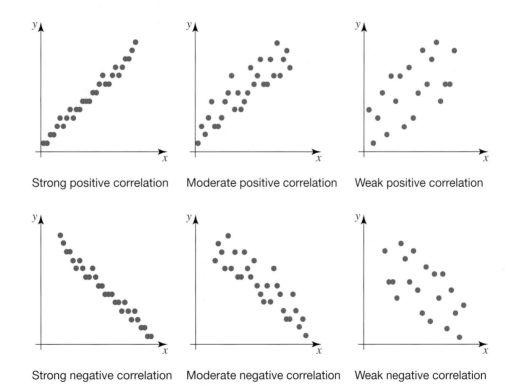

Strong positive correlation   Moderate positive correlation   Weak positive correlation

Strong negative correlation   Moderate negative correlation   Weak negative correlation

- Pearson's correlation coefficient, r, measures the strength of the linear association between two variables.

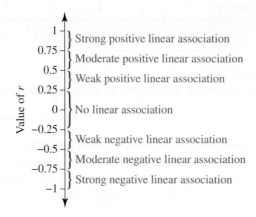

Making predictions

- A line of best fit can be drawn through the data points by hand or using technology.
- The line of best fit can be used to make predictions. Values can be read directly from the line of best fit on the graph, or the equation of the line of best fit can be determined and substitution used to make predictions.
- When the independent variable is equal to 0, the predicted value of the dependent variable is indicated by the y-intercept.
- For each increment of 1 unit of change in the independent variable, the average change in the dependent variable is indicated by the value of the gradient
- The reliability of making predictions relies heavily on the initial data. A larger sample and a strong correlation will make for a reliable prediction. In general, interpolation is more reliable than extrapolation.

Exercise 10.6 Review: exam practice

Simple familiar

1. **MC** Which of the following scatterplots best demonstrates a line of best fit?

A.

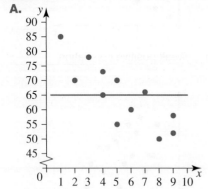

B.

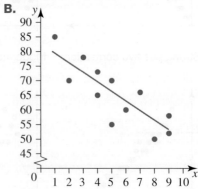

C.

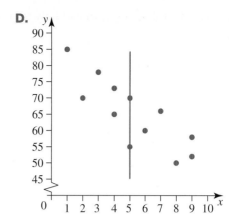

D.

2. **MC** The regression line equation for the graph is closest to which of the following?

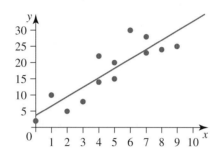

A. $y = 3.8 + 2.9x$

B. $y = -3.8 - 2.9x$

C. $y = -3.8 + 2.9x$

D. $y = 3.8 - 2.9x$

3. **MC** The type of correlation shown in the graph for question **2** would best be described as

A. weak, positive correlation.

B. moderate, positive correlation.

C. strong, positive correlation.

D. no correlation.

4. **MC** What type of correlation does an *r*-value of 0.64 indicate?

A. Strong, positive correlation

B. Strong, negative correlation

C. Moderate, positive correlation

D. Moderate, negative correlation

5. **MC** A gardener tracks a correlation coefficient of 0.79 between the growth rate of his trees and the amount of fertiliser used. What can the gardener conclude from this result?

A. An increase in tree growth increases the use of fertiliser.

B. An increase in the use of fertiliser increases the health of the trees.

C. The growth rate of the trees is influenced by the amount of fertiliser used.

D. The growth rate of the trees influences the quality of the fertiliser used.

6. **MC** When $y = 0.54 + 15.87x$, the value of y when $x = 2.5$ is which of the following?

A. 18.91

B. 40.215

C. 39.135

D. 6.888

7. **MC** Which of the following is most likely to be the graph for the regression line equation $y = 85 - 4x$?

A.
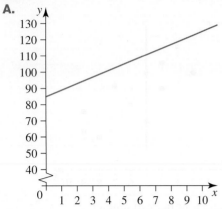

B.
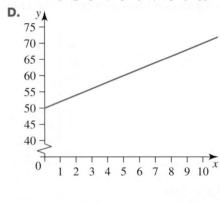

C.
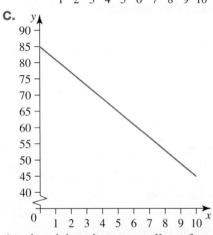

D.

8. An electrician charges a call out fee of $150 and an hourly rate of $80. The electrician uses the following equation to calculate his bill: $C = 80h + 150$ where C is the cost, in dollars, and h is the number of hours worked.
   a. Determine the cost if the electrician worked for 3 hours.
   b. Determine the cost if the electrician worked for 7 hours and 30 minutes.
   c. Determine the hours worked if the cost was $550.
   d. Determine the hours worked if the cost was $670.

9. **MC** For the following sample data set, which of the following is an example of interpolating data?

x	1	5	15	25
y	10	16	18	22

   A. Finding the value of x when $y = -7$ B. Finding the value of y when $x = 17$
   C. Finding the value of x when $y = 27$ D. Finding the value of y when $x = 37$

10. **MC** For the data set from question 9, what is the regression line equation?
   A. $y = 10 + x$ B. $y = 0.435 + 11.456x$
   C. $y = 0.876 + 0.936x$ D. $y = 11.496 + 0.435x$

11. For each of the following graphs, describe the strength of correlation between the independent and dependent variables.

a.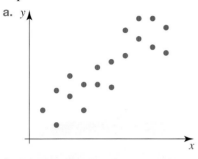

b.

c.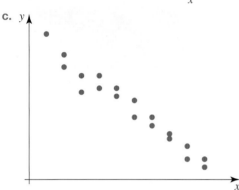

12. For each of the graphs in question 11, draw a line of best fit where possible.

Complex familiar

13. Identify the independent and dependent variable for each of the following scenarios.

a. In a junior Science class, students plot the time taken to boil various quantities of water.

b. Extra buses are ordered to transport a number of students to the school athletics carnival.

14. Use the following data to complete this question.

x	10	9	8	7	6	5	4	3	2	1
y	6	10	4	11	13	18	15	19	21	26

a. Plot the data on a scatterplot.
b. Comment on the direction and strength of the data.
c. Determine Pearson's correlation coefficient.
d. Use you answer from part c to further discuss the association between the variables.

15. A driver tracks how many kilometres (km) they travelled against the amount of fuel they used in litres (L) and the data is shown in the table.

Distance travelled (km)	38	66	95	167	181	207	275	459
Fuel used (L)	3.2	5.6	7.2	12.8	14.1	16.9	21.4	37.5

a. State the dependent and independent variables.
b. Sketch a scatterplot of distance travelled against fuel used.
c. Draw a line of best fit.
d. State the equation of the line of best fit in terms of the variables. Give values correct to three decimal places.

e. Using your equation in part **d**, estimate the fuel used when travelling 50 km.

f. Explain the reliability of your answer in part **e**.

16. Data on 15 people's shoe size and the length of their hair was collected.

Shoe size	Length of hair (cm)
6	9
8	14
7	12
8	1
9	7
6	8
7	5
12	22
8	15
9	8
10	18
12	4
7	5
9	9
11	3

a. Draw a scatterplot of this data.

b. Determine the equation of the line of best fit.

c. Determine Pearson's correlation coefficient.

d. What conclusions could you draw from this data?

Complex unfamiliar

17. During an interview investigating the link between the sales of healthy snack foods (functional foods) and the increasing consumer demand for these products, an advertising expert made the following comment:

There is a correlation but it's not causation ... our increasing need for healthy food and our laziness has resulted in mass innovation of functional foods.

Explain why he might have stated there is no causative link between the sales of healthy foods and laziness.

18. An independent agency test-drove a random sample of current model vehicles and measured their fuel-tank capacity against the average fuel consumption. Along with the following scatterplot, a regression equation of $y = 0.1119x + 0.6968$ was established.

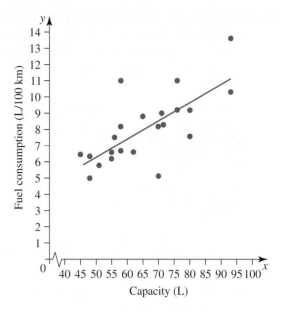

a. Identify the independent variable in this situation.

b. Rewrite the equation in terms of the independent and dependent variables.

c. It is often said that smaller vehicles are more economical. Determine correct to two decimal places the fuel consumption of a vehicle that had a 40-litre fuel tank.

d. Is your answer to part c an example of interpolation or extrapolation? Explain your response.

e. Calculate, correct to the nearest whole number, the tank size of a vehicle that had a fuel consumption rate of 10.2 L per 100 km.

f. Pearson's correlation coefficient for this data is 0.516. How can you use this value to evaluate the reliability of your data?

g. List at least two other factors that could influence the data.

19. The weight of top-brand runners was tracked against the recommended retail price, and the results were recorded in the following scatterplot.

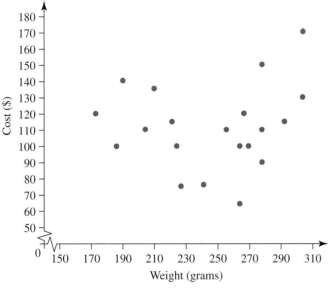

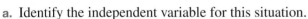

a. Identify the independent variable for this situation.

b. How would you describe the association between these two variables?

c. Identify two external factors that could explain the distribution of the data points.

20. The Bureau of Meteorology records data such as maximum temperatures and solar exposure on a daily and monthly basis. The following data table, for the Botanical Gardens in Melbourne, shows the monthly average amount of solar energy that fall on a horizontal surface and the monthly average maximum temperature. (*Note:* The data values have been rounded to the nearest whole number.)

Month	Jan.	Feb.	Mar.	Apr.	May	Jun.	Jul.	Aug.	Sep.	Oct.	Nov.	Dec.
Average solar exposure (MJ)	25	21	17	11	8	6	7	10	13	18	21	24
Average max daily temp. (°C)	43	41	34	33	24	19	24	24	28	32	25	40

a. Identify the independent and dependent variables for this situation.
b. Using technology, display the data in the form of a scatterplot.
c. Describe the trend of the data.
d. Calculate Pearson's correlation coefficient for this data. What does this value tell you about the reliability of the data?
e. Draw the regression line for this data and write its equation in terms of the variables.
f. Using your equation, calculate the amount of solar exposure for a monthly maximum temperature of 37 °C.
g. Extrapolate the data to find the average maximum temperature expected for a month that recorded an average solar exposure of 3 MJ.
h. Explain why part g is an example of extrapolation.

Answers

10 Scatterplots and lines of best fit

Exercise 10.2 Interpreting scatterplots

1.

	Independent	Dependent
a.	Number of hours	Test results
b.	Rainfall	Attendance
c.	Hours in gym	Visits to the doctor
d.	Length of essay	Memory taken
e.	Cost of care	Attendance
f.	Age of property	Cost of property
g.	Number of applicants	Cut-off OP score
h.	Running speed	Heart rate

2.
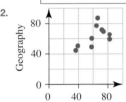

3.
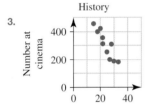

4.
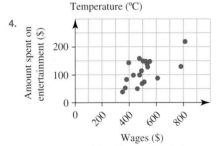

5. Moderate positive linear association
6. Moderate positive linear association
7. a. Yes — positive association
   b. Yes — positive association
   c. Yes — positive association
   d. Yes — negative association
   e. Yes — positive association
   f. Yes — negative association
   g. No — no association
8. a. Weak, negative association of linear form
   b. Moderate, negative association of linear form
   c. Moderate, positive association of linear form
   d. Strong, positive association of linear form
   e. No association
   f. Non-linear association

9. B
10.
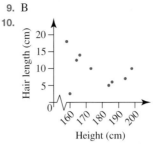

No relationship

11.
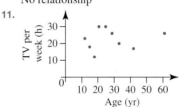

No relationship

12.
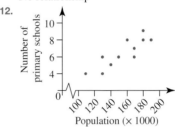

Moderate positive association of linear form.

13.
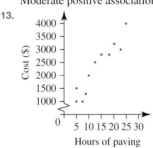

Strong positive association of linear form.

Exercise 10.3 The line of best fit

1.
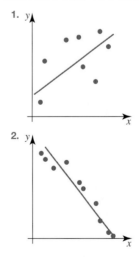

2.

3. a.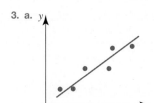

b.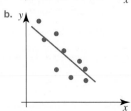

c.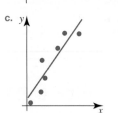

d.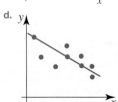

e.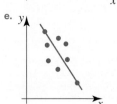

4.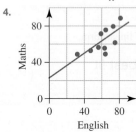

5. a.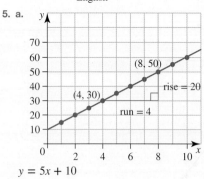

$y = 5x + 10$

b.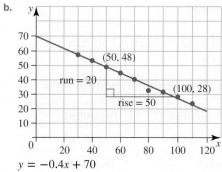

$y = -0.4x + 70$

c.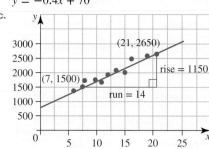

$y = 82x + 750$

6. C

7. a.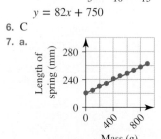

b. Manually, $L = 0.05M + 220$

8. a.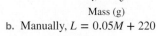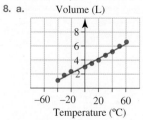

b. $V = 0.05T + 3.2$

9. a.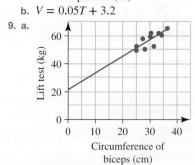

b. $S = 1.3B + 21$

10. $y = 1.25x - 33.75$

11. a. $y = 0.57x + 86.86$

---

b.

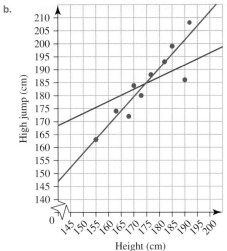

Height (cm)

c. Nidya's line of best fit is not a good representation of the data. In this instance having only two points of data to create the line of best fit was not sufficient.

12. a.

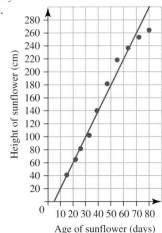

Age of sunflower (days)

Xavier's line is closer to the values above the line than those below it, and there are more values below the line than above it, so this is not a great line of best fit.

b. $y = 4.4x - 28$

c.

Age of sunflower (days)

Patricia's line is more appropriate as the data points lie on either side of the line and the total distance of the points from the line appears to be minimal.

d. $y = 4x - 22$

e. The line of best fit does not approximate the height for values that appear outside the parameters of the data set and the y-intercept lies well outside these parameters.

13. a. The increase in price of 4 cents for every additional person the venue holds.

 b. The price of a ticket if a venue has no capacity.

 c. No, as the smallest venues would still have some capacity.

14. a. Lines of best fit will vary, but should split the data points on either side of the line with the total distance to the line from the points minimised.

 b. Crime rate $= 0.07 \times$ population density $- 52.13$

 c. The amount of crime in a suburb with 0 people.

 d. No, if there are 0 people in a suburb there should be no crime.

Exercise 10.4 Making predictions using the line of best fit

1. a. $17.40 **b.** $8.40 **c.** 20 km **d.** 8.5 km

2. a. 159.7 cm **b.** 178 cm **c.** 31.15 cm **d.** 25.74 cm

3. a. 755 **b.** 295 **c.** 20 °C **d.** 38 °C

4. a. $8300

 b. $2500

 c. 14 000 km

 d. 35 500 km

5. a–b.

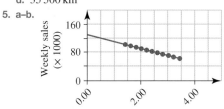

Price ($)

 c. $N = -20p + 135$

 d. i. 73 000

  ii. 43 000

 e. $2.75

 f. $1.75

6. a.

Distance (km)

 b. $A = 0.45d + 280$

 c. i. $582 **ii.** $1236

  iii. $741 **iv.** $1763

7. a. 174

 b. 60

 c. The estimate in part **a** is reliable as it was made using interpolation, it is located within the parameters of the original data set and appears consistent with the given data.

 The estimate in part **b** is unreliable as it was made using extrapolation and is located well outside the parameters of the original data set.

8. a. $112

 b. $305

c. All estimates outside the parameters of Georgio's original data set (400 km to 2000 km) will be unreliable, with estimates further away from the data set more unreliable than those closer to the data set.
Other factors that might affect the cost of flights include air taxes, fluctuating exchange rates and the choice of airlines for various flight paths.

9. a. $y = -17.31x + 116.38$
   b. For each increase in 1 L of lung capacity, the swimmers will take less time to swim 25 metres.
   c. i. 61 seconds
      ii. 40.2 seconds
      iii. 24.6 seconds
   d. As Mariana has only two data points and we have no idea of how typical these are of the data set, the equation for the line of best fit and the estimates established from it are all very unreliable.

10. a. 76
    b. 65
    c. Part **a** looks at data within the original data set range, while part **b** asks to predict data outside of the original data set range of 0–125 new customers each hour.

11. a.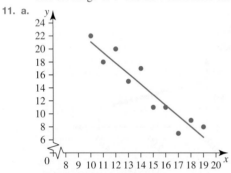
    b. $y = 37.703 - 1.648x$
    c. -0.201
    d. It is assumed the data will continue to behave in the same manner as the data originally supplied.

12. a. 130
    b. $3.6\,°C$
    c. The location of the fire, air temperature, proximity to water, and so on.

13. a.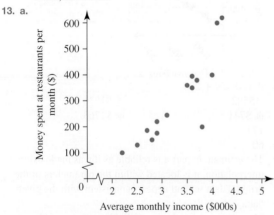
    Average monthly income ($000s)
    b. $R = -459.8 + 229.5I$
    c. $687.70
    d. Part **c** is asking you to predict data outside of the original data set range.
    e. To the nearest $10, $3160. This is an example of interpolation.

Exercise 10.5 Causality and correlation

1. Correlation measures the possible association between two variables. The correlation coefficient (r) measures the strength of this association. A strong correlation does not imply causation. Causation is a measure of how much of the change in one variable is caused by the change in the other variable.

2. B

3. a.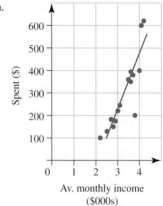
   b. Money spent $= 229.5 \times$ monthly income $- 460$
   c. $r = 0.8777$
   d. There is a strong positive linear association between the average monthly income and money spent at restaurants per month.

4. a.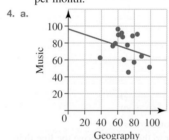
   b. Music mark $= -0.3 \times$ Geography mark $+ 95$
   c. $r = -0.28$
   d. There is a weak negative linear association between the Geography marks and the Music marks.

5. a.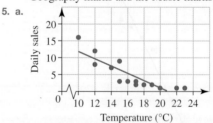
   b. Daily sales $= -1.07 \times$ temperature $+ 22.5$
   c. $r = -0.8621$
   d. There seems to be a strong negative association between daily sales and temperature. However the graph is not very linear.

6. $r = 0.8947$, which indicates a strong positive linear association.

7. a. Indicates a moderate positive linear association
   b. Indicates a strong negative linear association
   c. Indicates a weak to no negative linear association
   d. Indicates a weak positive linear association

8. a.

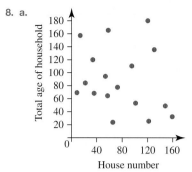

b. The data points appear random, indicating no correlation.

c. $r = -0.2135$

d. There is no association between the house number and the age of the household.

9. a.

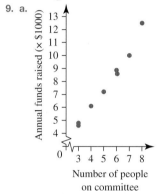

b. There is a strong (almost perfect) positive linear correlation between the number of people on the committee and the annual amount of funds raised.

c. $r = 0.99$

d. Causation cannot be established on a strong correlation alone.

10. It cannot be concluded that diabetes causes deterioration of the hip joint. The common cause could be age.

Exercise 10.6 Review: exam practice

1. B
2. A
3. B
4. C
5. C
6. B
7. C
8. a. $390   b. $750
   c. 5 hours   d. 6.5 hours
9. B
10. D
11. a. Moderate positive correlation
    b. No correlation
    c. Strong negative correlation

12. a.

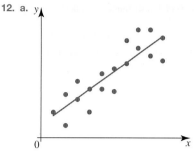

b. No line of best fit possible.

c.

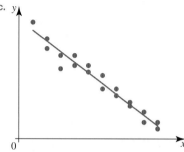

13. a. Independent variable = quantity of water
         Dependent variable = time

b. Independent variable = number of students
   Dependent variable = number of buses required

14. a.

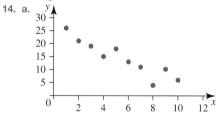

b. Strong negative correlation

c. $r = -0.9329$

d. Pearson's correlation coefficient confirms a strong association between the two variables.

15. a. Independent variable: Distance travelled
         Dependent variable: Fuel used

b.

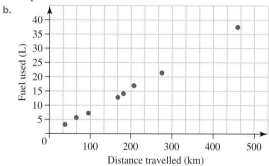

c.

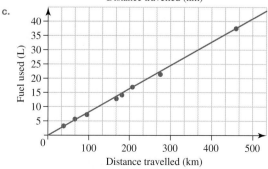

d. Fuel used (L) = 0.081 × distance travelled

e. 4.05 L

f. The estimate in part **e** is reliable since the prediction uses interpolation. This is when making a prediction inside the range of the supplied data.

16. a.

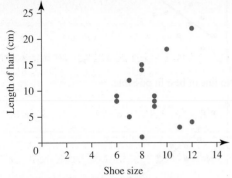

b. $y = 4.0386 + 0.6157x$

c. $r = 0.2055$

d. Based on the graph as well as Pearson's correlation coefficient, there is no association between the two variables. Therefore, no solid conclusions can be made to suggest a change in a person's shoe size will affect the length of their hair.

17. While there appears to be a link between the laziness of people and the increase in the sales of healthy foods, there are also many other factors, other than laziness. Based on this observation alone, the cause of an increase in healthy foods cannot be concluded to be due to laziness.

18. a. Fuel tank capacity

b. Average fuel consumption = 0.6968 + 0.1119 × fuel tank capacity

c. 5.17

d. Extrapolation

e. 85 L

f. This value indicates a moderate association between the variables. Therefore the data can be used; however, other factors should be considered.

g. Various possible answers, e.g. the manner in which a person drives the vehicle, weather conditions, road condition

19. a. Weight (grams)

b. No correlation

c. Various possible answers, e.g. popularity of the shoe, desired profits

20. a. Independent variable = average solar exposure
Dependent variable = max daily temperature

b.

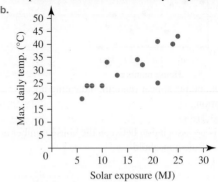

c. Strong positive correlation

d. $r = 0.8242$
This value indicates a strong association between the two variables.

e.

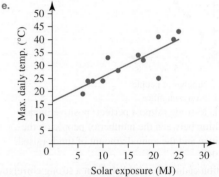

Max daily temperature = 16.232 + 0.9515 × average solar exposure

f. 22

g. 19

h. An x-value of 3 MJ is outside the original data set.

11 Simulations and simple probability

11.1 Overview

CONTENT

In this chapter, students will learn to:
- perform simulations of probability experiments using technology
- recognise that the repetition of chance events is likely to produce different results
- identify relative frequency as probability
- identify factors that could complicate the simulation of real-world events [complex]
- construct a sample space for an experiment
- use a sample space to determine the probability of outcomes for an experiment
- use arrays or tree diagrams to determine the outcomes and the probabilities for experiments.

Fully worked solutions for this chapter are available in the Resources section of your eBookPLUS at www.jacplus.com.au.

11.2 Theoretical probability and sample spaces

We come across probabilities in the media all the time; for example, 'There is a 50% chance it will rain tomorrow' and 'The Broncos have a $\frac{1}{10}$ chance to win the next Premiership'.

Being able to calculate the probability of an event allows us to determine how likely it is to occur. Probabilities are used in weather forecasting, in calculating the price of insurance premiums and in determining the likelihood of winning games. It pays to know your probability!

11.2.1 The probability scale

Probability is a measure of the likely occurrence of an event.

The probability of an event occurring is measured with a scale ranging from and including 0 (**impossible** event) to 1 (**certain** event).

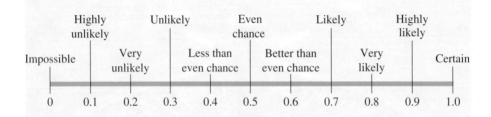

Probability can be written as a decimal number, fraction or percentage. For example, if the probability is even chance, this may be written as 0.5, $\frac{1}{2}$ or 50%.

The probability of an event (A) occurring can be denoted by P(A).

This means that, $0 \le P(A) \le 1$ with P(A) = 0 if A is an impossibility and P(A) = 1 if A is a certainty.

11.2.2 Sample spaces

An **outcome** is a particular result of an experiment.

A **sample space** is a list of all possible outcomes.

If an outcome appears more than once, it is counted only once and written once in the sample space.

A sample space is often listed as a set of outcomes. For example, the sample space for the spinner at right is {pink, blue, green, orange}; the sample space includes all of the colours on the spinner.

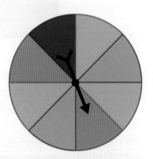

WORKED EXAMPLE 1

For each of the following experiments, list the sample space to show the possible outcomes.
a. **A coin is flipped.**
b. **An arrow is shot at a target where there is a blue area, a white area, a red area and a black area.**
c. **A circular spinner with seven sectors that are labelled 1 to 7 is spun.**

THINK	WRITE
a. A coin has two sides: head and tail. List each possible outcome.	a. When a coin is flipped, the two possible outcomes are head and tail. Therefore the sample space is {head, tail}.

---

b. There are four possible areas that the arrow could hit on the target: blue, white, red and black. List each possible outcome.

b. When an arrow is shot at the target, the four possible outcomes are blue, white, red and black.
Therefore the sample space is {blue, white, red, black}.

c. There are seven different sectors on the spinner so there are seven possible outcomes. List each possible outcome.

c. When the circular spinner is spun, the seven possible outcomes are 1, 2, 3, 4, 5, 6 and 7.
Therefore the sample space is {1, 2, 3, 4, 5, 6, 7}.

11.2.3 Calculating theoretical probability

A **favourable outcome** is an outcome that you want or are looking for.

The theoretical probability of an event can be calculated using the following probability formula.

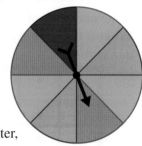

$$P(event) = \frac{number\ of\ favourable\ outcomes}{total\ number\ of\ outcomes}$$

Probabilities can be expressed as fractions, decimals or percentages. In this chapter, unless otherwise specified, we will express probabilities as fractions.

Equally likely outcomes have the same chance of occurring.

- For the spinner shown, $P(blue) = \frac{3}{8}$ because there are three favourable outcomes (blue sectors) out of eight possible outcomes (total number of equal sectors on the spinner).
- Orange and green are equally likely outcomes as they both occupy two of the eight sectors on the spinner. $P(orange) = P(green) = \frac{2}{8}$ or $\frac{1}{4}$.
- Pink is the least likely outcome as it occupies only one of the eight sectors on the spinner.
  $P(pink) = \frac{1}{8}$. Pink has the smallest value for its theoretical probability.

WORKED EXAMPLE 2

A standard six-sided die is rolled. What is the probability of rolling:
a. a 3
b. an odd number?

THINK

a. 1. The number 3 occurs once on a die.

2. The possible outcomes are 1, 2, 3, 4, 5 and 6.

WRITE

a. Number of favourable outcomes = 1

Number of outcomes = 6

3. Calculate the probability of rolling a 3.

$$P(3) = \frac{\text{number of favourable outcomes}}{\text{total number of outcomes}}$$
$$= \frac{1}{6}$$

b. 1. The odd numbers are 1, 3 and 5. There are six numbers on a die.

b. Number of favourable outcomes = 3

Number of outcomes = 6

2. Calculate the probability of rolling an odd number. Write the fraction in simplest form.

$$P(\text{odd number}) = \frac{\text{number of favourable outcomes}}{\text{total number of outcomes}}$$
$$= \frac{3}{6}$$

The probability of rolling an odd number is $\frac{1}{2}$.

WORKED EXAMPLE 3

A normal pack of 52 playing cards is well shuffled and one card is drawn.
What is the probability of drawing:

a. a club   **b.** an ace   **c.** a red queen?

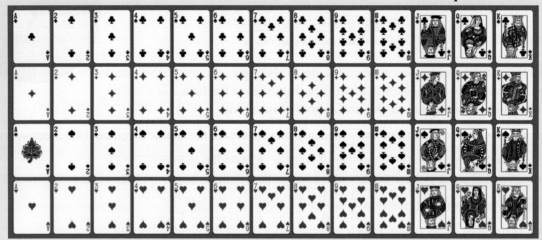

THINK

a. 1. There are 13 clubs in a deck of cards.

2. There are 52 cards in the deck, making 52 possible outcomes.

3. Calculate the probability of drawing a club, making sure to simplify the fraction.

b. 1. There are 4 aces in a deck of cards

2. There are 52 cards in the deck, making 52 possible outcomes.

WRITE

a. Number of favourable outcomes = 13

Number of outcomes = 52

$$P(\text{club}) = \frac{\text{number of favourable outcomes}}{\text{total number of outcomes}}$$
$$= \frac{13}{52}$$
$$= \frac{1}{4}$$

b. Number of favourable outcomes = 4

Number of outcomes = 52

---

3. Calculate the probability of drawing an ace. Write the fraction in simplest form.

$$P(ace) = \frac{\text{number of favourable outcomes}}{\text{total number of outcomes}}$$

$$= \frac{4}{52}$$

$$= \frac{1}{13}$$

c. 1. There are 2 red queens in a deck of cards

c. Number of favourable outcomes = 2

2. There are 52 cards in the deck, making 52 possible outcomes.

Number of outcomes = 52

3. Calculate the probability of drawing a red queen. Write the fraction in simplest form.

$$P(red\ queen) = \frac{\text{number of favourable outcomes}}{\text{total number of outcomes}}$$

$$= \frac{2}{52}$$

$$= \frac{1}{26}$$

11.2.4 The complement of an event

The probability of the **complement** of an event is the probability that the event discussed will not happen.

For example, if an event is 'it will rain today' then the complement of this is 'it will not rain today'. The other way to interpret the complement is to recognise it as the opposite of the event.

If the probability of an event is denoted by P(R), then the probability of the complement is denoted by P(R′).

In order to calculate P(R′) it is necessary to understand that the sum of the probability of an event and its complement is equal to 1:

$$P(R) + P(R') = 1$$

This equation can be manipulated so it is expressed in terms of the complement:

$$P(R') = 1 - P(R)$$

For example, if the probability of rain today is $\frac{7}{10}$, then:

$$P(R') = 1 - \frac{7}{10}$$

$$= \frac{3}{10}$$

This means there is a 30% chance that it will not rain.

Eli's Maths teacher told him that he had a **70% chance of passing** the final exam, based on Eli's prior exam scores.
a. What is the complement of this event?
b. What is the probability of the complement of this event? Give your answer as a percentage.

THINK	WRITE
a. The complement is if the discussed event will not happen.	a. The complement to Eli passing the exam is not passing the exam.
b. 1. Use the formula to find the complement.	b. $P(\text{passing}) = 70\% = \dfrac{7}{10}$ $P(\text{not passing}) = 1 - P(\text{passing})$ $= 1 - \dfrac{7}{10}$ $= \dfrac{3}{10}$
2. Convert the answer into a percentage by multiplying by 100.	$\dfrac{3}{10} \times 100 = \dfrac{300}{10}$ $= 30\%$

Exercise 11.2 Theoretical probability and sample spaces

Note: Unless otherwise stated, express probabilities as fractions.

1. Match the words with one of the numbers between 0 and 1 that are given. Choose the number depending on what sort of chance the word indicates, between *impossible* and *certain*. You may use a number more than once. The numbers to choose from are 1, 0.75, 0.5, 0.25 and 0.
   a. Certain
   b. Likely
   c. Unlikely
   d. Probable
   e. Improbable
   f. Definite
   g. Impossible
   h. Slim chance
   i. Sure thing
   j. Doubtful
   k. Fifty–fifty
   l. More than likely

2. List two events that have a probability of:
   a. 0
   b. 0.5
   c. 1
   d. 0.75
   e. 0.25.

3. **WE1** For each of the following experiments, list the sample space to show the total possible outcomes.
   a. A standard 6-sided die is rolled.
   b. A marble is randomly selected from a bag containing 4 green, 2 yellow and 3 blue marbles.
   c. A letter is selected from the letters of the alphabet.

4. **WE2** A standard 6-sided die is rolled. What is the probability of rolling:
   a. a 5
   b. a number less than 3
   c. an even number?

5. The letters of the word *MATHEMATICS* are each written on a small piece of card and placed in a bag. If one card is selected from the bag, what is the probability that it is:
   a. a vowel
   b. a consonant
   c. the letter M
   d. the letter C?

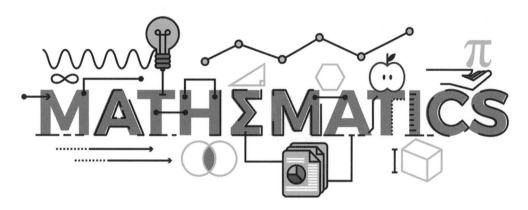

6. A spinner of 8 equally likely numbers is spun. The spinner is numbered 1–8.

   a. List the sample space.
   b. What is the probability, expressed as a percentage, of obtaining:
      i. an odd number
      ii. a number that is less than 6
      iii. the number 9
      iv. a number that is at most 8?

7. Answer these questions for each of the following spinners.
   i. Is there an equal chance of landing on each colour? Explain.
   ii. List all the possible outcomes.
   iii. Find the probability of each outcome.

   a.
   b.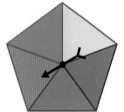
   c.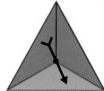

8. All the jelly beans in the photograph are placed in a bag for a simple probability experiment.
   a. Which colour jelly bean is most likely to be selected from the bag? Explain.
   b. Which colour jelly bean is least likely to be selected from the bag? Explain.
   c. Find the probability of selecting each coloured jelly bean from the bag.

9. **MC** What is the probability of obtaining a number less than 5 on the spinner shown?

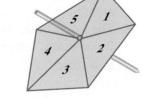

A. 1

B. 0

C. $\dfrac{1}{5}$

D. $\dfrac{4}{5}$

10. **WE3** A normal pack of 52 playing cards is well shuffled and one card is drawn. What is the probability of drawing:

a. a heart

b. a red card

c. an ace of diamonds

d. a picture card (jack, queen or king)?

11. Draw spinners with the following probabilities.

a. $P(\text{blue}) = \dfrac{1}{3}$ and $P(\text{white}) = \dfrac{2}{3}$

b. $P(\text{blue}) = \dfrac{1}{2}, P(\text{white}) = \dfrac{1}{4}, P(\text{green}) = \dfrac{1}{8}$ and $P(\text{red}) = \dfrac{1}{8}$

c. $P(\text{blue}) = 0.75$ and $P(\text{white}) = 0.25$

12. What is the complement of each of these events?

a. Owning a dog

b. Flipping a coin and getting a tail

c. Sleeping

d. Attending school

e. Passing your learner knowledge test

f. Owning a mobile phone

13. **WE4** Bol's Physics teacher told him that he had a 20% chance of passing the final exam, based on Bol's prior exam scores.

a. What is the complement of this event?

b. What is the probability of the complement of this event? Give your answer as a percentage.

14. There are three different colours of flowers in a basket: red, yellow, and purple. Assuming that you pick one flower from the basket, the complement of picking a red flower is picking a purple flower. True or false? Explain your answer.

15. If two dice are rolled, what are the probabilities that the sum of the rolled numbers is:

a. 5

b. not 5

c. 6 or 7

d. not 6 or 7

e. 2

f. not 2?

11.3 Multi-stage experiments (arrays and tree diagrams)

11.3.1 Arrays

When two events occur to form a single result, an **array** can be used to display the sample space.

Consider an experiment in which a red die and a blue die are rolled, and the two numbers that land uppermost are recorded. The array shows the sample space for that experiment. There are 36 possible outcomes in this sample space.

		Blue die					
		1	**2**	**3**	**4**	**5**	**6**
Red die	**1**	(1, 1)	(1, 2)	(1, 3)	(1, 4)	(1, 5)	(1, 6)
	2	(2, 1)	(2, 2)	(2, 3)	(2, 4)	(2, 5)	(2, 6)
	3	(3, 1)	(3, 2)	(3, 3)	(3, 4)	(3, 5)	(3, 6)
	4	(4, 1)	(4, 2)	(4, 3)	(4, 4)	(4, 5)	(4, 6)
	5	(5, 1)	(5, 2)	(5, 3)	(5, 4)	(5, 5)	(5, 6)
	6	(6, 1)	(6, 2)	(6, 3)	(6, 4)	(6, 5)	(6, 6)

An array is also useful when there are two events, but a different number of outcomes for each event. For example, the array shows the sample space for an experiment where a coin is flipped and a die is rolled.

		Die					
		1	**2**	**3**	**4**	**5**	**6**
Coin	**H**	(H, 1)	(H, 2)	(H, 3)	(H, 4)	(H, 5)	(H, 6)
	T	(T, 1)	(T, 2)	(T, 3)	(T, 4)	(T, 5)	(T, 6)

WORKED EXAMPLE 5

The two spinners shown are spun and the numbers that are chosen on each spinner are recorded. What is the probability of spinning at least one 1?

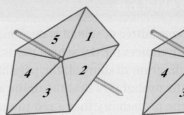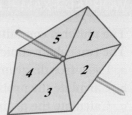

THINK

WRITE

1. Draw the array for the sample space. Highlight or circle the outcomes that have at least one 1, as shown in green. There are 9 outcomes with at least one 1.

Spinner 2

		1	2	3	4	5
Spinner 1	1	(1, 1)	(1, 2)	(1, 3)	(1, 4)	(1, 5)
	2	(2, 1)	(2, 2)	(2, 3)	(2, 4)	(2, 5)
	3	(3, 1)	(3, 2)	(3, 3)	(3, 4)	(3, 5)
	4	(4, 1)	(4, 2)	(4, 3)	(4, 4)	(4, 5)
	5	(5, 1)	(5, 2)	(5, 3)	(5, 4)	(5, 5)

2.	Calculate the probability using the formula. There are 25 outcomes in the sample space.	$P(\text{at least one 1}) = \dfrac{\text{number of favourable outcomes}}{\text{total number of outcomes}}$
		$= \dfrac{9}{25}$

11.3.2 Tree diagrams

Tree diagrams are used to list all possible outcomes of two or more events. The branches show the possible links between one outcome and the next outcome. The tree diagram at right represents the sample space for flipping two coins.

When all outcomes are equally likely, probabilities can be calculated from tree diagrams. For example, in the tree diagram shown, the probability of obtaining one tail and one head can be calculated:

$$P(\text{one H and one T}) = \dfrac{\text{number of outcomes with one H and one T}}{\text{total number of outcomes}}$$

$$= \dfrac{2}{4}$$

$$= \dfrac{1}{2}$$

Each event is treated as if it were a separate event, even if events occur at the same time. For example, the example tree diagram applies to experiments where:
- two coins are flipped together
- the same coin is flipped twice.

WORKED EXAMPLE 6

The uniform committee at school is deciding on a new colour combination for the school uniform. The two colour choices for the school pants, shirt and jumper are red and yellow.
a. Use a tree diagram to show all the possible combinations.
b. Calculate the probability that the uniform will consist of only one colour.
c. Calculate the probability that a red jumper will be part of the uniform.
d. Calculate the probability that the uniform will be red pants, a yellow shirt and a red jumper.

THINK

WRITE

a. The three events are the colour of the pants, the colour of the shirt and the colour of the jumper. There are two choices for each of these.

a.

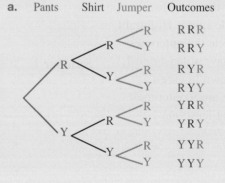

---

b. For the uniform to consist of only one colour, all pieces need to be red or all pieces need to be yellow. There is 1 outcome with all red and 1 outcome with all yellow. There are 8 outcomes altogether.

b. $P(\text{all one colour}) = \dfrac{2}{8}$

$$= \dfrac{1}{4}$$

c. Four of the outcomes have a red jumper.

c. $P(\text{red jumper}) = \dfrac{4}{8}$

$$= \dfrac{1}{2}$$

d. Calculate the probability of each option and multiply them together.

d. $P(\text{red pants}) = \dfrac{1}{2}$

$$P(\text{yellow shirt}) = \dfrac{1}{2}$$

$$P(\text{red jumper}) = \dfrac{1}{2}$$

$$P(\text{RYR}) = \dfrac{1}{2} \times \dfrac{1}{2} \times \dfrac{1}{2} = \dfrac{1}{8}$$

Exercise 11.3 Multi-stage experiments (arrays and tree diagrams)

Note: Unless otherwise stated, express probabilities as fractions.

1. A spinner is divided into 5 sections (red, blue, green, yellow and orange). A six-sided die is also rolled. Draw the array for one spin of the spinner and one roll of the die.

2. **WE5** The two spinners shown are spun and the numbers that are chosen on each spinner is recorded. What is the probability of spinning exactly one 5?
   You must show your working using an array.

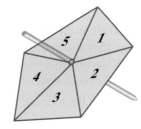 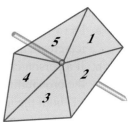

3. **MC** A six-sided die and a coin are tossed at the same time. What is the sample space?
   A. $\{1, 2, 3, 4, 5, 6, 7, 8\}$
   B. $\{1, 2, 3, 4, 5, 6, H, T\}$
   C. $\{(1, H), (2, H), (3, H), (4, H), (5, H), (6, H)\}$
   D. $\{(1, H), (2, H), (3, H), (4, H), (5, H), (6, H), (1, T), (2, T), (3, T), (4, T), (5, T), (6, T)\}$

4. In the game of Pretzel, the two spinners shown are spun and each contestant places the two body parts spun in contact and holds them there until the next two body parts are spun. If you fall over, you are out. The winner is the last person standing.

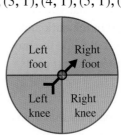 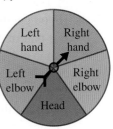

   When the spinners are spun, what is the probability, expressed as a percentage, that the contestants will need to place:
   a. their right foot on their head
   b. their left elbow on a knee
   c. their head on a foot
   d. their left hand on a body part lower than their belly button?

5. A restaurant has the menu shown.
   a. Use a tree diagram to show how many different combinations are possible for a meal with one selection at each course.
   b. If a person selects a dinner at random, what is the probability that:
      i. they have soup for entrée
      ii. they have spring rolls and
      iii. curry
      iv. they don't have ice-cream for dessert
      v. they have soup and pasta but not ice-cream?

Entrée

Pumpkin soup
Mini spring rolls

Main course

Pasta bolognaise
Grilled fish
Beef curry

Dessert

Lemon pie
Ice-cream

6. **WE6** The uniform committee at school is deciding on a new colour combination for the school uniform.

   The two colour choices for the school hat, bag and hoodie are blue and white.
   a. Use a tree diagram to show all the possible combinations.
   b. Calculate the probability that the uniform will consist of only one colour.
   c. Calculate the probability that a white hat will be part of the uniform.
   d. Calculate the probability that a white hat, blue bag and blue hoodie will be part of the uniform.

7. For a fundraiser, your class is selling single-scoop ice-cream cones at lunchtime. There is a choice of vanilla and chocolate ice-cream, topped with chocolate curls, sprinkles or melted chocolate or no topping at all.
   a. Draw a tree diagram to show all the possible combinations of ice-cream and topping that you could have from the stall.
   b. If a person chooses an ice-cream at random, calculate the probability that:
      i. both the ice-cream and the topping are chocolate
      ii. it is a vanilla ice-cream with sprinkles
      iii. there is no topping on the ice-cream.

8. In your drawers at home, there are two white T-shirts, a green T-shirt and a red T-shirt. There is also a pair of black pants and a pair of khaki pants.
   a. Draw a tree diagram to show all the possible combinations of T-shirts and pants that you could wear.
   b. If you get dressed in the dark and put on one T-shirt and one pair of pants, calculate the probability that you put on the red T-shirt and khaki pants.

9. In the last Science test, your friend guessed the answers to three true/false questions.
   a. Use a tree diagram to show all the different answer combinations for the three questions.
   b. Calculate the probability that your friend:
      i. got all three answers correct
      ii. got exactly two correct answers
      iii. got no correct answers.

10. Assuming that the chance of a baby being a boy or a girl is the same:
    a. Calculate the probability that a family with three children has:
        i. all boys
        ii. two girls and one boy
        iii. three children of the same gender
        iv. at least one girl
        v. two children of the same gender.
    b. If the family was expecting another baby, what is the probability that the new baby will be a boy?
    c. If the family already has three boys, what is the probability that the new baby will be a boy?
    d. If the family has three girls, what is the probability that the new baby will be a boy?
    e. How likely is the combination of children shown in the photo?

11. A fair coin is flipped 3 times. Calculate the probability of obtaining:
    a. at least two heads or at least two tails
    b. exactly two tails.

12. **MC** A fair coin is flipped 3 times. What is the probability of obtaining at most 1 tail?

    A. $\dfrac{1}{3}$ B. $\dfrac{1}{2}$

    C. $\dfrac{3}{8}$ D. $\dfrac{1}{8}$

13. A spinner of 7 equally likely numbers (numbered 1–7) is spun twice and the two numbers are added. Calculate the probability of obtaining a total of:
    a. 2 or 14   b. 9   c. at least 12.

14. Two dice are rolled and the product of the two numbers is found. Calculate the probability that the product of the two numbers is:
    a. an odd number
    b. a prime number
    c. more than 1
    d. at most 36.

15. A fair die is rolled and a fair coin flipped. Calculate the probability of obtaining:
    a. an even number and a head
    b. a tail from the coin
    c. a prime number from the die
    d. a number less than 5 and a head.

16. a. Use an array to display the sample space for an experiment where two dice are rolled and the sum of the two numbers appearing is found.
    b. List any patterns you found in the sample space.

c. Copy and complete the following table.

Sum of two dice	2	3	4	5	6	7	8	9	10	11	12
Probability											

d. List any patterns that you notice in the probabilities you found in the table.

17. In a new game, two dice are rolled and the difference between the two dice is noted. If the difference is larger than 2, you win; otherwise you lose.
    a. Use an array to find the sample space for the game.
    b. Calculate the probability of winning (having a difference greater than 2).
    c. Calculate the probability of losing (having a difference of 2 or less).
    d. A fair game is one in which the chances of winning are the same as the chances of losing. Is this game fair?

18. The game rock, paper, scissors is sometimes used to make a decision between two people. It uses the three different hand signs shown in the photo at right. On the count of 3, each player displays one of the hand signs. If both players display the same hand sign then the game is a tie. The photos show all other possible results.

    Rock   Paper   Scissors

    a. Draw an array to show all the possible combinations for a single game of rock, paper, scissors.

    Rock breaks scissors

    Rock wins
    Paper covers rock

    Person 2

		Rock	Paper	Scissors
Person 1	**Rock**			
	Paper			
	Scissors			

    Paper wins
    Scissors cut paper

    Calculate the probability that:
    b.   i. rock wins
        ii. paper wins
       iii. scissors wins.
    c. Is rock, paper, scissors a fair game?
    d. A friend suggested that you add dynamite to the game with the following rules.
    • Dynamite blows up the rock and wins.
    • Dynamite has its fuse cut by scissors and loses.
    • Dynamite sets the paper on fire and wins.
        i. Modify your array in part **a** to include dynamite.
       ii. Does each hand sign have an equal chance of winning?
      iii. Support your answer to part **ii** using the array and probabilities.

    Scissors wins

    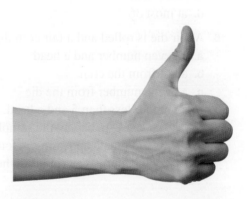

19. There are three different ways to go from school to the shops. There are two different ways to go from the shops to the library. There is only one way to go from the library to home.

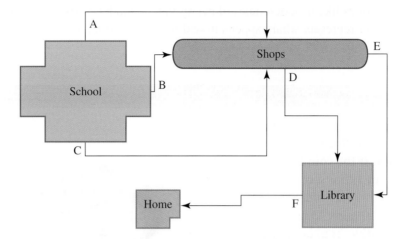

This afternoon you need to travel from school to home via the shops and the library.
a. Use a tree diagram to calculate the number of different routes you could use on your journey. (*Hint:* Use the letters A, B and C to represent the different routes from school to the shops and D and E to represent the different routes from the shops to the library. Use F to represent the route from the library to home.)
b. Would the number of outcomes be different if we omitted the last leg of the journey from the library to home? Why?
20. A die is rolled and 2 coins are flipped. Display the sample space using an appropriate method.

11.4 Relative frequency and simulations

11.4.1 Relative frequency

Often in real life it is not possible to calculate a theoretical probability for an event. In these situations, an experiment is carried out.

An experiment that is repeated a number of times is called a **trial** and becomes part of a larger experiment. For example, a die is rolled and the number is recorded. This is an experiment. If the experiment is repeated another 49 times, the first roll of the die is now referred to as the first trial in an experiment where the die is rolled 50 times.

A **successful trial** is a trial where the result is the outcome that you wanted.

The **relative frequency**, sometimes called the empirical or experimental probability, of an event is calculated using the following formula:

$$\text{relative frequency} = \frac{\text{number of successful trials}}{\text{total number of trials}}$$

The relative frequency of an event for a very large number of trials gives an indication of the value of the theoretical probability.

As the number of trials increases, the relative frequency of an event will gradually become closer in value to the theoretical probability. This is known as the 'law of large numbers'.

If the number of trials is large enough, the relative frequency can be used as an estimate of probability.

11.4.2 Simulations

Sometimes it is not possible to conduct trials of a real experiment because it is too expensive, too difficult or impracticable. In situations like this, outcomes of events can be modelled using devices such as spinners, dice or coins to simulate or represent what happens in real life. These simulated experiments are called **simulations**.

WORKED EXAMPLE 7

To simulate whether a baby is born male or female, a coin is flipped. If the coin lands heads up, the baby is a boy. If the coin lands tails up, the baby is a girl. Student 1 flips a coin 10 times and obtains 3 heads and 7 tails, while Student 2 flips a coin 100 times and obtains 43 heads and 57 tails.

a. Determine the relative frequency of females in both cases. Give your answers as decimals.

b. Compare the results from the two simulations and determine which is a better estimate of the true probability.

THINK	WRITE
a. 1. A baby born female is a successful trial. For Student 1: in the simulation, females are tails. There were 7 tails, so 7 females were born and there were 7 successful trials out of 10 trials. Write the formula for relative frequency and substitute the results.	**a.** Student 1: $$\text{Relative frequency} = \frac{\text{number of successful trials}}{\text{total number of trials}}$$ $$= \frac{7}{10}$$ $$= 0.7$$
2. For Student 2: there were 57 tails, so 57 females were born and there were 57 successful trials out of 100 trials. Write the formula for relative frequency and substitute the results.	Student 2: $$\text{Relative frequency} = \frac{\text{number of successful trials}}{\text{total number of trials}}$$ $$= \frac{57}{100}$$ $$= 0.57$$
b. 1. Compare the results of the two simulations. Theoretically, there should be an equal number of males and females born, so we would expect the results to be close to 0.5.	**b.** Student Relative frequency 1 0.7 2 0.57
2. Give your answer.	The true probability is likely to be close to 0.5, so Student 2 achieved the better estimate of probability.

As you can see from Worked example 7, the greater the number of trials, the better the estimate of probability achieved.

11.4.3 Using technology to perform simulations

Simulating experiments using manual devices such as dice and spinners can take a lot of time. A more efficient method of collecting results is to use a list of randomly generated numbers.

Random number generators can generate a series of numbers between two given values; for example, decimals between 0 and 0.9999. The following table shows some Excel formulae that generate random numbers.

Formula	Output
=RAND()	A random decimal from 0 to 0.9999
=INT(6^*RAND()+1)	Integers between 1 and 6 that can be used to simulate rolling a die. Both of these formulae can be modified to generate the type of numbers that you require.
=RANDBETWEEN(1,6)	Generates random numbers between 1 and 6.

WORKED EXAMPLE 8

a. **Use a random number generator to simulate the number of chocolate chips in 50 biscuits, with a maximum of 80 chocolate chips in each.**
b. **Calculate the probability that there will be more than 30 chocolate chips in a randomly chosen biscuit.**

THINK

a. Use a random number generator to generate 50 numbers between 0 and 80.
Use either of the following formulas in Excel.
=INT(80*RAND()+1)
or
=RANDBETWEEN(1,80)

b. 1. Count the number of biscuits with more than 30 chocolate chips.
There are 34 in the example shown.

WRITE

a.

8	55	8	62	80	50	42	80	57	39
15	7	64	73	47	12	74	74	16	42
41	22	50	33	68	72	64	16	6	72
70	72	52	48	14	22	59	48	65	34
67	62	72	59	10	30	13	7	40	18

b.

8	55	8	62	80	50	42	80	57	39
15	7	64	73	47	12	74	74	16	42
41	22	50	33	68	72	64	16	6	34
70	72	52	48	14	22	59	48	65	34
67	62	72	59	10	30	13	7	40	18

2. The probability is the number of successful outcomes (34) divided by the number of trials (50).

$$P(\text{more than 30 chocolate chips}) = \frac{34}{50}$$
$$= \frac{17}{25}$$

11.4.4 Factors that could complicate simulations

Simulations are used to reproduce real-life situations when it is not feasible to implement the actual test.

The reasons for not conducting an actual test can include financial or safety constraints, or just the overall complexity of running the test.

Most simulations are conducted using computers and various other mechanical devices. For example, airbag manufacturers run airbag tests using cars, but instead of using real humans they use mannequins. These simulations are fairly accurate, but they have their limitations. One problem with the airbag simulation is the difficulty of modelling the interaction of the airbag with the seatbelt. There are numerous

other factors that can affect the success of the airbag in an accident. Some of these factors are easy to simulate, but others are not yet known or are too difficult to simulate.

WORKED EXAMPLE 9

Six runners are to compete in a race.
a. Explain how simulation could be used to determine the winner.
b. Are there any limitations in your answer?

THINK	WRITE
a. There are six runners, so six numbers are needed to represent them. A die is the obvious choice. Assign numbers 1 to 6 to each runner.	a. Each runner could be assigned a number on a die, which could be rolled to determine the winner of the race. When the die is rolled, it simulates a race. The number that lands uppermost is the winner of the race.
b. In this simulation, each runner had an equally likely chance of winning each game.	b. The answer depends on each runner having an equally likely chance of winning the race. This is rarely the case.

11.4.5 Using relative frequencies: polls

Statisticians sometimes want to compare the attitudes of certain groups, with frequency tables used to visually show the statistical differences in the groups' attitudes. Frequency tables typically show results obtained from taking a random sample from a large population, or from gathering all the data from an entire population when the population is small.

When calculating relative frequencies, six steps are followed.
1. Gather the required statistics.
2. Determine the total number of people polled.

---

3. Determine each group's attitude in relation to the total number of people polled and write in fractional form; for example:

- Of the 100 people polled, 30 liked the caravan park $= \dfrac{30}{100}$.

- Of the 100 people polled, 70 did not like the caravan park $= \dfrac{70}{100}$.

4. Convert the fraction to a decimal.

$$\dfrac{30}{100} = 0.3$$

5. Convert the decimal to a percentage.

$$\dfrac{30}{100} = 0.3$$
$$= 30\%$$

6. Check that all relative frequencies add to 100%.
   Note: When the decimals or percentages are rounded off, the figure may be slightly below or above 100%.

WORKED EXAMPLE 10

In a recent election, there were 4 candidates. During a telephone poll conducted a week before the election, 500 randomly selected voters were asked to indicate their preferences for the 4 candidates. The results are shown in the table.

Candidate	1	2	3	4
Number of voters who preferred this candidate	115	168	145	72

Estimate each candidate's probability of winning the election, giving your answers as decimals.

THINK

1. The probability of a candidate winning is the number of votes they received divided by the total number of votes.

2. The closer the relative frequency is to 1, the higher the probability of success.

WRITE

$P(\text{candidate 1 wins}) = \dfrac{115}{500} = 0.230$

$P(\text{candidate 2 wins}) = \dfrac{168}{500} = 0.336$

$P(\text{candidate 3 wins}) = \dfrac{145}{500} = 0.290$

$P(\text{candidate 4 wins}) = \dfrac{72}{500} = 0.144$

Candidate 2 is the favourite to win.

11.4.6 Calculating the number of successful trials

Sometimes, a relative frequency is known and may be used to calculate the number of individuals who responded a certain way or the number of successful trials.

Number of successful trials = relative frequency × total number of trials

For example, you may have determined the relative frequency of individuals that vaccinate their children is 88% or 0.88. In a population of 300 individuals, it can be assumed that approximately 264 individuals vaccinated their children (0.88×300).

WORKED EXAMPLE 11

In 2017, the Australian Marriage Law Postal Survey was conducted to gauge support for legalising same-sex marriage in Australia.

The table shows the breakdown of votes per State as a decimal frequency.

State/Territory	Yes	No
New South Wales	0.578	0.422
Victoria	0.649	0.351
Queensland	0.607	0.393
South Australia	0.625	0.375
Western Australia	0.637	0.363
Northern Territory	0.606	0.394
Australian Capital Territory	0.740	0.260

5000 individuals from Queensland were randomly selected and asked for their vote.
a. Approximately how many will state they answered 'yes'?
b. Approximately how many will state they answered 'no'?

THINK	WRITE
a. 1. Look on the chart and find the row with Queensland and those who stated 'yes'.	a. 0.607
2. Multiply 5000 (number of individuals) by the result from step **1**. The answer represents the number of Queenslanders who answered 'yes'.	$0.607 \times 5000 = 3035$
3. Since a fractional result is not valid in this context, round to the nearest whole number.	Approximately 3035 of the individuals answered 'yes'.
b. 1. The number of Queenslander questions was 5000. We know 3485 voted 'yes', so we subtract the answer to part **a** from 5000.	b. $5000 - 3035 = 1965$
2. Write the answer.	Approximately 1965 of the 5000 individuals answered 'no'.

Exercise 11.4 Relative frequency and simulations

Note: Unless otherwise stated, express probabilities as fractions.

1. **WE7** To simulate whether or not the weather will be suitable for sailing, a coin is flipped. If the coin lands heads up, the weather is perfect. If the coin lands tails up, the weather is not suitable.

   Mandy flips a coin 20 times and obtains 13 heads and 7 tails. Sophia flips a coin 100 times and obtains 47 heads and 53 tails.

   a. Determine the relative frequency of perfect weather days in both cases.

   b. Compare the results from the two simulations and determine which is a better estimate of the true probability of the experiment.

2. A box of 50 batteries was tested and three were found to be flat.

   a. What was the relative frequency of a flat battery?

   b. In a box of 500 batteries, how many batteries would you expect to be flat?

3. The results of a class experiment that involved rolling a standard six-sided die 300 times are shown in the table.

Number on die	1	2	3	4	5	6
Number of times rolled	42	50	61	37	52	58

   Use the results to calculate the experimental probability of rolling:

   a. the number 5   b. an odd number   c. an even number.

4. In his last 20 games, a basketball player sank 17 free throws and missed 11. Estimate the probability of sinking his next free throw.

5. **MC** A student rolls a pair of dice 10 times and records the number of times the total is 7. She repeats the experiment numerous times and her results are as follows.

$$1, 5, 3, 2, 1, 0, 2, 3, 4, 3, 2, 1, 0, 1, 2, 6, 2, 0, 1, 2$$

   What is the estimated probability of getting a total of 7 when 2 dice are tossed?

   A. $\dfrac{7}{12}$ **B.** $\dfrac{41}{100}$ **C.** $\dfrac{41}{200}$ **D.** $\dfrac{41}{400}$

6. a. **WE8** Use a random number generator to simulate the number of walnuts in a 40 different carrot cakes, with a maximum of 60 in each cake.

   b. Calculate the probability that there will be more than 40 walnuts in a randomly chosen cake.

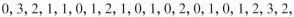

7. The gender of babies in a set of triplets is simulated by flipping 3 coins. If a coin lands tails up, the baby is a boy. If a coin lands heads up, the baby is a girl. In the simulation, the trial is repeated 40 times and the following results show the number of heads obtained in each trial.

   0, 3, 2, 1, 1, 0, 1, 2, 1, 0, 1, 0, 2, 0, 1, 0, 1, 2, 3, 2,
   1, 3, 0, 2, 1, 2, 0, 3, 1, 3, 0, 1, 0, 1, 3, 2, 2, 1, 2, 1

   a. Calculate the probability, as a percentage, that exactly one of the babies in a set of triplets is female.

   b. Calculate the probability, as a percentage, that more than one of the babies in the set of triplets is female.

8. **WE9** There are 24 drivers lining up at the start of a Formula 1 race.
   a. Explain how simulation could be used to determine the winner.
   b. Are there any limitations in your answer?

9. **WE10** A school canteen asked 150 random Year 12 students to indicate their favourite food before they asked the entire school to place their vote. The food with the most votes was going to be made half price for a week.

   The results from the Year 12 students are shown in the table.

Meal	Tally
Hamburger	45
Fish and chips	31
Macaroni and cheese	30
Lamb souvlaki	25
BBQ pork ribs	19

Estimate each food's probability in the overall vote, giving your answers as decimals.

10. If a computer manufacturer wanted to simulate the probability of their laptops having a fault, would it be appropriate to use a random number generator to select random computers on the assembly line? Why or why not?

11. A random number generator was used to select how many students will receive a free lunch on any given day. The maximum number of students who can receive a free lunch is 10 and the minimum is 0.
    a. If the test is run for 30 days, what is the theoretical probability that:
        i. 8 or more students are selected on any given day
        ii. 0 students are selected on any given day
        iii. 10 students are selected on any given day?
    b. Use a random number generator to simulate the free-lunch program over 30 days, and compare the results to those from part a.
    c. Are the results from the random number generator the same as your predictions? Why or why not?

12. Determine if the following simulations are easy or difficult to undertake. If your response is 'difficult', explain why.
    a. Predicting how a helicopter pilot will react in an emergency
    b. Estimating how many times you will win when spinning a roulette wheel
    c. Predicting how many kilometres a new car will travel before any major mechanical problems
    d. Estimating how many car accidents you will have in your lifetime
    e. Estimating the number of defective mobile phones that one manufacturer will produce over 2 months

13. **WE11** In 2018, a survey was conducted to gauge support for changing the age of gaining a provisional licence from a minimum of 17 years old to a minimum of 16 years old.

    The table shows the breakdown of votes by age as a decimal frequency.

Age	Yes	No
16–20	0.871	0.129
21–25	0.764	0.236
26–30	0.641	0.359
31–35	0.443	0.557
36–40	0.398	0.602
41–45	0.287	0.713
46–50	0.213	0.787
51–55	0.122	0.878

2800 individuals aged 16–20 were randomly selected and asked for their vote.
    a. Approximately how many will state they answered 'yes'?
    b. Approximately how many will state they answered 'no'?

14. a. Five thousand new domestic university students are selected to take part in a study. Use the data in the chart to predict how many students will not come back after their first year of university for the 4-year period of 2015–18.

b. Based on the domestic trend for students not coming back after the first year, do you believe that your estimate in part **a** will be higher or lower than the actual percentage? Explain your answer.

Completion rates of higher education students — cohort analysis, 2005–2014					
National total (domestic students)	Year	Completed (in any year)	Still enrolled at the end of the 4-year cohort period	Re-enrolled, but dropped out	Never came back after the first year
	2005	47.3%	30.2%	11.5%	11.0%
	2006	46.7%	32.7%	10.6%	10.0%
	2007	45.9%	33.8%	10.4%	9.9%
	2008	46.7%	33.5%	10.9%	8.9%
	2009	46.1%	34.3%	11.1%	8.5%
	2010	45.1%	34.7%	11.2%	9.0%
	2011	45.0%	34.5%	11.7%	8.8%

Source: Department of Education and Training, *Completion Rates of Higher Education Students — Cohort Analysis,* 2005–2014

15. The results of a coin-flipping experiment are shown.

Outcome	Frequency
Heads	38
Tails	62
Total	100

a. What are the relative frequencies, expressed as a percentage, of:
  i. heads   **ii.** tails?
b. If the experiment was repeated, would you expect the same results? Explain your answer.
c. If the experiment was repeated 100 times, what do you expect to happen to the relative frequencies of the outcomes?

16. *Note:* You will need a 6-sided die for this question. The latest promotion by a chocolate bar manufacturer claims 'Every third one is free'. If there is a special message inside the wrapper, you get a refund on the purchase price. The manufacturer has put the special message in every third wrapper. When the chocolate bars are packaged into boxes, they are mixed randomly. Use the numbers 3 and 6 on a die to represent getting a free chocolate bar.
a. Carry out 50 trial purchases of chocolate bars.
b. Calculate the relative frequency of a free chocolate bar.
c. Compare your results with the students around you.
d. Did other students get different results? Explain why.
e. In 100 trials, how many free chocolate bars would you expect to receive?

17. a. How many times, to the nearest whole number, would you expect to roll a 4 if you rolled a fair die:
  i. 10 times   **ii.** 60 times   **iii.** 1000 times?
b. If you actually carried out these experiments, would you get the exact number that you predicted? Explain your answer.

18. A simulation of the ages of a class of Year 12 students using random numbers is shown.

$$16 \quad 17 \quad 16 \quad 17 \quad 18 \quad 18$$

$$16 \quad 17 \quad 16 \quad 16 \quad 16 \quad 17$$

$$16 \quad 17 \quad 17 \quad 17 \quad 17 \quad 18$$

$$18 \quad 17 \quad 18 \quad 17 \quad 16 \quad 17$$

   a. Calculate the relative frequency of each age.
   b. Carry out the same simulation using your random number generator.
   c. Compare the results of the two simulations and comment on any differences or similarities.

19. A player suspects that a die is loaded (unfair), so she rolls the die 600 times to find out. The results of the experiment are shown in the table.

Number on die	1	2	3	4	5	6
Number of times rolled	63	90	96	118	95	138

   a. Predict the number of times each number should appear if the die were a fair die.
   b. Explain whether you think the die is loaded.

11.5 Review: exam practice

11.5.1 Summary: Simulations and simple probability

Theoretical probability and sample spaces
- Probability is measure of the likely occurrence of an event.
- Probability is measured on a scale from 0 (impossible) to 1 (certain).
- Probability can be written as a decimal, fraction or a percentage.
- The probability of an event (A) occuring can be denoted by P(A).
- An outcome is a particular result of an experiment.
- A sample space is a list of all possible outcomes; often listed as a set of values {}.
- A favourable outcomes is an outcome you want or are looking for.
- Theoretical probability can be calculated by:

$$P(\text{event}) = \frac{\text{number of favourable outcomes}}{\text{total number of outcomes}}$$

- Equally likely outcomes have the same chance of occuring.
- The complement of an event is the probability that the event discussed will *not* happen.
- The sum of the probability of an event and its complement is equal to 1.

$$P(R) + P\left(R'\right) = 1$$

Multi-stage experiments (arrays and tree diagrams)
- An array can be used to display the sample space when two events occur to form a single result.
- Tree diagrams list all possible outcomes of two or more events. The branches show the possible links between one outcomes and the next outcome.
- Each event in a tree diagram is treated as if it were a separate event, even if the events occur at the same time.

Relative frequencies and simulations

- If a theoretical probability cannot be calculated, an experiment can be repeated a number of times (a trial) as part of a larger experiment.
- A successful trial is where the result is the desired outcome.
- Relative frequency, or empirical or experimental probability, can be determined by the formula:

$$\text{relative frequency} = \frac{\text{number of successful trials}}{\text{total number of trials}}$$

- As the number of trials increases, the relative frequency of an event will gradually become closer in value to the theoretical probability. If enough trials are carried out, relative frequency can be used as an estimate of probability.
- If an experiment cannot be carried out, outcomes of events can be modelled to simulate or represent what happens in real life.
- Random number generators can generate a series of numbers between two given values.
- Many factors can complicate simulations; modelling real life events can be very complex.

Exercise 11.5 Review: exam practice

Note: Unless otherwise stated, express probabilities as fractions.

Simple familiar

1. What is the sample space for the following scenarios?
   a. Rolling an 8-sided die numbered 1–8
   b. Drawing an 'a' from a bag containing the letters of the alphabet
   c. Drawing the winning ticket in a raffle with 100 tickets sold
2. Choose which of the following events are 'most likely' and 'very unlikely'.
   a. A cyclone in Victoria
   b. The temperature rising above 30 degrees Celsius on a summer day in Darwin
   c. Snow in Cairns
   d. Getting struck by lightning during a storm in Brisbane
3. Create three events that meet the following criteria.
   a. Less than even chance of occurring
   b. Greater than even chance of occurring
4. Two fair dice are rolled and the product of the numbers is recorded.
   a. Construct a table to illustrate the sample space. List the sample space for the experiment.
   b. What is the most likely outcome?
   c. What is the least likely outcome?
5. Two dice are rolled and the difference of the numbers is recorded.
   a. Construct a table to illustrate the sample space. List the sample space for the experiment.
   b. What is the most likely outcome?
   c. What is the least likely outcome?
6. a. What is the probability of rolling a 3 on a 6-sided die?
   b. What is the complement of the event in part a?
   c. What is the probability of the complement occurring?
7. a. If you toss two fair dice, what is the probability that the two numbers you roll are the same?
   b. What is the probability of the complement of this event?
   c. Would it be easier to calculate the probability of the complement before calculating the original event? Explain your answer.

8. The following table shows what people do with their old mobile phones in Australia.

Kept it	Shared it	Recycled it	Sold it	Lost it
48%	27%	15%	8%	2%

Source: Data from 2015 Deloitte Touche Tohmatsu, *Mobile Consumer Survey — The Australian Cut*

What is the probability of NOT selling or NOT losing your mobile phone in Australia? Give your answer as a percentage.

9. In the game of backgammon, two dice are rolled together. The resulting numbers can be used separately in two separate moves or their total can be used for one move.

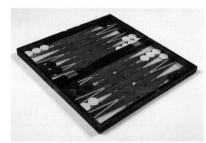

   a. List the sample space of rolling two dice simultaneously.
   b. Calculate the probability of 'getting a 4 on a die or a total of 4 on both dice'.
   c. Calculate the probability of NOT 'getting a 4 on a die or a total of 4 on both dice'.
   d. Calculate the probability of 'getting a 4 or getting a 5 on a die or a total of 5 on both dice'.

10. A soft-drink chemist is trying to create a new drink with three different ingredients. The chemist wants to experiment with the mixing order of the three key ingredients. Draw a tree diagram to show the sample space and determine how many combinations are possible.

11. A car dealer offers these package options on one of their new models.
   • Interior: leather or suede
   • Transmission: manual or automatic
   • Front seats: bucket, captain or regular
      a. How many different configurations of options are available?
      b. What is the theoretical probability a random customer chooses a bucket seat in their package?

12. A fair coin is flipped and then an 8-sided die is rolled.
   a. List all the possible outcomes in an array.
   b. What is the probability of getting a head and an 8?
   c. What is the probability of getting an even number or a head?

Complex familiar

13. At a fair, people were asked to taste and state their preference for two new brands of soft drink: brand A and brand B.
   a. List all the possible outcomes using a tree diagram.
   b. What is the probability, expressed as a percentage, that a randomly selected person liked:
      i. brand A but not brand B
      ii. brand A and brand B
      iii. brand A?

14. **a.** How many times, to the nearest whole number, would you expect to roll a 6 if you rolled a fair die:

    i. 10 times

    ii. 150 times

    iii. 5000 times?

  b. If you actually did the experiments in part **a**, would you get the exact number that you predicted? Explain your answer.

15. You are about to sit a Mathematics examination that contains 40 multiple-choice questions. You didn't have time to study so you are going to choose the answers completely randomly. There are four choices for each question.

Explain how you could use random numbers to select your answers for you.

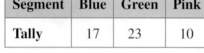

16. The spinner shown was spun 50 times and the outcome each time was recorded in the table.

Segment	Blue	Green	Pink
Tally	17	23	10

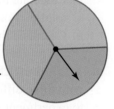

  a. List the sample space.

  b. Given the experimental results, determine the relative frequency for each segment.

  c. The theoretical probability of the spinner landing on any particular segment with one spin is $\frac{1}{3}$. How could the experiment be changed to give a better estimate of the true probabilities?

Complex unfamiliar

17. Jose has to call his girlfriend, but he cannot remember her full number. He knows the first eight digits, but can't recall the final two digits.

  a. What is the probability that he correctly guesses either one of the two final digits (but not both)?

  b. What is the probability that he correctly guesses both of the two final digits?

  c. What is the probability that he incorrectly guesses both of the final two digits?

18. A friend offers you a chance to roll a single die three times, and if you roll a 3 on any one of the rolls he will pay for your lunch. Otherwise, you will pay for his lunch. Should you play his game? Explain your answer.

19. There are five different possible hair colours: red, brunette, brown, grey and blonde. Assume that each hair colour is equally likely. If there are 5 people in a room, what is the probability that at least 2 have the same coloured hair? Give your answer as a percentage correct to two decimal places. (*Hint:* Use the complement to solve this problem.)

20. Sonya knows her credit card number for online shopping. However, she has forgotten the three-digit CVV number on the back of the card, and wants to purchase a dress before the sale ends.

  a. What is the probability she correctly guesses all three digits?

  b. What is the probability she incorrectly guesses all three digits?

  c. What is the probability that she correctly guesses at least one digit?

  d. She remembers that her CVV number does not have any repeated digits.

What is the probability of her correctly guessing all the numbers now she has the information?

Answers

11 Simulations and simple probability

Exercise 11.2 Theoretical probability and sample spaces

1. a. 1   b. 0.75   c. 0.25
   d. 0.75   e. 0.25   f. 1
   g. 0   h. 0.25   i. 1
   j. 0.25   k. 0.5   l. 0.75

2. Answers will vary. Sample responses can be found in the worked solutions in the online resources.

3. a. $\{1, 2, 3, 4, 5, 6\}$
   b. $\{\text{green, yellow, blue}\}$
   c. $\{a, b, c, d, e, f, g, h, i, j, k, l, m, n, o, p, q, r, s, t, u, v, w, x, y, z\}$

4. a. $\dfrac{1}{6}$ b. $\dfrac{1}{3}$ c. $\dfrac{1}{2}$

5. a. $\dfrac{4}{11}$ b. $\dfrac{7}{11}$
   c. $\dfrac{2}{11}$ d. $\dfrac{1}{11}$

6. a. $\{1, 2, 3, 4, 5, 6, 7, 8\}$
   b. i. 50%   ii. 62.5%
      iii. 0%   iv. 100%

7. a. i. No, the colours are not represented equally.
      ii. Blue, green, yellow, red
      iii. $P(\text{blue}) = \dfrac{1}{3}$, $P(\text{green}) = \dfrac{1}{6}$, $P(\text{red}) = \dfrac{1}{3}$,
          $P(\text{yellow}) = \dfrac{1}{6}$

   b. i. Yes, as there are equal numbers of the same colour.
      ii. Blue, green, yellow, pink, orange
      iii. $P(\text{blue}) = \dfrac{1}{5}$, $P(\text{green}) = \dfrac{1}{5}$, $P(\text{pink}) = \dfrac{1}{5}$,
          $P(\text{yellow}) = \dfrac{1}{5}$, $P(\text{orange}) = \dfrac{1}{5}$

   c. i. No, the colours are not represented equally.
      ii. Pink, blue
      iii. $P(\text{blue}) = \dfrac{1}{3}$, $P(\text{pink}) = \dfrac{2}{3}$

8. a. Blue. There are more blue jelly beans (6) than beans of any other colour.
   b. Yellow. There are fewer yellow jelly beans (2) than beans of any other colour.
   c. $P(\text{blue}) = \dfrac{6}{17}$, $P(\text{green}) = \dfrac{5}{17}$, $P(\text{red}) = \dfrac{4}{17}$,
      $P(\text{yellow}) = \dfrac{2}{17}$

9. D

10. a. $\dfrac{1}{4}$ b. $\dfrac{1}{2}$
    c. $\dfrac{1}{52}$ d. $\dfrac{3}{13}$

11. a.

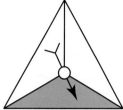

b.

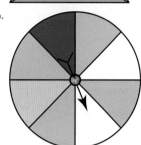

c.

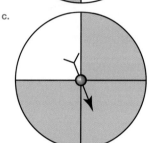

12. a. Not owning a dog
    b. Flipping a coin and getting a head
    c. Being awake
    d. Truancy
    e. Failing your learner knowledge test
    f. Not owning a mobile phone

13. a. Failing the final exam
    b. 80%

14. False. The complement of picking a red flower is not picking a red flower; therefore, the complement of picking a red flower is picking a purple or yellow flower.

15. a. $\dfrac{1}{9}$ b. $\dfrac{8}{9}$ c. $\dfrac{11}{36}$
    d. $\dfrac{25}{36}$ e. $\dfrac{1}{36}$ f. $\dfrac{35}{36}$

Exercise 11.3 Multi-stage experiments (arrays and tree diagrams)

1. * See the table at the bottom of the page.
2. * See the table at the bottom of the page.

$$P(\text{exactly one } 5) = \frac{8}{25}$$

3. D

4. a. 5%   b. 10%
   c. 10%   d. 20%

5. a.

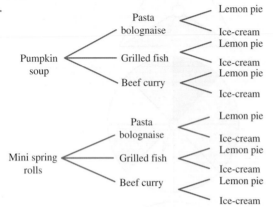

b. i. $\dfrac{1}{2}$

  ii. $\dfrac{1}{6}$

  iii. $\dfrac{1}{2}$

  iv. $\dfrac{1}{12}$

6. a.

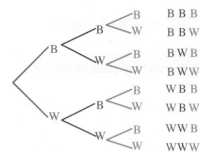

Hat	Bag	Hoodie	Outcomes
B	B	B	B B B
B	B	W	B B W
B	W	B	B W B
B	W	W	B W W
W	B	B	W B B
W	B	W	W B W
W	W	B	W W B
W	W	W	W W W

b. $\dfrac{1}{4}$

c. $\dfrac{1}{2}$

d. $\dfrac{1}{8}$

7. a.

b. i. $\dfrac{1}{4}$

  ii. $\dfrac{1}{8}$

  iii. $\dfrac{1}{4}$

---

*1.

	Die					
	1	**2**	**3**	**4**	**5**	**6**
Red	$(R,1)$	$(R,2)$	$(R,3)$	$(R,4)$	$(R,5)$	$(R,6)$
Blue	$(B,1)$	$(B,2)$	$(B,3)$	$(B,4)$	$(B,5)$	$(B,6)$
Green	$(G,1)$	$(G,2)$	$(G,3)$	$(G,4)$	$(G,5)$	$(G,6)$
Yellow	$(Y,1)$	$(Y,2)$	$(Y,3)$	$(Y,4)$	$(Y,5)$	$(Y,6)$
Orange	$(O,1)$	$(O,2)$	$(O,3)$	$(O,4)$	$(O,5)$	$(O,6)$

(Spinner — row label on left)

*2.

	Spinner 2				
	1	**2**	**3**	**4**	**5**
1	$(1,1)$	$(1,2)$	$(1,3)$	$(1,4)$	$(1,5)$
2	$(2,1)$	$(2,2)$	$(2,3)$	$(2,4)$	$(2,5)$
3	$(3,1)$	$(3,2)$	$(3,3)$	$(3,4)$	$(3,5)$
4	$(4,1)$	$(4,2)$	$(4,3)$	$(4,4)$	$(4,5)$
5	$(5,1)$	$(5,2)$	$(5,3)$	$(5,4)$	$(5,5)$

(Spinner 1 — row label on left)

8. a.

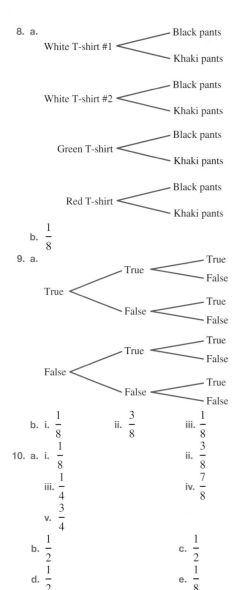

b. $\dfrac{1}{8}$

9. a.

True
— True
  — True
  — False
— False
  — True
  — False

False
— True
  — True
  — False
— False
  — True
  — False

b. i. $\dfrac{1}{8}$ **ii.** $\dfrac{3}{8}$ **iii.** $\dfrac{1}{8}$

10. a. i. $\dfrac{1}{8}$ **ii.** $\dfrac{3}{8}$

iii. $\dfrac{1}{4}$ **iv.** $\dfrac{7}{8}$

v. $\dfrac{3}{4}$

b. $\dfrac{1}{2}$ **c.** $\dfrac{1}{2}$

d. $\dfrac{1}{2}$ **e.** $\dfrac{1}{8}$

11. a. 1   **b.** $\dfrac{3}{8}$

12. B

13. a. $\dfrac{2}{49}$ **b.** $\dfrac{6}{49}$ **c.** $\dfrac{6}{49}$

14. a. $\dfrac{1}{4}$ **b.** $\dfrac{1}{6}$

c. $\dfrac{35}{36}$ **d.** 1

15. a. $\dfrac{1}{4}$ **b.** $\dfrac{1}{2}$

c. $\dfrac{1}{2}$ **d.** $\dfrac{1}{3}$

16. a.

+	1	2	3	4	5	6
1	2	3	4	5	6	7
2	3	4	5	6	7	8
3	4	5	6	7	8	9
4	5	6	7	8	9	10
5	6	7	8	9	10	11
6	7	8	9	10	11	12

b. The diagonals in the sample space are repeating numbers. Also, 7 is the outcome that shows up the most, and each outcome further away from 7 has a reduced chance of occurring.

c. * See the table at the bottom of the page.

d. The probabilities on either side of the 7 decrease at the same rate.

17. a.

−	1	2	3	4	5	6
1	0	1	2	3	4	5
2	1	0	1	2	3	4
3	2	1	0	1	2	3
4	3	2	1	0	1	2
5	4	3	2	1	0	1
6	5	4	3	2	1	0

b. $\dfrac{1}{3}$ **c.** $\dfrac{2}{3}$ **d.** No

18. a.

	Person 2		
	Rock	**Paper**	**Scissors**
Rock	RR	RP	RS
Paper	PR	PP	PS
Scissors	SR	SP	SS

(Person 1 labels the rows)

---

*16.c.

Sum of two dice	2	3	4	5	6	7	8	9	10	11	12
Probability	$\dfrac{1}{36}$	$\dfrac{1}{18}$	$\dfrac{1}{12}$	$\dfrac{1}{9}$	$\dfrac{5}{36}$	$\dfrac{1}{6}$	$\dfrac{5}{36}$	$\dfrac{1}{9}$	$\dfrac{1}{12}$	$\dfrac{1}{18}$	$\dfrac{1}{36}$

b. i. $\dfrac{2}{9}$

ii. $\dfrac{2}{9}$

iii. $\dfrac{2}{9}$

c. Yes

d. i. * See the table at the bottom of the page.

ii. No

iii. $P(\text{Rock wins}) = \dfrac{2}{16} = \dfrac{1}{8}$

$P(\text{Paper wins}) = \dfrac{2}{16} = \dfrac{1}{8}$

$P(\text{Scissors win}) = \dfrac{4}{16} = \dfrac{1}{4}$

$P(\text{Dynamite wins}) = \dfrac{4}{16} = \dfrac{1}{4}$

Scissors and dynamite are twice as likely to win as rock and paper.

19. a.

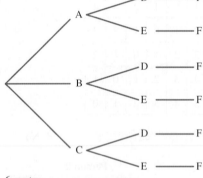

6 routes

b. No. There are still six paths, even if we omit home from the journey. Since there is only one choice to go from the library to home, the outcomes do not change.

20.

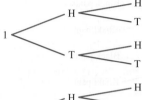

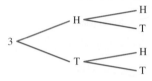

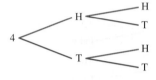

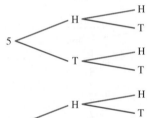

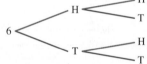

Exercise 11.4 Relative frequency and simulations

1. a. Mandy's simulation: $\dfrac{13}{20}$

Sophia's simulation: $\dfrac{47}{100}$

b. Mandy's simulation shows a higher probability of perfect weather days. Sophia's simulation is likely to be more reliable as there is a higher number of coin tosses.

2. a. $\dfrac{3}{50}$ b. 30 batteries

3. a. $\dfrac{13}{75}$ b. $\dfrac{31}{60}$ c. $\dfrac{29}{60}$

*18. d. i.

	Person 2			
	Rock	**Paper**	**Scissors**	**Dynamite**
Rock	RR	RP	RS	RD
Paper	PR	PP	PS	PD
Scissors	SR	SP	SS	SD
Dynamite	DR	DP	DS	DD

(Person 1 labels the rows)

4. $\dfrac{17}{28}$

5. C

6. Answers will vary. Sample responses can be found in the worked solutions in the online resources. You can use =RANDBETWEEEN(1, 60) in Excel to generate a random number.

7. a. 35%   **b.** 40%

8. a. Each driver could be assigned a number from 1 to 24. A random number generator could then be used to simulate the race. The number that is generated is the winner of the race.

b. The answer depends on each driver having an equally likely chance of winning the race. This is rarely the case.

9. a. P(fish and chips) $= \dfrac{45}{150} = 0.30$

b. P(fish and chips) $= \dfrac{31}{150} = 0.21$

c. P(macaroni and cheese) $= \dfrac{30}{150} = 0.20$

d. P(lamb souvlaki) $= \dfrac{25}{150} = 0.17$

e. P(BBQ pork ribs) $= \dfrac{19}{150} = 0.13$

10. Yes. It is too expensive to check every computer for every possible fault. A random number generator will allow the manufacturer to gain insight into the probability of a computer having a fault.

11. a. i. $\dfrac{3}{11}$

    ii. $\dfrac{1}{11}$

    iii. $\dfrac{1}{11}$

b. Answers will vary. Sample responses can be found in the worked solutions in the online resources.

c. Answers will vary. Accuracy can be improved by using the random number generator to test more days. As more data is obtained, the results will become closer and closer to the theoretical probability.

12. Answers will vary. Sample responses can be found in the worked solutions in the online resources.

13. a. 2439   **b.** 361

14. a. 440 domestic students (using most recent data)

b. Based on the trend observed from 2005 to 2011, the number of students who never came back after the first year has steadily decreased from 11.0% to 8.8%. Hence, the estimate in part **a** may be too high.

15. a. i. 38%

    ii. 62%

b. No. If the experiment was repeated, it is unlikely that the results would be the same as or similar to the previous experiment.

c. According to the law of large numbers, as the number of trials increases, the gap between the experimental probability and the theoretical probability decreases.

16. a–c. Answers will vary. Sample responses can be found in the worked solutions in the online resources.

d. Yes. Since it is a random experiment, and only 50 experiments are executed, everyone should have different results.

e. Approximately 33

17. a. i. 2

    ii. 10

    iii. 167

b. It is unlikely that you would achieve the theoretical probability during an experiment. However, the more times the experiment is repeated, the closer the results will be to theoretical probability. To obtain the theoretical probability consistently would require that the experiment was run an infinite amount of times.

18. a. $P(16) = \dfrac{1}{3}$

    $P(17) = \dfrac{11}{24}$

    $P(18) = \dfrac{5}{24}$

b. Answers will vary. Sample responses can be found in the worked solutions in the online resource.

c. Answers will vary. Sample responses can be found in the worked solutions in the online resource.

19. a. 100 times

b. There are not enough trials to decide whether the die is loaded, although the trend suggests that it might be. There is a large difference between the frequencies of 1 and 6. (The numbers 1 and 6 are on opposite sides of a standard die, so a weight on the 1 side would tend to make 6 show up more often.)

Exercise 11.5 Review: exam practice

1. a. {1, 2, 3, 4, 5, 6, 7, 8}

b. {a, b, c, d, e, f, g, h, i, j, k, l, m, n, o, p, q, r, s, t, u, v, w, x, y, z}

c. {winning ticket, losing tickets}

2. a. Very unlikely   **b.** Most likely

c. Very unlikely   **d.** Very unlikely

3. Answers will vary. Sample responses can be found in the worked solutions in the online resources.

4. a.

×	1	2	3	4	5	6
1	1	2	3	4	5	6
2	2	4	6	8	10	12
3	3	6	9	12	15	18
4	4	8	12	16	20	24
5	5	10	15	20	25	30
6	6	12	18	24	30	36

Sample space is {1, 2, 3, 4, 5, 6, 8, 9, 10, 12, 15, 16, 18, 20, 24, 25, 30, 36}

b. 6 or 12

c. 1, 9, 16, 25 or 36

5. a.

−	1	2	3	4	5	6
1	0	1	2	3	4	5
2	1	0	1	2	3	4
3	2	1	0	1	2	3
4	3	2	1	0	1	2
5	4	3	2	1	0	1
6	5	4	3	2	1	0

Sample space is $\{0, 1, 2, 3, 4, 5\}$

b. 1

c. 5

6. a. $\dfrac{1}{6}$

b. The complement is the set $\{1, 2, 4, 5, 6\}$. or not rolling a 3.

c. $\dfrac{5}{6}$

7. a. $\dfrac{1}{6}$ **b.** $\dfrac{5}{6}$

c. No. The probability of the original event is a smaller probability, and thus requires less counting.

8. 90%

9. a.

Die 2

	1	**2**	**3**	**4**	**5**	**6**
1	(1, 1)	(1, 2)	(1, 3)	(1, 4)	(1, 5)	(1, 6)
2	(2, 1)	(2, 2)	(2, 3)	(2, 4)	(2, 5)	(2, 6)
3	(3, 1)	(3, 2)	(3, 3)	(3, 4)	(3, 5)	(3, 6)
4	(4, 1)	(4, 2)	(4, 3)	(4, 4)	(4, 5)	(4, 6)
5	(5, 1)	(5, 2)	(5, 3)	(5, 4)	(5, 5)	(5, 6)
6	(6, 1)	(6, 2)	(6, 3)	(6, 4)	(6, 5)	(6, 6)

(left label: **Die 1**)

b. $\dfrac{7}{18}$ **c.** $\dfrac{11}{18}$ **d.** $\dfrac{2}{3}$

10.

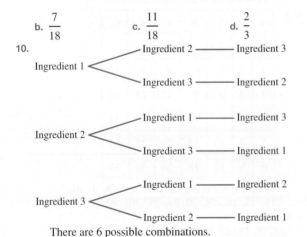

There are 6 possible combinations.

11. a. 12 **b.** $\dfrac{1}{3}$

12. a.

	1	**2**	**3**	**4**	**5**	**6**	**7**	**8**
H	H1	H2	H3	H4	H5	H6	H7	H8
T	T1	T2	T3	T4	T5	T6	T7	T8

b. $\dfrac{1}{16}$

c. $\dfrac{3}{4}$

13. a.

Like brand — A
Dislike brand — A
Like brand B Dislike brand B Like brand B Dislike brand B

b. i. 25% **ii.** 25% **iii.** 50%

14. a. i. 1 **ii.** 25 **iii.** 833

b. No, these are estimated numbers calculated from the theoretical probabilities. As the experiment is carried out more times, the experimental probabilities will approach the theoretical probabilities. There is a slight chance in any experiment that you may obtain the theoretical probability, but it is unlikely.

15. You could run your random number generator using the numbers 1- 4. These numbers would represent A, B, C and D respectively in the multiple-choice questions. You would run this simulation 40 times.

16. a. Blue, green, pink

$$P(\text{blue}) = \dfrac{17}{50}$$

b. $P(\text{green}) = \dfrac{23}{50}$

$$P(\text{pink}) = \dfrac{1}{5}$$

c. Run the experiment more times (e.g. 10 000 times).

17. a. $\dfrac{9}{50}$ **b.** $\dfrac{1}{100}$ **c.** $\dfrac{81}{100}$

18. No. The theoretical probability of rolling a 3 in three rolls is only $\dfrac{91}{216}$.

19. 96.16%

20. a. $\dfrac{1}{1000}$ **b.** $\dfrac{999}{1000}$

c. $\dfrac{3}{10}$ **d.** $\dfrac{1}{720}$

12 Loans and compound interest

12.1 Overview

LEARNING SEQUENCE

12.1 Overview
12.2 Simple interest
12.3 Simple interest as a recurrence relation
12.4 Compound interest
12.5 Comparing simple and compound interest
12.6 Review: exam practice

CONTENT

In this chapter, students will learn to:

- review the principles of simple interest through substitution of given values for other pronumerals into a mathematical formula to find the value of the subject of the formula
- understand the concept of compound interest as a recurrence relation
- consider similar problems involving compounding [complex]
- use technology (online calculator) to calculate the future value of a compound interest loan or investment and the total interest paid or earned
- use technology (spreadsheet) to calculate the future value of a compound interest loan or investment and the total interest paid or earned [complex]
- use technology (online calculator) to compare, numerically and graphically, the growth of simple interest and compound interest loans and investments
- use technology (spreadsheet) to compare, numerically and graphically, the growth of simple interest and compound interest loans and investments [complex]
- use technology (online calculator) to investigate the effect of the interest rate and the number of compounding periods on the future value of a loan or investment
- use technology (spreadsheet) to investigate the effect of the interest rate and the number of compounding periods on the future value of a loan or investment [complex].

Fully worked solutions for this chapter are available in the Resources section of your eBookPLUS at www.jacplus.com.au.

12.2 Simple interest

12.2.1 Introducing simple interest

When you lend money for a certain period of time (a **term deposit**) to a bank, building society or other financial institution, you expect to be rewarded by eventually getting your money back, plus an extra amount commonly known as **interest** (I).

Similarly, if you borrow money from any institution by taking out a loan or mortgage, you must pay back the original sum plus interest.

The following examples deal with **simple interest**, that is, interest which is paid only on the original sum of money invested or borrowed.

The interest earned from an investment depends on the **rate of interest**, the time period (**term**) for which it is invested and the amount (**principal**) invested. This relationship can be expressed formally by a mathematical equation.

Simple interest

The formula used to calculate simple interest is given by:

$$I = P \times i \times n \quad \text{or} \quad I = Pin$$

where:

I = simple interest ($)
P = principal ($) — that is, the sum of money borrowed or invested
i = interest rate as a decimal per time period
n = number of time periods — that is, the period of time for which the sum of money is to be borrowed or invested.

Note: The simple interest formula is also commonly written as $I = \dfrac{PrT}{100}$.

When we invest money, at the end of the time period we collect our investment (principal) and also the interest it has earned.

The sum of the principal, P, and the interest, I, is called the *total amount* and is denoted by the symbol A.

Total amount

The formula used to calculate the total amount is given by:

$$A = P + I$$

where:

A = total amount at the end of the term ($)
P = principal ($)
I = simple interest ($).

WORKED EXAMPLE 1

Calculate the amount of simple interest, I, earned and the total amount, A, at the end of the term, if:

a. **$12 000 is invested for 5 years at 9.5% p.a.**
b. **$2500 is invested for 3 months at 4.5% p.a.**

THINK	WRITE
a. 1. Write down the formula for simple interest.	**a.** $I = Pin$
2. Write down the known values of the variables.	$P = \$12\,000$ $i = \dfrac{9.5}{100}$ $= 0.095$ $n = 5$ years
3. Substitute the values into the given formula.	$I = 12\,000 \times 0.095 \times 5$
4. Evaluate.	$= 5700$
5. Answer the question and include the appropriate unit.	The amount of simple interest earned is \$5700.
6. Write down the formula for the total amount.	$A = P + I$
7. Substitute the values for P and I.	$= 12\,000 + 5700$
8. Evaluate.	$= 17\,700$
9. Answer the question and include the appropriate unit.	The total amount at the end of the term is \$17\,700.
b. 1. Write down the formula for simple interest.	**b.** $I = Pin$
2. Write down the known values of the variables. *Note: n must be expressed in years, so divide 3 months by 12.*	$P = \$2500$ $i = \dfrac{4.5}{100}$ $= 0.045$ $n = 3$ months $= \dfrac{3}{12}$ years $= 0.25$ years
3. Substitute the values into the given formula.	$I = 2500 \times 0.045 \times 0.25$
4. Evaluate and round off the answer to two decimal places.	$= 28.125$ $= 28.13$
5. Answer the question and include the appropriate unit.	The amount of simple interest earned is \$28.13.
6. Write down the formula for the total amount.	$A = P + I$
7. Substitute the values for P and I.	$= 2500 + 28.13$
8. Evaluate.	$= 2528.13$
9. Answer the question and include the appropriate unit.	The total amount at the end of the term is \$2528.13.

After comparing investment options from a
variety of institutions, Lynda and Jason decided
to invest their $18 000 in state government bonds
at 7.75% p.a. The investment is for 5 years and the
interest is paid biannually (twice per year). Calculate
how much interest:

a. they receive in every payment

b. will be received in total.

THINK	WRITE

a. 1. Write down the formula for simple interest.

a. $I = Pin$

2. Write down the known values of the variables.
Note: n must be expressed in years so divide
6 months by 12.

$P = \$18\,000$

$i = \dfrac{7.75}{100}$

$= 0.0775$

$n = 6$ months

$= \dfrac{6}{12}$ years

$= 0.5$ years

3. Substitute the values into the given formula.

$I = 18\,000 \times 0.0775 \times 0.5$

4. Evaluate.

$= 697.5$

5. Answer the question and include the appropriate
unit.

Lynda and Jason receive $697.50 in
interest every 6 months.

b. **Method 1**

1. Write down the formula for simple interest.

b. $I = Pin$

2. Write down the known values of the variables.

$P = \$18\,000$

$i = 0.0775$

$n = 5$ years

3. Substitute the values into the given formula.

$I = 18\,000 \times 0.0775 \times 5$

4. Evaluate.

$= 6975$

5. Answer the question and include the appropriate
unit.

Lynda and Jason will receive a total of
$6975 in interest.

b. **Method 2**

1. Multiply the interest received in each 6-month
period by the number of 6-month periods in
5 years; that is, multiply $697.50 by 10.

Interest obtained every 6 months
$= \$697.50$
Number of payments to be received
$= 10$
Total interest received $= \$697.50 \times 10$
$\qquad\qquad\qquad\qquad = \$6975$

2. Answer the question.

Lynda and Jason will receive a total of
$6975 in interest.

12.2.2 Transposing the simple interest formula

In many cases we may wish to calculate the principal, interest rate or time period of a loan. In these situations it is necessary to rearrange or transpose the simple interest formula. This can be done before or after substitution.

Transposed simple interest formula

It may be easier to use the transposed formula when finding P, i or n.

> **Simple interest formula transpositions**
>
> To calculate the interest rate, use $i = \dfrac{I}{Pn}$
>
> To calculate the period of the loan or investment, use
> $n = \dfrac{I}{Pi}$
>
> To calculate the principal, use $P = \dfrac{I}{in}$

WORKED EXAMPLE 3

Calculate the interest rate offered when \$720 is invested for 36 months and earns \$205.20 simple interest. Express rates in % per annum.

THINK	WRITE
Method: transpose before substitution	
1. Write the transposed simple interest formula for i.	$i = \dfrac{I}{Pn}$
2. List the values of P, I and n. n must be expressed in years.	$P = \$720$ $I = \$205.20$ $n = 36 \text{ months} = 3 \text{ years}$
3. Substitute into the formula.	$i = \dfrac{205.20}{720 \times 3}$
4. Evaluate on a calculator.	$\dfrac{205.20}{720 \times 3} = 0.095$
5. An interest rate is normally displayed as a percentage. Convert this decimal to a percentage by multiplying by 100.	$0.095 \times 100 = 9.5\%$
6. Write your answer.	The interest rate offered was 9.5% per annum.

WORKED EXAMPLE 4

Calculate the period of time, to the nearest year, for an investment of \$255 at simple interest of 8.5% p.a. to earn \$86.70 in interest.

THINK	WRITE
Method: transpose before substitution	
1. Write the transposed simple interest formula for n.	$n = \dfrac{I}{Pi}$

2. List the values of P, I and i.	$P = \$255$
	$I = \$86.70$
	$i = \dfrac{8.5}{100} = 0.085$
3. Substitute into the formula.	$n = \dfrac{86.70}{255 \times 0.085}$
4. Evaluate on a calculator.	$\dfrac{86.70}{255 \times 0.085} = 4$
5. Write your answer.	The period of the investment was 4 years.

WORKED EXAMPLE 5

Calculate the principal invested for an investment offering simple interest of 9% p.a., earning $215 interest over 4 year.

THINK	WRITE
Method: transpose after substitution	
1. Write the simple interest formula.	$I = Pin$
2. List the values of I, i and n.	$I = \$215$
	$i = \dfrac{9}{100} = 0.09$
	$n = 4$ years
3. Substitute into the formula.	$I = P \times i \times n$
	$215 = P \times 0.09 \times 4$
4. Evaluate 0.09×4 using a calculator to simplify the equation.	$215 = P \times 0.36$
5. Make P the subject by dividing both sides by 0.36.	$P = \dfrac{215}{0.36}$
6. Use a calculator to evaluate.	$P = 597.22$
7. Write your answer.	The amount invested was $597.22.

on Resources

📄 **Digital document** SpreadSHEET Simple interest (doc-9498)

Exercise 12.2 Simple interest

1. **WE1** Calculate the amount of simple interest, I, earned and the total amount, A, at the end of the term for each of the following.
   - a. $680 for 4 years at 5% p.a.
   - b. $210 for 3 years at 9% p.a.
   - c. $415 for 5 years at 7% p.a.
   - d. $460 at 12% p.a. for 2 years
   - e. $1020 at $12\frac{1}{2}$% p.a. for 2 years
   - f. $713 at $6\frac{3}{4}$% p.a. for 7 years

g. $821 at $7\frac{1}{4}$% p.a. for 3 years

h. 11.25% p.a. on $65 for 6 years

i. 6.15% p.a. on $21.25 for 9 years

j. 9.21% p.a. on $623.46 for 4 years

k. $13\frac{3}{4}$% p.a. on $791.35 for 5 years

2. **WE2** Sue and Harry invested $14 500 in state government bonds at 8.65% p.a. The investment is for 10 years and the interest is paid biannually (twice per year). Calculate how much interest:
   a. they receive every payment
   b. will be received in total.

3. **WE3** For each of the following, calculate the interest rate offered. Express rates in % per annum.
   a. Loan of $10 000, with a $2000 interest charge, for 2 years
   b. Investment of $5000, earning $1250 interest, for 4 years
   c. Loan of $150, with a $20 interest charge, for 2 months
   d. Investment of $1400, earning $178.50 interest, for 6 years
   e. Investment of $6250, earning $525 interest, for $2\frac{1}{2}$ years

4. **WE4** For each of the following, calculate the period of time (to the nearest month) for which the principal was invested or borrowed.
   a. Investment of $1000, at simple interest of 5% p.a., earning $50 interest
   b. Loan of $6000, at simple interest of 7% p.a., with an interest charge of $630
   c. Loan of $100, at simple interest of 24% p.a., with an interest charge of $6
   d. Investment of $23 000, at simple interest of $6\frac{1}{2}$% p.a., earning $10 465 interest
   e. Loan of $1 500 000, at simple interest of 1.5% p.a., with an interest charge of $1875

5. **WE5** For each of the following, calculate the principal invested.
   a. Simple interest of 5% p.a., earning $307 interest over 2 years
   b. Simple interest of 7% p.a., earning $1232 interest over 4 years
   c. Simple interest of 8% p.a., earning $651 interest over 18 months
   d. Simple interest of $5\frac{1}{2}$% p.a., earning $78 interest over 6 years
   e. Simple interest of 6.25% p.a., earning $625 interest over 4 years

6. Determine the interest earned on the following investments.
   a. $690 invested at 12% p.a. simple interest for 15 months
   b. $7500 invested for 3 years at 12% per year simple interest
   c. $25 000 invested for 13 weeks at 5.2% p.a. simple interest
   d. $250 invested at 21% p.a. for $2\frac{1}{2}$ years

7. Determine the amount to which each investment has grown after the investment periods shown in the following examples.
   a. $300 invested at 10% p.a. simple interest for 24 months
   b. $750 invested for 3 years at 12% p.a. simple interest
   c. $20 000 invested for 3 years and 6 months at 11% p.a. simple interest
   d. $15 invested at $6\frac{3}{4}$% p.a. for 2 years and 8 months
   e. $10.20 invested at $8\frac{1}{2}$% p.a. for 208 weeks

8. **MC** If John had $63 in his bank account and earned 9% p.a. over 3 years, what would be the simple interest earned?
   A. $5.67   **B.** $17.01   **C.** $22.68   **D.** $80.01

9. **MC** If $720 was invested in a fixed deposit account earning $6\frac{1}{2}\%$ p.a. for 5 years, what would be the interest earned at the end of 5 years?

A. $23.40   B. $216.00   C. $234.00   D. $954.00

10. **MC** Bodgee Bank advertised a special offer. If a person invests $150 for 2 years, the bank will pay 12% p.a. simple interest on the money. How much would the investor have earned at the expiry date?

A. $36   B. $48   C. $186   D. $300

11. **MC** Joanne asked Sally for a loan of $125 to buy new shoes. Sally agreed on the condition that Joanne paid it back in two years at 3% p.a. simple interest. What amount did Joanne pay Sally at the end of the two years?

A. $7.50   B. $125   C. $130.50   D. $132.50

12. **MC** Two banks pay simple interest on short-term deposits. Hales Bank pays 8% p.a. over 5 years and Countrybank pays 10% p.a. for 4 years. What is the difference between the two banks' final payout figure if $2000 was invested in each account?

A. $0   B. $150   C. $800   D. $1200

13. **MC** Joanne's accountant found that for the past 2 years she had earned a total of $420 interest in an account paying 6% p.a. simple interest. When she calculated how much she invested, what was the amount?

A. $50.40

B. $350

C. $3500

D. $5040

14. **MC** A loan of $1000 is taken over 5 years. The total amount repaid for this loan is $1800. What is the simple interest rate per year on this loan?

A. 5%   B. 8%   C. 9%   D. 16%

15. **MC** Jarrod decides to buy a motorbike at no deposit and no repayments for 3 years. He takes out a loan of $12 800 and is charged at 7.5% p.a. simple interest over the 3 years. What is the lump sum Jarrod has to pay in 3 years' time?

A. $2880   B. $9920   C. $13 760   D. $15 680

16. Jill and John decide to borrow money to improve their yacht, but cannot agree which loan is the better value. They would like to borrow $2550. Jill goes to the Big-4 Bank and finds that they will lend her the money at $11\frac{1}{3}\%$ p.a. simple interest for 3 years. John finds that the Friendly Building Society will lend the $2550 to them at 1% per month simple interest for the 3 years.

a. Which institution offers the better rate over the 3 years?

b. Explain why.

17. Lennie Cavan earned $576 in interest when she invested in a fund paying 9.5% p.a. simple interest for 4 years. How much did Lennie invest originally?

18. Lennie's sister Lisa also earned $576 interest at 9% p.a. simple interest, but she only had to invest it for 3 years. What was Lisa's initial investment?

19. Jack Kahn put some money away for 5 years in a bank account which paid $3\frac{3}{4}\%$ p.a. interest. He found from his bank statement that he had earned \$66. How much did Jack invest?

20. James needed to earn \$225. He invested \$2500 in an account earning simple interest at a rate of 4.5% p.a. How many months will it take James to achieve his aim?

21. Anna invested \$85 000 in Ski International debentures. She earns 7.25% p.a. which is paid quarterly for one year.
    a. Calculate how much interest:
        i. Anna receives quarterly
        ii. will be received in total, over a year.
    b. Would Anna receive the same amount of interest over a 3-year period if it were paid annually rather than quarterly?

22. Mrs Williams invested \$60 000 in government bonds at 7.49% p.a. with interest paid biannually (i.e. every 6 months).
    a. How much interest is she paid each 6 months?
    b. How much interest is she paid over $3\frac{1}{2}$ years?
    c. How long would the money need to be invested to earn a total of \$33 705 in interest?

12.3 Simple interest as a recurrence relation

The value of a simple interest loan or investment can also be calculated using recurrence relations. A **recurrence relation** is a mathematical rule that we can use to generate a sequence of numbers, where each term in the sequence can be found from the previous term. It has two parts:
1. a starting point — the value of one of the terms in the sequence
2. a rule that can be used to generate successive terms in the sequence.

In a simple interest loan or investment, the amount of interest paid or earned remains constant for each time period; hence, simple interest is a special case of linear growth in which the starting value is the amount borrowed or invested. The recurrence relation for simple interest is given by the following rule.

Recurrence relation for simple interest

The recurrence relation for the value of the simple interest loan or investment after n years is:

$$V_0 = \text{principal}, \quad V_{n+1} = V_n + d$$

where:

V_n represents the value of the investment after n time periods

d is the amount of simple interest earned each time period, which can be found using the formula $d = i \times V_0$

V_{n+1} represents the value of the investment after $n + 1$ time periods.

Note that in the formula $d = i \times V_0$, i represents the interest rate as a decimal per time period.

When using a recurrence relation to model a simple interest problem, it may be necessary to determine the total amount of interest earned. This can be found by using the following formula.

Total interest earned

Amount of simple interest = (amount of investment) − principal

$$I = A - P$$

It is important to clarify the differences between using a recurrence relation to model a simple interest problem and using the simple interest formula from Section 12.2.

	Calculating the amount of interest	Calculating the total amount of the loan or investment
Simple interest formula	$I = Pin$	$A = P + I$
Recurrence relation	$I = A - P$ or $I = V_0 \times i \times n$ where A is the value of the recurrence relation.	$V_{n+1} = V_n + d$, $V_0 = $ principal where $d = i \times V_0$.

WORKED EXAMPLE 6

$575 is invested in a simple interest account for 3 years at 4% p.a.
a. **Set up a recurrence relation to find the value of the investment after n years.**
b. **Use the recurrence relation to find the value of the investment at the end of each of the first 3 years.**

THINK

a. 1. Write the formula to calculate the amount of interest earned per period.

2. Substitute the values of V_0 and i.

3. Write the fomula for the recurrence relation that models simple interest.

4. Substitute the values of V_0 and d to give the recurrence relation.

b. 1. Since V_0 represents the initial amount (principal) of money invested, V_1 would represent the value of the investment after 1 year. To calculate V_1, substitute $n = 0$ into the recurrence relation.

2. Substitute the value of V_0 into this equation and complete the addition.

3. To determine the value of V_2, the value of the investment after 2 years, substitute $n = 1$ into the recurrence relation.

4. Substitute the value of V_1 into this equation and complete the addition.

5. To determine the value of V_3, the value of the investment after 3 years, substitute $n = 2$ into the recurrence relation.

6. Substitute the value of V_2 into this equation and complete the addition.

WRITE

a. $d = i \times V_0$

$d = \dfrac{4}{100} \times 575$
$= 23$

$V_0 = $ principal, $V_{n+1} = V_n + d$

$V_0 = 575$, $V_{n+1} = V_n + 23$

b. $V_{n+1} = V_n + 23$
$V_{0+1} = V_0 + 23$
$V_1 = V_0 + 23$

$V_1 = 575 + 23$
$= 598$

$V_{n+1} = V_n + 23$
$V_{1+1} = V_1 + 23$
$V_2 = V_1 + 23$

$V_2 = 598 + 23$
$= 621$

$V_{n+1} = V_n + 23$
$V_{2+1} = V_2 + 23$
$V_3 = V_2 + 23$

$V_3 = 621 + 23$
$= 644$

7. State the answers.	The value of the investment after 1 year is $598.
	The value of the investment after 2 years is $621.
	The value of the investment after 3 years is $644.

WORKED EXAMPLE 7

Sandra invests $420 in a fixed deposit account that pays 6% p.a. simple interest for 15 months.
a. How much is her investment worth after 15 months?
b. Represent the account balance for each of the 15 months graphically.

THINK	WRITE
a. 1. Write the simple interest formula.	**a.** $I = Pin$
2. To calculate the amount of interest earned over the 15 months, substitute the values into the formula and evaluate. Fifteen months converted into years is $\frac{15}{12}$.	$I = 420 \times \dfrac{6}{100} \times \dfrac{15}{12}$ $= 31.50$
3. Write the answer.	Interest = $31.50
4. To determine the total amount add the interest to the principal.	$A = P + I$ $= 420 + 31.50$ $= 451.50$
5. Write the answer.	Investment worth = $451.50
b. 1. Since the investment is simple interest, the value of the investment increases by the same amount each month. To determine the monthly increase, divide the total interest by the number of months.	**b.** Increase per month $= \dfrac{31.50}{15}$ $= \$2.10$
2. Draw a set of axes displaying only positive values. Label the horizontal axis as Time (months). Label the vertical axis as Investment ($). Plot the points starting at $(0, 420)$ and increasing by 2.10 each month to the final point of $(15, 451.50)$.	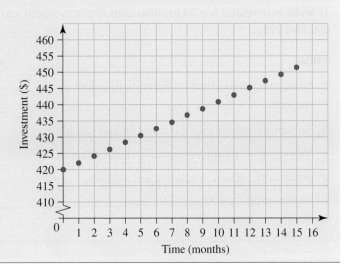

Finding V_0, i and n

We may not always want to calculate the interest. There are times that we may want to calculate the principal, interest rate or number of time periods. To do this, the simple interest formula needs to be transposed. The transposed version for each of these are as follows.

$$V_0 = \frac{I}{i \times n}$$

$$i = \frac{I}{V_0 \times n}$$

$$n = \frac{I}{V_0 \times i}$$

Note that V_0 and P are interchangeable.

WORKED EXAMPLE 8

A building society offers 8% p.a. simple interest on an investment. The investment earned $320 after 5 years. How much was initially invested?

THINK	WRITE
1. Write the simple interest formula to calculate the principal.	$V_0 = \dfrac{I}{i \times n}$
2. Substitute values into the formula and calculate.	$V_0 = \dfrac{320}{\frac{8}{100} \times 5}$
	$= \dfrac{320}{0.08 \times 5}$
	$= 800$
3. Write the answer.	The principal was $800.

WORKED EXAMPLE 9

If $840 is invested for 24 months and the invesment earns $126 in interest, calculate the yearly interest rate.

THINK	WRITE
1. Write the simple interest formula to calculate the rate.	$i = \dfrac{I}{V_0 \times n}$
2. Substitute values into the formula and calculate. Twenty-four months equals $\dfrac{24}{12}$ or (2) years.	$i = \dfrac{126}{840 \times \frac{24}{12}}$ $= 0.075$
3. Convert the decimal into a percentage by multiplying by 100.	$i = 0.075 \times 100$ $= 7.5$
4. Write the answer.	The yearly interest rate was 7.5%.

Exercise 12.3 Simple interest as a recurrence relation

1. **WE6** $688 is invested in a simple interest account for 4 years at 6.25% p.a.
   a. Set up a recurrence relation to calculate the value of the investment after n years.
   b. Use the recurrence relation in part a to determine the value of the investment after 4 years.

2. $728 is invested in a simple interest account for 3 years at 8.5% p.a.
   a. Set up a recurrence relation to calculate the value of the investment after n years.
   b. Use the recurrence relation in part a to determine the value of the investment after 3 years.

3. **WE7** Marina invests $1560 in a fixed deposit account that pays 10.4% p.a. simple interest for 18 months.
   a. How much is her investment worth after 18 months?
   b. Represent the account balance for each of the first 18 months graphically.

4. a. Determine the amount the investment has grown if $770 was invested at 9.2% p.a. for 2 years.
   b. Represent the balance of the account for each of the first 12 months.

5. **WE8** A buliding society offers 6% p.a. simple interest on an investment. The investment earned $724 after 3 years. How much was initially invested?

6. Find the principal that was invested if simple interest is 9% p.a. which earns $468 in interest over 4 years.

7. **WE9** If $4500 is invested for 4 years and the investment earns $870 in interest, calculate the yearly interest rate.

8. Calculate the annual interest rate offered for an investment of $600 that earns a total of $80 in interest over 6 months.

9. Calculate the period of time for which the principal was invested if $3800 was invested with a simple interest of 6% p.a. and $420 interest was earned.

10. Determine the value of the following investments by using a recurrence relation.
    a. $860 for 4 years at 4% p.a.
    b. $390 for 5 years at 7% p.a.

11. Calculate the interest charged or earned on the following loans and investments.
    a. $1100 loaned at 11% p.a. simple interest for 21 months
    b. $5750 invested for 2 years at 0.8% per month simple interest
    c. $31 500 borrowed for 26 weeks at 0.15% per week simple interest
    d. $5575 invested at 1.8% per month for 3.5 years

12. Determine the amount to which each investment has grown in each of the following cases.
    a. $500 invested at 12% p.a. simple interest for 24 months
    b. $680 invested for 2 years at 1.3% per month simple interest
    c. $18 000 invested for 4 years and 6 months at 9.5% p.a. simple interest

13. Jeremy invested the $850 he received for his birthday from his family members. He invested at 9.5% p.a. simple interest for a period of 4 years.
    a. What is Jeremy's investment worth after 4 years?
    b. Represent the balance at the end of each year graphically.

14. Joel and Shae decided to borrow money to put in a basketball area and new decking at their house. They cannot agree on who they should take the loan out with. They would like to borrow $25 000. Joel goes to Tigers Bank, which offers the money at 9.8% p.a. simple interest for 4 years. Shae finds that Twins Bank will lend the money to them at 0.9% per month simple interest for 4 years.
    a. Which option offers the better deal over 4 years?
    b. Explain why.

15. The value of a simple interest investment at the end of year 2 is $3136. At the end of year 3 the investment is worth $3512.32. Use a recurrence relation to work out how much was invested.

16. Abuk needed to earn $320 in one year. She invested $8000 in an account earning simple interest at a rate of 4.8% p.a. paid monthly. How many months will it take Abuk to achieve her goal?

12.4 Compound interest

Compound interest is a different type of interest. Simple interest depends on the initial amount invested; however, it is more common for interest to be calculated on the changing value throughout the time period of a loan or investment. This is known as compounding.

In compouding, the interest is added to the balance, and the next interest calculation is made on the new value. This means that the amount of interest received changes over time.

The final amount of a compound interest investment or loan depends on the rate of interest, the time period for which it is invested, and the amount invested. This relationship can be expressed formally by a mathematical equation.

> ### Final amount of compound interest investment or loan
> The formula used to determine the value of a compound interest investment or loan is given by the rule:
>
> $$A = P(1 + i)^n$$
>
> where:
> A = the final amount of a compound interest investment or loan ($)
> P = principal ($) — that is, the sum of money borrowed or invested
> i = interest rate as a decimal per time period
> n = number of compounding time periods.

12.4.1 Compound interest as a recurrence relation

Compound interest calculations can also be evaluated using a recurrence relation. These recurrence relations involve a **compounding factor**, R, which is the factor by which the value of an investment or loan changes each time period.

> ### Recurrence relation for value of compound interest investment or loan
> The recurrence relation for the value of a compound interest investment or loan after n compounding periods is given by:
>
> $$V_0 = \text{principal}, \quad V_{n+1} = RV_n$$
>
> where:
> V_n is the value of the investment after n compounding periods
> R is the compounding factor, with $R = (1 + i)$
> i is the interest rate as a decimal per compounding period.

12.4.2 Value after *n* time periods

Since recurrence relations use a step-by-step method to generate the final amounts of a compound interest investment or loan, it can become tedious to determine the final amount of an investment or loan after, say, 25 years. It is possible to derive a rule for calculating the final amount (any term in the sequence) directly.

Recurrence relation for the value of a compound interest loan or investment after n time periods

The value of a compound interest investment or loan after the nth compounding period is given by the rule:

$$V_0 = \text{starting value}, \quad V_n = R^n \times V_0$$

where:
V_n is the value of the investment or loan after n compounding periods
R is the compounding factor, with $R = (1 + i)$
i is the interest rate as a decimal per compounding period
n is the number of compounding periods.

The above rule can be interchanged with $A = P(1 + i)^n$. Can you see how these two rules are the same?

Note that the compound interest formula gives the final amount (*total value*) of an investment, not just the interest earned as in the simple interest formula.

Another rule can be used to find the total interest.

Total interest compounded

The total interest compounded after n time periods, I, is given by the rule:

$$I = \text{final amount} - \text{principal}$$
$$I = V_n - V_0$$

where
$V_0 = \text{principal (\$)}$.

If compound interest is used, the value of the investment at the end of each period grows by an increasing amount. Therefore, when plotted, the values of the investment at the end of each period form an exponential curve.

Now let us consider how the formula is used.

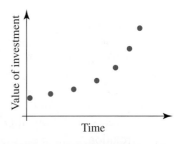

WORKED EXAMPLE 10

$5000 is invested for 4 years at 6.5% p.a. interest, compounded annually.
a. Generate the compound interest formula for this investment.
b. Calculate the amount in the balance after 4 years and the interest earned over this period.

THINK	WRITE
a. 1. Write the compound interest formula.	a. $A = P(1 + i)^n$
2. List the values of n, i and P.	$n = 4$
	$i = \dfrac{6.5}{100} = 0.065$
	$P = 5000$
3. Substitute the values into the formula.	$A = 5000(1 + 0.065)^4$

4. Complete the addition inside the brackets to simplify.

$A = 5000(1.065)^4$

b. 1. Use your calculator to evalute $5000(1.065)^4$ correct to two decimal places. This is the amount of the investment after 4 years.

b. $A = 5000(1.065)^4$
$= 6432.33$

2. To determine the amount of interest earned, subtract the principal from the balance.

$I = \text{final amount} - \text{principal}$
$= 6432.33 - 5000$
$= \$1432.33$

3. Write your answer.

The amount of interest earned is $1432.33 and the balance is $6432.33.

Alternative method — using a recurrence relation

a. 1. Write the recurrence relation for the value of a compound interest investment or loan.

a. $V_0 = \text{principal}, \ V_{n+1} = RV_n$

2. List the values of V_0 and R.

$V_0 = 5000$
$R = 1 + i$
$= 1 + \dfrac{6.5}{100}$
$= 1 + 0.065$
$= 1.065$

3. Substitute the values into the rule to write the recurrence relation.

$V_0 = 5000, \ V_{n+1} = 1.065V_n$

b. 1. Calculate the value of the investment after 1 year (V_1). Substitute $n = 0$ into the recurrence relation.

b. $V_{0+1} = 1.065V_0$
$V_1 = 1.065 \times V_0$
$= 1.065 \times 5000$
$= 5325$

2. Calculate the value of the investment after 2 years (V_2). Substitute $n = 1$ into the recurrence relation.

$V_{1+1} = 1.065V_1$
$V_2 = 1.065 \times V_1$
$= 1.065 \times 5325$
$= 5671.125$

3. Calculate the value of the investment after 3 years (V_3). Substitute $n = 2$ into the recurrence relation.

$V_{2+1} = 1.065V_2$
$V_3 = 1.065 \times V_2$
$= 1.065 \times 5671.125$
$= 6039.748125$

4. Calculate the value of the investment after 4 years (V_4). Substitute $n = 3$ into the recurrence relation.

$V_{3+1} = 1.065V_3$
$V_4 = 1.065 \times V_3$
$= 1.065 \times 6039.748125$
$= 6432.33...$

5. Write the answer.

The amount of the investment after 4 years is $6432.33.

12.4.3 Non-annual compounding

In Worked example 10, interest was compounded annually. However, in many cases the interest is compounded more often than once a year; for example, quarterly (every 3 months), monthly, weekly or daily. In these situations, n and i still have their usual meanings and we calculate them as follows.

Adjusting the rate and number of periods for non-annual compounding

Number of interest periods, n = number of years × number of interest periods per year

$$\text{Interest rate per period, } i = \frac{\text{interest rate per annum}}{\text{number of interest periods per year}}$$

WORKED EXAMPLE 11

If \$3200 is invested for 5 years at 6% p.a. interest compounded quarterly:
a. determine the number of interest-bearing periods, n
b. calculate the interest rate per period, i
c. calculate the balance of the account after 5 years
d. graphically represent the balance at the end of each quarter for 5 years and describe the shape of the graph.

THINK	WRITE
a. Calculate n.	a. $n = 5\,(\text{years}) \times 4\,(\text{quarters})$ $= 20$
b. Convert % p.a. to % per quarter to match the time over which the interest is calculated. Divide i% p.a. by the number of compounding periods per year, namely 4. Write as a decimal.	b. $i = \dfrac{6\% \text{ p.a.}}{4}$ $i = 1.5\%$ per quarter
c. 1. Write the compound interest formula. You could use the compound interest formula directly or use a recurrence relation. In this case, we will use the formula directly to save time.	c. $A = P\,(1 + i)^n$ or $V_0 = $ principal, $V_{n+1} = RV_n$
2. List the values of n, i and P.	$n = 20$ $i = \dfrac{1.5}{100} = 0.015$ $P = 3200$
3. Substitute into the formula.	$A = 3200(1 + 0.015)^{20}$
4. Simplify by completing the addition.	$= 3200(1.015)^{20}$
5. Use your calculator to evaluate $3200(1.015)^{20}$ correct to two decimal places.	$= \$4309.94$
6. Write your answer.	The balance of the account after 5 years is \$4309.94.
d. 1. Using a calculator, find the balance at the end of each quarter. This will involve using the formula $A = 3200\,(1.015)^n$ repeatedly for different values of n.	d. \$3200, \$3284, \$3296.72, \$3346.17 ... , \$4309.94

2. Plot these values on a set of axes. The first point is $(0, 3200)$, which represents the principal.

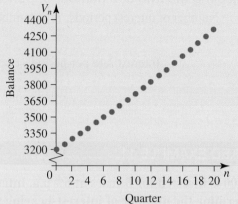

3. Comment on the shape of the graph.

The graph shows exponential growth as the interest is added at the end of each quarter and the following interest is calculated on the *new* balance.

The situation often arises where we require a certain amount of money by a future date. It may be to pay for a holiday or to finance the purchase of a car. It is then necessary to know what principal should be invested now so that it will increase in value to the desired final balance within the time available.

WORKED EXAMPLE 12

Determine the principal that will grow to \$4000 in 6 years, if interest is compounded quarterly at 6.5% p.a.

THINK	WRITE
1. Write the compound interest formula. Here we will use a recurrence relation rule. You could also use $A = P(1 + i)^n$.	$V_0 =$ starting value, $V_n = R^n \times V_0$
2. Calculate n (there are 4 quarters in a year).	$n = 6 \times 4$ $= 24$
3. Calculate i.	$i = \dfrac{6.5\%}{4}$ $= 1.625\%$
4. Calculate R.	$R = 1 + i$ $= 1 + \dfrac{1.625}{100}$ $= 1 + 0.01625$ $= 1.01625$
5. The symbol for the value of the investment after 6 years (24 compounding periods) is V_{24}. State its value.	$V_{24} = 4000$
6. Substitute V_{24} and R into the rule and simplify.	$V_{24} = R^{24} \times V_0$ $4000 = (1.01625)^{24} \times V_0$
7. Transpose to isolate V_0.	$V_0 = \dfrac{4000}{(1.016\,25)^{24}}$

8. Use a calculator to evaluate $\dfrac{4000}{(1.016\,25)^{24}}$ correct to two decimal places.

$= 2716.73$

9. Write a summary statement.

The principal would need to be $2716.73.

12.4.4 Using a spreadsheet to calculate compound interest

Calculating compound interest over a long period of time requires numerous repetitive calculations. Spreadsheets are a useful tool to perform these calculations.

WORKED EXAMPLE 13

A bank offers interest of 7.5% per annum compounded yearly on investments. $5000 is invested for 10 years. Using a spreadsheet, determine the value of the investment after the 10 years.

THINK

1. Open a spreadsheet labelling column A as 'n, year' and column B as 'value, $'. Place these labels in row 1 as shown. In column A (starting in cell A2), input the numbers from 0 to 10. We have placed all the values of n (from year 0 to year 10) in column A. Column B will display the value of the investment for its corresponding year number.

2. Write down the formula for compound interest.

3. Calculate $1 + i$.

4. Substitute this value and the value of P into the compound interest formula. We are now ready to input the compound interest formula into the spreadsheet.

5. Use the cursor to click into cell B2. Type the following: =5000*(1.075)^A2 followed by Enter. This should automatically input the value of 5000 in cell B2.

WRITE

	A	B
1	n, year	value, $
2	0	
3	1	
4	2	
5	3	
	⋮	
12	10	

$A = P(1 + i)^n$

$1 + i = 1 + \dfrac{7.5}{100}$

$= 1 + 0.075$

$= 1.075$

$A = 5000\,(1.075)^n$

	A	B
1	n, year	value, $
2	0	5000
3	1	

6. Copy the calculation for the next 10 years by using the fill-down function (or drag the bottom right corner down the cells). Complete column B to show 10 years.

	A	B
1	n, year	value, $
2	0	5000
3	1	5375
4	2	5778.125
5	3	6211.484
6	4	6677.346
7	5	7178.147
8	6	7716.508
9	7	8295.246
10	8	8917.389
11	9	9586.193
12	10	10305.16

7. Answer the question.

After 10 years, the value of the investment is $10 305.16.

---

on Resources

Digital document SpreadSHEET Compound interest (doc-9603)

Exercise 12.4 Compound interest

1. **WE10** $2500 is invested for 5 years at 7.5% p.a. compounding annually.
   a. Generate the compound interest formula for this investment.
   b. Calculate the amount in the balance after 5 years and the interest earned over this period.
2. Use the compound interest formula, $V_n = R^n \times V_0$, to calculate the amount, V_n, for each of the following.
   a. $V_0 = \$500, n = 2, i = 8\%$ p.a.
   b. $V_0 = \$1000, n = 4, i = 13\%$ p.a.
   c. $V_0 = \$3600, n = 3, i = 7.5\%$ p.a.
   d. $V_0 = \$2915, n = 5, i = 5.25\%$ p.a.
3. Using a recurrence relation, determine:
   i. the balance
   ii. the interest earned (interest compounded annually)
      a. if $2000 is invested for 1 year at 7.5% p.a.
      b. if $2000 is invested for 2 years at 7.5% p.a.
      c. if $2000 is invested for 6 years at 7.5% p.a.

4. Determine the number of time periods, n, if interest is compounded:
   a. annually for 5 years
   b. quarterly for 5 years
   c. biannually for 4 years
   d. monthly for 6 years
   e. 6-monthly for $4\frac{1}{2}$ years
   f. quarterly for 3 years and 9 months.

5. Calculate the interest rate per period, i, if the annual rate is:
   a. 6% and interest is compounded quarterly
   b. 4% and interest is compounded half-yearly
   c. 18% and interest is compounded monthly
   d. 7% and interest is compounded quarterly.

6. Determine the amount in the account and interest earned after $6750 is invested for 7 years at 5.25% p.a. interest compounded annually.

7. **WE11** If $4200 is invested for 3 years at 7% p.a. interest compounded quarterly:
   a. determine the number of interest-bearing periods, n
   b. calculate the interest rate per period, i
   c. calculate the balance of the account after 3 years
   d. graphically represent the balance at the end of each quarter for 3 years and describe the shape of the graph.

8. If $7500 is invested for 2 years at 5.5% p.a., interest compounded monthly:
   a. determine the number of interest-bearing periods, n
   b. calculate the interest rate per period, i
   c. calculate the balance of the account after 2 years
   d. graphically represent the balance at the end of each month for 2 years and describe the shape of the graph.

9. **WE12** Determine the principal that will grow to $5000 in 5 years, if interest is added quarterly at 7.5% p.a.

10. Determine the principal that will grow to $6300 in 7 years, if interest is added monthly at 5.5% p.a.

11. Assume $1500 is invested for 2 years into an account paying 8% p.a. Calculate the balance if:
    a. interest is compounded yearly
    b. interest is compounded quarterly
    c. interest is compounded monthly
    d. interest is compounded daily.
    e. Compare your answers to parts a–d.

12. Use the recurrence relation $V_{n+1} = 1.045 V_n$ to answer the following questions.
    a. If the balance in an account after 1 year is $2612.50, what will the balance be after 3 years?
    b. If the balance in an account after 2 years is $4368.10, what will the balance be after 5 years?
    c. If the balance in an account after 2 years is $6552.15, what was the initial investment?

13. Determine the amount that accrues in an account which pays compound interest at a nominal rate of:
    a. 7% p.a. if $2600 is invested for 3 years (compounded monthly)
    b. 8% p.a. if $3500 is invested for 4 years (compounded monthly)
    c. 11% p.a. if $960 is invested for $5\frac{1}{2}$ years (compounded fortnightly)
    d. 7.3% p.a. if $2370 is invested for 5 years (compounded weekly)
    e. 15.25% p.a. if $4605 is invested for 2 years (compounded daily).

14. **MC** The greatest return is likely to be made if interest is compounded
    A. annually.
    B. quarterly.
    C. monthly.
    D. fortnightly.

15. **MC** If $12 000 is invested for $4\frac{1}{2}$ years at 6.75% p.a., compounded fortnightly, the amount of interest that would accrue would be closest to
    A. $3600.
    B. $4200.
    C. $5000.
    D. $12 100.

16. Use the compound interest formula to find the principal, V_0, for each of the following.

   a. $V_n = \$5000$, $i = 0.09$, $n = 4$
   b. $V_n = \$2600$, $i = 0.082$, $n = 3$
   c. $V_n = \$3550$, $i = 0.015$, $n = 12$
   d. $V_n = \$6661.15$, $i = 0.008$, $n = 36$

17. Determine the principal that will grow to:

   a. \$3000 in 4 years, if interest is compounded 6-monthly at 9.5% p.a.

   b. \$2000 in 3 years, if interest is compounded quarterly at 9% p.a.

   c. \$5600 in $5\frac{1}{4}$ years, if interest is compounded quarterly at 8.7% p.a.

   d. \$10 000 in $4\frac{1}{4}$ years, if interest is compounded monthly at 15% p.a.

18. Calculate the interest accrued in each case in question 17.

19. **WE13** A bank offers 2.90% per annum compounded yearly on investments. \$25 000 is invested for 5 years. Using a spreadsheet, determine the value of the investment after the 5 years.

12.5 Comparing simple and compound interest

12.5.1 Simple interest or compound interest?

Earlier in this chapter we have looked at simple interest (interest calculated on the principal) and compound interest (interest calculated on the principal and the interest). An example of the differences between simple and compound interest is shown.

Simple interest	Compound interest
Initial principal, $P = \$1000$	Initial principal, $P = \$1000$
Rate of interest, $i = 10\%$	Rate of interest, $i = 10\%$
Interest for Year 1 10% of \$1000, $I_1 = \$100$	Interest for Year 1 10% of \$1000, $I_1 = \$100$
Principal at the beginning of Year 2 $P_2 = \$1000$	Principal at the beginning of Year 2 $P_2 = \$1000 + \$100$ $= \$1100$
Interest for Year 2 10% of \$1000, $I_2 = \$100$	Interest for Year 2 10% of \$1100, $I_2 = \$110$
Principal at the beginning of Year 3 $P_3 = \$1000$	Principal at the beginning of Year 3 $P_3 = \$1100 + \$110$ $= \$1210$
Interest for Year 3 10% of \$1000, $I_3 = \$100$	Interest for Year 3 10% of \$1210, $I_3 = \$121$
Principal at the beginning of Year 4 $P_4 = \$1000$	Principal at the beginning of Year 4 $P_4 = \$1210 + \$121$ $= \$1331$

Interest for Year 4 10% of 1000, $I_4 = \$100$	Interest for Year 4 10% of 1331, $I_4 = \$133.10$
Principal at the beginning of Year 5 $P_5 = \$1000$	Principal at the beginning of Year 5 $P_5 = \$1331 + \$133.10$ $= \$1464.10$
Interest for Year 5 10% of 1000, $I_5 = \$100$	Interest for Year 5 10% of 1464.10, $I_5 = \$146.41$
The simple interest earned over a 5-year period is $500.	The compound interest earned over a 5-year period is $610.51.

This table has illustrated how, as the principal for compound interest increases periodically (i.e. $P = \$1000, \$1100, \$1210, \$1331, \$1464.10...,$) so does the interest (i.e. $I = \$100, \$110, \$121, \$133.10, \$146.41...,$), while for simple interest both the principal and the interest earned remain constant, that is, $P = \$1000$ and $I = \$100$. The difference of $110.51 between the compound interest and simple interest earned in a 5-year period represents the interest earned on added interest.

If we were to place the set of data obtained in two separate tables and represent each set graphically — as the total amount of the investment, A, versus the year of the investment, n — we would see that the simple interest investment grows at a constant rate while the compound interest investment grows exponentially.

1. The simple interest investment is represented by a straight line.

n	A
0	1000
1	1100
2	1200
3	1300
4	1400
5	1500

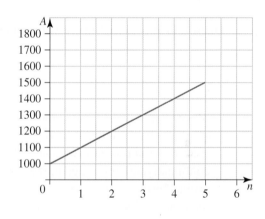

2. The compound interest investment is represented by a curve.

n	A
0	1000
1	1100
2	1210
3	1331
4	1464.10
5	1610.51

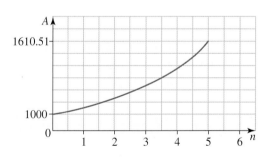

Graphing simple interest functions

Suppose that we invest $10 000 at 5% p.a. simple interest. The table below shows the amount of interest that we will receive after various lengths of time.

No. of years	1	2	3	4	5
Interest	$500	$1000	$1500	$2000	$2500

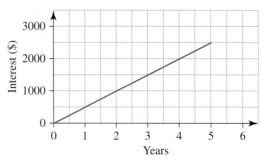

The amount of interest earned can be graphed by the **linear function** at right.

Note that the gradient of this graph is 500, which is the amount of one year's interest, or 5% of the principal. This means that for every 1 unit increase on the x-axis (1 year in time) the y-axis (Interest) increases by 500 units ($500).

WORKED EXAMPLE 14

Leilay invests $6000 into a saving account earning simple interest at a rate of 4% p.a.

a. Complete the table below to calculate the future value of the investment at the end of each year.

No. of years	1	2	3	4	5
Future value					

b. Draw a graph of the future value against the number of years the money is invested.

THINK

a. 1. Calculate the interest earned after 1 year.

   2. Since simple interest earns the same amount of interest each year ($240 in this case) we can continually add this value to the original principal to determine the future value.

   3. Complete the table.

WRITE

a. $I = Pin$

$I = 6000 \times 0.04 \times 1 = 240$

Original principal = $6000
Future value after 1 year = $6000 + $240 = $6240
Future value after 2 years = $6240 + $240 = $6480
Future value after 3 years = $6480 + $240 = $6720
...

No. of years	1	2	3	4	5
Future value	$6240	$6480	$6720	$6960	$7200

b. Draw the graph with Years on the horizontal axis and Future value on the vertical axis.

b.

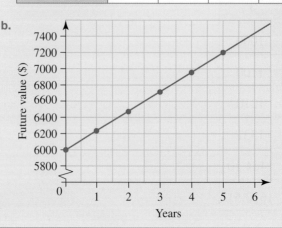

We are able to compare the interest that is earned by an investment at varying interest rates by graphing the interest earned at varying rates on the one set of axes.

Aston has \$12 000 to invest. Three different banks offer simple interest rates of 4%, 5% and 6%.
a. Complete the table below to show the interest that he could earn over 5 years.

No. of years	1	2	3	4	5
Interest (4%)					
Interest (5%)					
Interest (6%)					

b. Show this information in graph form.

THINK

a. 1. Use the simple interest formula to calculate the interest earned on \$12 000 at 4% p.a. for 1, 2, 3, 4 and 5 years.

2. Use the simple interest formula to calculate the interest earned on \$12 000 at 5% p.a. for 1, 2, 3, 4 and 5 years.

3. Use the simple interest formula to calculate the interest earned on \$12 000 at 6% p.a. for 1, 2, 3, 4 and 5 years to complete the table.

WRITE

a. $I = Pin$
$I = 12000 \times 0.04 \times 1 = 480$
$I = 12000 \times 0.04 \times 2 = 960$
...

$I = Pin$
$I = 12000 \times 0.05 \times 1 = 600$
$I = 12000 \times 0.05 \times 2 = 1200$
...

$I = Pin$
$I = 12000 \times 0.06 \times 1 = 720$
$I = 12000 \times 0.06 \times 2 = 1440$
...

No. of years	1	2	3	4	5
Interest (4%)	\$480	\$960	\$1440	\$1920	\$2400
Interest (5%)	\$600	\$1200	\$1800	\$2400	\$3000
Interest (6%)	\$720	\$1440	\$2160	\$2880	\$3600

b. Draw a line graph for each investment.

b.

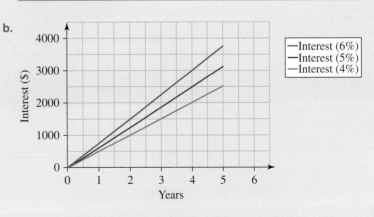

Graphing compound interest functions

Earlier we drew graphs of the simple interest earned by various simple interest investments. We found these graphs were linear because the interest earned in each interest period was the same.

With compound interest the interest earned in each interest period increases, and so when we graph the future value of the investment, an exponential graph results. The shape of the graph is a smooth curve which gets progressively steeper. We can use the compound interest formula to complete tables that will then allow us to graph a compound interest function.

WORKED EXAMPLE 16

Olivia invests $5000 at 5% p.a., with interest compounded annually.
a. Complete the table below to show the future value at the end of each year.

No. of years	0	1	2	3	4	5
Future value						

b. Draw a graph of the future value of the investment against the number of years.

THINK

a. 1. Use the compound interest formula to calculate final amount at the end of every year for 5 years.

  2. Complete the table rounding all values correct to two decimal places.

b. Draw a graph by drawing a smooth curve between the marked points.

WRITE

a. $A = P(1 + i)^n$

$A = 5000(1 + 0.05)^1 = 5250$

$A = 5000(1 + 0.05)^2 = 5512.50$

$A = 5000(1 + 0.05)^3 = 5788.125$

...

No. of years	0	1	2	3	4	5
Future value	$5000	$5250	$5512.50	$5788.13	$6077.53	$6381.41

b.

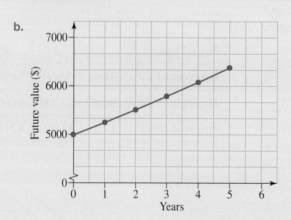

To graph the interest earned, the principal must be subtracted from the future value of the investment. As with simple interest, such graphs can be used to compare investments.

WORKED EXAMPLE 17

Paul is to invest $2000 at 5% p.a., 6% p.a. and 7% p.a., compounded annually.

a. **Copy and complete the table below to calculate the amount of interest of each investment at the end of each year.**

No. of years	1	2	3	4	5
Interest (5%)					
Interest (6%)					
Interest (7%)					

b. **Draw a graph that will allow the amount of interest earned to be compared.**

THINK

a. 1. Use the compound interest formula $A = P(1 + i)^n$ to determine the value of each investment over a 5-year period. For example, to determine the value of the 5% p.a. investment after 1 year evaluate $A = 2000(1 + 0.05)^1$.

2. To calculate the amount of interest earned for the investments at the end of each year, subtract the principal ($2000) from all values. We do this since the formula for calculating the amount of interest is $I = A - P$.

WRITE

a.

No. of years	1	2	3	4	5
Interest (5%)	$2100	$2205	$2315.25	$2431.01	$2552.56
Interest (6%)	$2120	2247.20	$2382.03	$2524.95	$2676.45
Interest (7%)	$2140	$2289.80	$2450.09	$2621.59	$2805.10

No. of years	1	2	3	4	5
Interest (5%)	$100	$205	$315.25	$431.01	$552.56
Interest (6%)	$120	$247.20	$382.03	$524.95	$676.45
Interest (7%)	$140	$289.80	$450.09	$621.59	$805.10

b. Draw each graph by joining the points with a smooth curve.

b.

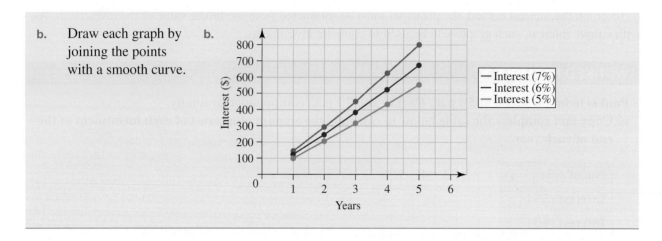

Comparing simple and compound interest functions using a spreadsheet

Your grandfather left you $20 000 in his will. You have no need to use the money at this stage, so you are looking at investing it for approximately 12 years. Your research has narrowed down your options to 4.25% p.a. simple interest or 3.6% p.a. interest compounding yearly. At this stage, you do not anticipate having to withdraw your money in the short term; however, it may become necessary to do so.

Let us investigate to determine which would be the better option if you were forced to withdraw your money at any period of time within 12 years.

The spreadsheet and graphs we are aiming to produce appear as follows.

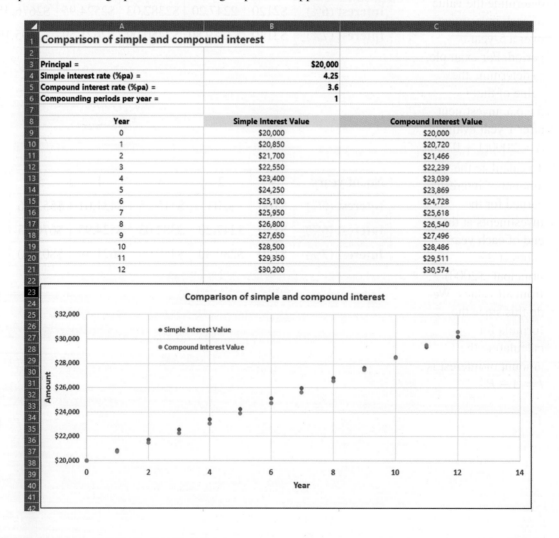

1. Enter the spreadsheet heading shown in cell A1.
2. Enter the side headings in cells A3 to A6.
3. In cell B3 enter the value of 20 000 then format it to currency with zero decimal places.
4. Enter the numeric values shown in cells B4, B5 and B6.
5. In row 8, enter the column headings shown.
6. In cell A9, enter the value 0.
7. In cell A10, enter the formula =A9+1. The value of 1 should appear. Copy this formula down from cell A11 to A21.
8. The formula for the simple interest value is:

$$\text{principal} + \text{interest} = \text{principal} + \text{principal} * \text{rate} * \text{time}/100$$

   Enter the formula =$B$3+$B$3*$B$4*A9/100 in cell position B9, then copy this formula down from B9 to B21. Format these cells to currency with zero decimal places. Check the values that appear agree with those on the spreadsheet displayed.
9. The formula for the compound interest value for cell position C9 is:

$$= \$B\$3 * (1 + \$B\$5/(100 * \$B\$6)) \wedge A9$$

   Enter this value, then copy it down from C9 to C21. Format these cells to currency with zero decimal places. Check the values with those on the spreadsheet displayed.
10. Use the graphing facility of your spreadsheet to produce a graph similar to the one shown.
11. From the table and the shape of the graphs it is obvious that a critical point occurs somewhere between the 10-year and 11-year marks. Modify your spreadsheet by inserting rows between these two years. Enter part-year values, such as 10.2, 10.5 and so on, in column A. Copy the formulas in columns B and C to complete the entries. Continue to investigate until you can find a fairly exact value for the time when these two graphs cross.
12. Write a paragraph summarising the results of your spreadsheet and graphs. Describe which option would be the better one, considering that you may be forced to withdraw your money at any time within the 12 years. Support any conclusions by referring to your spreadsheet and graphs.

12.5.2 Using an online calculator to investigate a compound interest loan or investment

In the real-world there are many types of technologies that can be used to calculate the final amount, the interest rate, the number of compounding periods, the payment or even the principal of a compound interest loan or invesment. In this section, an online calculator involving the time value of money (TVM solver) will be used to investigate compound interest problems. The one we will use can be found at:

http://www.fncalculator.com/financialcalculator?type=tvmCalculator

A picture of this financial solver is shown at right.

To use a Finance Solver, you need to know the meaning of each of its inputs. These are defined as follows:
- *Present Value* is the current (present) value of the loan or investment.
- *Payment* is the amount paid at each time period.
- *Future Value* is the value of the investment or loan after a specified number of compounding periods.
- *Annual Rate (%)* is the annual interest rate.
- *Periods* is the total number of compounding periods.

Mode	● End ○ Beginning	
Present Value		PV
Payment		PMT
Future Value		FV
Annual Rate (%)		Rate
Periods		Periods
Compounding	Monthly ▼	
	Reset	

- *Compounding* is the number of times the interest is compounded per year.
- *Mode* is used to indicate whether the interest is compounded at the end or at the beginning of the time period. Leave this set at END.

Note: If there is no mention of a payment or repayment being made, leave the cell blank.

When using a financial solver, you must be very careful in the way you enter information because it needs to know which way the money is flowing. It does this by following a *sign convention*.

Sign conventions for the flow of money

- If you receive money, or someone owes you money, we treat this as a positive (+ve).
- If you pay out money or you owe someone money, we treat this as a negative (−ve).

Let us now analyse some problems using a Financial Solver.

Analysing the effect of changing the interest rate

When people shop around for the best loan, there are many factors that need to be considered; however, the one that most people look at is the interest rate charged on the loan. In this section we will look at the effect of the interest rate on the future value of the loan or investment.

To help us investigate the effect of the interest rate on a loan or investment we will use an online calculator.

WORKED EXAMPLE 18

Kasey invested \$5000 over a period of 5 years. How much more would she earn investing at 5.5% interest compounding monthly compared to 5% interest compounding monthly?

THINK

1. Consider the 5% interest investment.
   Since Kasey has given the money to the bank, we input this as a negative since she no longer has possession of the money.
   Input the following values into the online calculator:
   Present Value: −5000
   Payment: leave blank
   Rate: 5
   Periods: 60
   (5 years × 12 each year)
   Compounding: Monthly
   To find the future value click on the FV button.

2. Write the answer.

WRITE

Mode	⦿ End ◯ Beginning	
Present Value	-5,000	PV
Payment		PMT
Future Value	6,416.79	FV
Annual Rate (%)	5	Rate
Periods	60	Periods
Compounding	Monthly ▾	

The future value at 5% is \$6416.79.

3. Consider the 5.5% interest investment.
   Input the following values into the online calculator:
   Present Value: −5000
   Payment: leave blank
   Rate: 5.5
   Periods: 60
      (5 years × 12 each year)
   Compounding: Monthly
   To find the future value click on the FV button.

Mode	● End ○ Beginning	
Present Value	-5,000	PV
Payment		PMT
Future Value	6,578.52	FV
Annual Rate (%)	5.5	Rate
Periods	60	Periods
Compounding	Monthly ▼	

4. Write the answer.

The future value at 5.5% is $6578.52.

5. Calculate the difference by subtraction.

The difference = 6578.52 − 6416.79
               = 161.73

6. Write the answer.

By increasing the interest rate to 5.5% from 5%, the investment increased by $161.73 over the 5-year period.

Note: In the situation in Worked example 18, we observe that an increase in the interest rate has resulted in an increase in the future value. As this is an investment the increased future value works in Kasey's favour; however, if this was a $5000 loan from the bank an increase in the interest rate would generally not be favourable to the person borrowing the money.

Analysing the effect of changing the number of compounding periods

To help us investigate the effect of the number of compounding periods we will use an online calculator.

WORKED EXAMPLE 19

Hugh wants to invest his $7500 for a period of 7 years. He believes he is better to invest his money with more compounding periods at 9.5% p.a. How much more would Hugh earn on his investment if he had it compound fortnightly compared to monthly.

THINK

WRITE

1. Consider compounding monthly.
   Since Hugh has given the money to the bank, we input this as a negative value.
   Input the following values into the online calculator:
   Present Value: −7500
   Payment: leave blank
   Rate: 9.5
   Periods: 84
      (7 years × 12 each year)
   Compounding: Monthly
   To find the future value click on the FV button.

Mode	● End ○ Beginning	
Present Value	-7,500	PV
Payment		PMT
Future Value	14,545.54	FV
Annual Rate (%)	9.5	Rate
Periods	84	Periods
Compounding	Monthly ▼	

2. Write the answer.

The future value of the investment with monthly compounding is $14 545.54.

3. Consider compounding fornightly.
Input the following values into the online calculator:
Present Value: −7500
Payment: leave blank
Rate: 9.5
Periods: 182
    (7 years × 26 each year)
Compounding: Bi-Weekly
To find the future value click on the FV button.

Mode	⦿ End ○ Beginning	
Present Value	-7,500	PV
Payment		PMT
Future Value	14,566.01	FV
Annual Rate (%)	9.5	Rate
Periods	182	Periods
Compounding	Bi-Weekly ▾	

4. Write the answer.

The future value of the investment with fortnightly compounding is $14 566.01.

5. Calculate the financial benefit of fortnightly compounding compared to monthly compounding.

Difference = 14 566.01 − 14 545.54
       = 20.47

6. Write the answer.

Hugh would earn $20.47 more in the investment if he chooses fortnightly compounding.

Note: In the situation in Worked example 19, we observe that an increase in the frequency of compounding has resulted in an increase in the future value. Again, as this is an investment, the increased future value works in Hugh's favour. Discuss with your classmates what the effect of increasing the frequency of compounding would have if you borrowed money.

Analysing the effect of changing the interest rate and number of compounding periods

We will now investigate the effect of changing the interest rate and the number of compounding periods, using an online calculator.

WORKED EXAMPLE 20

From Worked example 19, Hugh now has two options to invest his $7500 for 7 years:
- **Option A: At 12% compounding monthly**
- **Option B: At 11.8% compounding daily**

What option should Hugh take? Justify your answer.

THINK

1. Consider option A.
Since Hugh has given the money to the bank, we input this as a negative value.
Input the following values into the online calculator:
Present Value: −7500
Payment: leave blank
Rate: 12
Periods: 84
    (7 years × 12 each year)
Compounding: Monthly
To find the future value click on the FV button.

WRITE

Mode	⦿ End ○ Beginning	
Present Value	-7,500	PV
Payment		PMT
Future Value	17,300.42	FV
Annual Rate (%)	12	Rate
Periods	84	Periods
Compounding	Monthly ▾	

2. Write the answer.

The future value of the investment with 12% monthly compounding is $17 300.42.

3. Consider option B.
Input the following values into the online calculator:
Present Value: −7500
Payment: leave blank
Rate: 11.8
Periods: 2555
    (7 years × 365 each year)
Compounding: Daily
To find the future value click on the FV button.

4. Write the answer.

The future value of the investment with 11.8% daily compounding is $17 128.94.

5. Write the answer with justification.

Hugh would be better choosing Option A since his investment would be $171.48 greater after 7 years of investment.

Note: When the interest rate and the frequency of compounding are changed simultaneously, the overall effect on the future value may not be obvious before any calculations are carried out. Generally speaking, an investment with a lower rate would yield less interest. In some cases we may need to complete the calculations first to see the effect on the future value.

Exercise 12.5 Comparing simple and compound interest

1. **WE14** An amount of $8000 is invested into a savings account earning simple interest at a rate of 5% p.a.

   a. Copy and complete the table below to calculate the future value of the investment at the end of each year.

No. of years	1	2	3	4	5
Future value					

   b. Draw a graph of the future value against the number of years the money is invested.

2. **WE15** Adam has $2500 to invest. Three different banks offer simple interest rates of 1.60%, 1.80% and 2% per annum.

   a. Complete the table below to show the interest that Adam could earn over 5 years.

No. of years	1	2	3	4	5
Interest (1.60%)					
Interest (1.80%)					
Interest (2%)					

   b. Show this information in graph form.

3. **WE16** Monique invests $12 000 at 8% p.a., with interest compounded annually.
   a. Copy and complete the table below to show the future value at the end of each year.

No. of years	1	2	3	4	5
Future value					

   b. Draw a graph of the future value of the investment against the number of years.
4. Draw a graph to represent the future value of the following investments over the first five years.
   a. $15 000 at 7% p.a. with interest compounded annually
   b. $2000 at 10% p.a. with interest compounded annually
5. A graph is drawn to show the future value of an investment of $2000 at 6% p.a. with interest compounding six-monthly.
   a. Complete the table below.

Years	0.5	1	1.5	2	2.5	3	3.5	4	4.5	5
Future value										

   b. Use the table to draw the graph.
6. An amount of $1200 is invested at 4% p.a. with interest compounding quarterly.
   a. Graph the future value of the investment at the end of each year for 10 years.
   b. Graph the compound interest earned by the investment at the end of each year.
7. **WE17** James has $8000 to invest at either 4% p.a., 6% p.a. or 8% p.a. compounding annually.
   a. Complete the table below to calculate the amount of interest of each investment at the end of each year.

No. of years	1	2	3	4	5
Interest (4%)					
Interest (6%)					
Interest (8%)					

   b. Draw a graph that will allow the amount of interest earned to be compared.
8. Petra has $4000 to invest at 6% p.a.
   a. Complete the table below to show the future value of the investment at the end of each year, if interest is compounded:
      i. annually
      ii. six-monthly.

No. of years	1	2	3	4	5
Annually					
Six-monthly					

   b. Show this information in graphical form.
9. **WE18** Donna invested $3500 over a period of 4 years. How much more would she earn investing at 6.5% interest compounding monthly compared to 6% interest compounding monthly?

10. **WE19** Sam wants to invest his $12 400 for a period of 8 years. He believes he is better to invest his money with more compounding periods at 7.25% p.a. How much more would Sam earn on his investment if he had it compound fortnightly compared to annually?

11. **WE20** Tamara has two options to invest her $10 000 for 9 years:
    Option A: At 8% compounding biannually (twice per year)
    Option B: At 7.8% compounding daily
    What option should Tamara take? Justify your answers.

12. Kerry invests $100 000 at 8% p.a. for a one-year term. For such large investments interest is compounded daily.
    a. Calculate the daily percentage interest rate, correct to four decimal places.
    b. Calculate the compounded value of Kerry's investment on maturity.
    c. Calculate the amount of interest earned on this investment.
    d. Calculate the extra amount of interest earned, compared with the interest calculated at a simple interest rate.

13. Simon invests $4000 for 3 years at 6% p.a. simple interest. Monica also invests $4000 for 3 years but her interest rate is 5.6% p.a., with interest compounded quarterly.

    a. Calculate the value of Simon's investment on maturity.
    b. Show that the compounded value of Monica's investment is greater than Simon's investment.
    c. Explain why Monica's investment is worth more than Simon's, despite receiving a lower rate of interest.

14. a. A bank offers interest at 7% per annum, compounded yearly. If a customer puts $900 in the bank, how much will be there after 3 years?
    b. Another bank offers interest at 8% per annum, compounded yearly. If a customer puts $14 000 in the bank, how much will be there after 4 years?

15. A bank offers a term deposit for 3 years at an interest rate of 8% p.a. with a compounding period of 6 months. What would be the end value of a $5000 investment under these conditions?

16. A bank offers 8% per annum compounded yearly on investments. Assume $2000 is invested for 5 years. Using a spreadsheet, determine the value of the investment after the 5 years.

17. A bank offers compound interest on its savings account of 4% p.a. compounding yearly. A boy opens an account with $1125.
    a. Using a spreadsheet, calculate the value of the boy's investment for the first four years.
    b. Using your values from part a, construct a compound interest graph.

18. A sum of $1000 is invested at 10.5% p.a. interest. Using a spreadsheet, calculate the value of the investment for the following years and types of interest.

	Principal	Annual interest rate	Time (years)	Future value (Simple interest)	Future value (Compound interest with yearly compounding)
a.	$1000	10.5%	3		
b.	$1000	10.5%	6		
c.	$1000	10.5%	9		
d.	$1000	10.5%	12		

19. A sum is invested into a bank account that is compounded annually. The graph shows the value of the investment over 5 years.
   a. What was the sum of money invested?
   b. Calculate the compound interest rate for this investment.

20. Using a spreadsheet, determine how long it would take to 'double your money' with compound interest at the following annual rates. State your answers in years. Is there a pattern to your answers?
   a. 5%   b. 7%   c. 10%   d. 13%

21. There are two options available for a $4500 investment.
       Option 1: simple interest at 6.25% per year
       Option 2: compound interest 6% per annum, compounded biannually
   To calculate the interest on the investment for Option 2 after the first year, a student performed the following calculation:

$$\text{Option 2: investment value} = 4500 + \left(4500 \times \frac{6 \times 1}{100} \times 1\right)$$

   a. Explain why the rate, i, should be 3%.
   b. Using a spreadsheet, determine the value of the investment for both Option 1 and Option 2 after three years.
   c. Using a spreadsheet, determine how many years it will take for Option 2 to be the better option.

12.6 Review: exam practice

12.6.1 Loans and compound interest: summary

Simple interest
- Simple interest is given by $I = Pin$
   where:
   I = simple interest ($)
   P = principal ($) — that is, the sum of money borrowed or invested

i = interest rate as a decimal per time period

n = number of time periods — that is, the period of time for which the sum of money is to be borrowed or invested.

- The total amount is given by $A = P + I$.
- When calculating simple interest, the interest earned is the same for each time period.
- To calculate the principal, use $P = \dfrac{I}{in}$.
- To calculate the interest rate, use $i = \dfrac{I}{Pn}$.
- To calculate the period of the loan or investment, use $n = \dfrac{I}{Pi}$.

Simple interest as a recurrence relation

- The recurrence relation for the value of the simple interest loan or investment after n years is $V_0 = $ principal, $V_{n+1} = V_n + d$
  where:
  V_n represents the value of the investment after n time periods
  d is the amount of simple interest earned each time period, which can be found using the formula $d = i \times V_0$
  V_{n+1} represents the value of the investment after $n + 1$ time periods.
- Amount of simple interest = (amount of investment) − principal
- When a simple interest function is graphed, it gives rise to a linear graph.

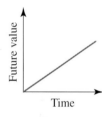

Compound interest

- The formula used to determine the final amount of a compound interest investment or loan is given by the rule: $A = P(1 + i)^n$
  where:
  A = the final amount of a compound interest investment or loan ($)
  P = principal ($) — that is, the sum of money borrowed or invested
  i = interest rate as a decimal per time period
  n = number of compounding time periods.

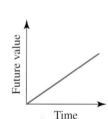

- The recurrence relation for the value of a compound interest investment or loan after n compounding periods is given by: $V_0 = $ principal, $V_{n+1} = RV_n$
  where:
  V_n is the value of the investment after n compounding periods
  R is the compounding factor, with $R = (1 + i)$
  i is the interest rate as a decimal per compounding period.
- To determine the amount of interest accrued on a compound interest investment or loan use:
  $I = $ final amount − principal.
- The value of a compound interest invesment or loan after the nth compounding period can be found by using: $V_0 = $ starting value, $V_n = R^n \times V_0$.
- For non-annual compounding the value of n and i will need to be adjusted using the following formulas before substituting them into the compound interest formula.

$$\text{Number of interest periods, } n = \text{number of years} \times \text{number of interest periods per year}$$

$$\text{Interest rate per period, } i = \frac{\text{interest rate per annum}}{\text{number of interest periods per year}}$$

- When a compound interest function is graphed, it gives rise to an exponential curve.

Using a financial solver

- A financial solver is a useful tool to easily solve problems relating to compound interest loans and investments. To use a financial solver, you need to know the meaning of its inputs. These are defined as follows:
    - *Present Value* is the current (present) value of the loan or investment.
    - *Payment* is the amount paid at each time period.
    - *Future Value* is the value of the investment or loan after a specified number of compounding periods.
    - *Annual Rate (%)* is the annual interest rate.
    - *Periods* is the total number of compounding periods.
    - *Compounding* is the number of times the interest is compounded per year.
    - *Mode* is used to indicate whether the interest is compounded at the end or at the beginning of the time period.
- When using a financial solver, you must be very careful in the way you enter information because it needs to know which way the money is flowing. It does this by following a sign convention.
    - If you receive money, or someone owes you money, we treat this as a positive.
    - If you pay out money or you owe someone money, we treat this as a negative.

Exercise 12.6 Review: exam practice

Simple familiar

1. Calculate the simple interest earned on an investment of $5000 at 4% p.a. for 5 years.
2. Calculate the simple interest earned on each of the following investments.
    a. $3600 at 9% p.a. for 4 years
    b. $23 500 at 6% p.a. for 2 years
    c. $840 at 2.5% p.a. for 2 years
    d. $1350 at 0.2% p.a. for 18 months
    e. $45 820 at 4.75% p.a. for $3\frac{1}{2}$ years
3. Dion invests $32 500 in a debenture paying 5.6% simple interest for 4 years.
    a. Calculate the interest earned by Dion.
    b. Calculate the total value of Dion's investment after 4 years.
    c. If the debenture paid Dion in quarterly instalments, calculate the value of each interest payment.
4. **MC** Two banks pay simple interest on short-term deposits. Bank A pays 6% p.a. over 4 years and Bank B pays 6.5% p.a. for $3\frac{1}{2}$ years. What is the difference between the two banks' final payout figures if $5000 was invested in each account?
    A. $0   **B.** $62.50   **C.** $1137.50   **D.** $1200
5. **MC** Keira invested $360 in a bank for 3 years at 8% simple interest each year. At the end of the 3 years, what total amount will she receive?
    A. $28.80   **B.** $86.40   **C.** $388.80   **D.** $446.40
6. **MC** Philip borrowed $7000 and intended to pay it back in 4 years. The terms of the loan indicated Philip was to pay 9% p.a. simple interest. What was the interest Philip paid on the loan?
    A. $630   **B.** $2520   **C.** $9520   **D.** $9881
7. Michelle invests $15 000 for a period of 4 years. Calculate the interest rate, given that Michelle earned a total of $3900 in simple interest.
8. Bradley invests $23 500 at 4.6% p.a. If he earned $1351.25 in simple interest, calculate the length of time for which the money was invested.

9. **MC** A loan of $5000 is taken over 5 years. The simple interest is calculated monthly. The interest bill on this loan is $1125. What is the simple interest rate per year on this loan?

   A. 3% **B.** 3.75% **C.** 4% **D.** $4\frac{1}{2}$%

10. **MC** What principal invested in an investment bond will accumulate $2015 in simple interest after 6 months invested at $6\frac{1}{2}$% p.a.?

    A. $6000 **B.** $50 000 **C.** $60 000 **D.** $62 000

11. **MC** A loan of $10 000 is taken over 10 years. The total interest bill on this loan is $2000. What is the simple interest rate per year on this loan?

    A. 1.5% **B.** 2% **C.** 3% **D.** $4\frac{1}{2}$%

12. An amount of $7500 is to be invested at 6% p.a. simple interest.
    a. Copy and complete the table below to calculate the interest over 5 years.

No. of years	1	2	3	4	5
Interest					

    b. Draw a graph of the interest earned against the length of the investment.
    c. What is the gradient of the linear graph drawn?
    d. Use your graph to find the amount of interest that would have been earned after 10 years.

Complex familiar

13. Vicky invests $2400 at 5% p.a. for 3 years with interest compounded annually. Calculate the compounded value of the investment at the end of the term.
14. Barry has an investment with a present value of $4500. The investment is made at 6% p.a. with interest compounded biannually. Calculate the future value of the investment in 4 years.
15. Calculate the compounded value of each of the following investments.
    a. $3000 at 7% p.a. for 4 years with interest compounded annually
    b. $9400 at 10% p.a. for 3 years with interest compounded biannually
    c. $11 400 at 8% p.a. for 3 years with interest compounded quarterly
    d. $21 450 at 7.2% p.a. for 18 months with interest compounded biannually
    e. $5000 at 2.6% p.a. for $2\frac{1}{2}$ years with interest compounded quarterly
16. Dermott invested $11 500 at 3.2% p.a. for 2 years with interest compounded quarterly. Calculate the total amount of interest earned on this investment.

Complex unfamiliar

17. Kim and Glenn each invest $7500 for a period of 5 years.
    a. Kim invests her money at 9.9% p.a. with interest compounded annually. Calculate the compounded value of Kim's investment.
    b. Glenn invests his money at 9.6% p.a. with interest compounded quarterly. Calculate the compounded value of Glenn's investment.
    c. Explain why Glenn's investment has a greater compounded value than Kim's.
18. $20 000 is to be invested at 4% p.a. with interest compounded annually.
    a. Copy and complete the table below to calculate the future value at the end of each year.

No. of years	1	2	3	4	5
Future value					

    b. Draw a graph of the interest earned against the length of the investment.
    c. Use your graph to determine the future value of the investment after 10 years.

19. Christof can choose from two investment accounts. One returns 4.1% p.a. compounding monthly while the other returns 4% p.a. compounding daily. Which investment account should he choose?

20. A building society advertises investment accounts at the following rates:
   - 3.875% p.a. compounding daily
   - 3.895% p.a. compounding monthly
   - 3.9% p.a. compounding quarterly.

   Which account should an investor choose? Justify your answer by providing evidence for your choice.

Answers

12 Loans and compound interest

Exercise 12.2 Simple interest

1. a. $136.00, $816
   b. $56.70, $266.70
   c. $145.25, $560.25
   d. $110.40, $570.40
   e. $255, $1275
   f. $336.89, $1049.89
   g. $178.57, $999.57
   h. $43.88, $108.88
   i. $11.76, $33.01
   j. $229.68, $853.14
   k. $544.05, $1335.40
2. a. $627.13
   b. $12 542.50
3. a. 10% p.a.
   b. 6.25% p.a.
   c. 80% p.a.
   d. 2.125% p.a. or $2\frac{1}{8}$% p.a.
   e. 3.36% p.a.
4. a. 1 year
   b. 18 months
   c. 3 months
   d. 7 years
   e. 1 month
5. a. $3070
   b. $4400
   c. $5425
   d. $236.36
   e. $2500
6. a. $103.50
   b. $2700
   c. $325
   d. 131.25
7. a. $360
   b. $1020
   c. $27 700
   d. $17.70
   e. $13.67
8. B
9. C
10. A
11. D
12. A
13. C
14. D
15. D
16. a. The Big-4 Bank offers the better rate.
    b. The Big-4 Bank charges $11\frac{1}{3}$% p.a. for a loan while The Friendly Building Society charges 12% (= 12 × 1% per month).
17. $1515.79
18. $2133.33
19. $352
20. 24 months
21. a. i. $1540.63
       ii. $6162.50
    b. Yes
22. a. $2247
    b. $15 729
    c. $7\frac{1}{2}$ years

Exercise 12.3 Simple interest as a recurrence relation

1. a. $A_{n+1} = A_n + 43, A_0 = 688$
   b. $860
2. a. $B_{n+1} = B_n + 61.88, B_0 = 728$
   b. $913.64
3. a. $1803.36

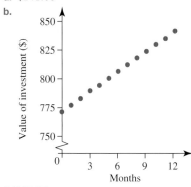

b.

4. a. $141.68

   b.

5. $4022.22
6. $1300
7. 4.83%
8. 26.66%
9. 1.84 years
10. a. $997.60
    b. $526.50
11. a. $211.75
    b. $1104
    c. $1228.50
    d. $4214.70
12. a. $620
    b. $892.16
    c. $25 695
13. a. $1173
    b.

14. a. Tigers Bank
    b. The interest on the Tigers Bank loan is $9800 compared to Twins Bank at $10 800, so Tigers Bank is the best option since it requires payment of $1000 less interest.
15. $2383.36
16. 10 months

Exercise 12.4 Compound interest

1. a. $A = 2500 (1.075)^5$
   b. $A = \$3589.07, I = \$1089.07$
2. a. $583.20
   b. $1630.47
   c. $4472.27
   d. $3764.86
3. a. i. $2150
      ii. $150
   b. i. $2311.25
      ii. $311.25

c. i. $3086.60
   ii. $1086.60

4. a. 5   b. 20
   c. 8   d. 72
   e. 9   f. 15

5. a. 1.5%   b. 2%
   c. 1.5%   d. 1.75%

6. $9657.36

7. a. 12
   b. 1.75%
   c. $5172.05
   d.

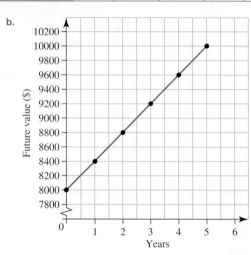

Despite it being difficult to see on the graph, the graph is exponential.

8. a. 24
   b. 0.4583%
   c. $8369.92
   d.

Despite it being difficult to see on the graph, the graph is exponential.

9. $3448.40

10. $4290.73

11. a. $1749.60
    b. $1757.49
    c. $1759.33
    d. $1760.24
    e. The balance increases as the compounding periods become more frequent.

12. a. $2852.92   b. $4984.72   c. $6000

13. a. $605.60   b. $1314.84   c. $795.77
    d. $1043.10   e. $1641.82

14. D

15. B

16. a. $3542.13   b. $2052.54
    c. $2969.18   d. $5000

17. a. $2069.61   b. $1531.33
    c. $3564.10   d. $5307.05

18. a. $930.39   b. $468.67
    c. $2035.90   d. $4692.95

19. $28 841.44

Exercise 12.5 Comparing simple and compound interest

1. a.

No. of years	1	2	3	4	5
Future value	$8400	$8800	$9200	$9600	$10 000

b.

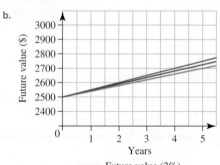

2. a.

No. of years	1	2	3	4	5
Interest (1.60%)	$2540	$2580	$2620	$2660	$2700
Interest (1.80%)	$2545	$2590	$2635	$2680	$2725
Interest (2%)	$2550	$2600	$2650	$2700	$2750

b.

—— Future value (2%)
—— Future value (1.80%)
—— Future value (1.60%)

3. a.

No. of years	1	2	3	4	5
Future value	$12 960	$13 997	$15 117	$16 326	$17 632

b.

4. a.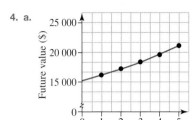

b.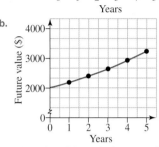

b.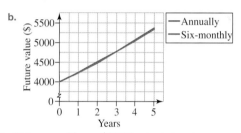

5. a. * See the table at the bottom of the page.

b.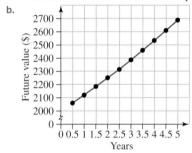

6. a.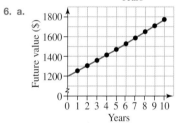

b.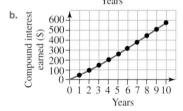

8. a.

No. of years	1	2	3	4	5
Annually	$4240	$4494	$4764	$5050	$5353
Six-monthly	$4244	$4502	$4776	$5067	$5376

b.

9. Donna would earn $89.36 extra interest.
10. Sam would earn $422 extra on his investment.
11. Option A
12. **a.** 0.0219%
   b. $108 320.72
   c. $8320.72
   d. $320.72
13. **a.** $4720
   b. $4726.24
   c. Compounding interest
14. **a.** $1102.54
   b. $19 046.85
15. $6326.59

7. a.

No. of years	1	2	3	4	5
Interest (4%)	$320	$653	$999	$1359	$1733
Interest (6%)	$480	$989	$1528	$2100	$2706
Interest (8%)	$640	$1331	$2078	$2884	$3755

---

*5. a.

Years	0.5	1	1.5	2	2.5	3	3.5	4	4.5	5
FV	$2060	$2122	$2185	$2251	$2319	$2388	$2460	$2534	$2610	$2688

16.

	A	B	C
1	t	Value	
2	0	$2000.00	
3	1	$2160.00	= (B2 * 0.08) + B2
4	2	$2332.80	
5	3	$2519.42	
6	4	$2720.98	
7	5	$2938.66	

After 5 years the investment is worth $2938.66.

17. a.

	A	B	C
1	t	Value	
2	0	$1125.00	
3	1	$1170.00	= (B2 * 0.04) + B2
4	2	$1216.80	
5	3	$1265.47	
6	4	$1316.09	

After 4 years the investment is worth $1316.09.

b.

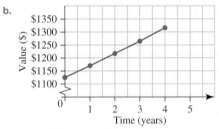

18. * See the table at the bottom of the page.

19. a. $12 000

   b. 5% p.a.

20. a. At 5% compound interest, the investment is doubled in 15 years.

   b. At 7% compound interest, the investment is doubled in 11 years.

   c. At 10% compound interest, the investment is doubled in 8 years.

   d. At 13% compound interest, the investment is doubled in 6 years.

As the rate of interest increases, the length of time taken to double the investment decreases.

21. a. Investment is compounded half-yearly. 6% is for the full year, therefore i is half that value: $\dfrac{6}{2} = 3$.

   b. Option 1: $5343.75
   Option 2: $5373.24

   c. After 2 years, Option 2 will be the better option with a value of $5064.79, which is more than Option 1 ($5062.50).

Exercise 12.6 Review: exam practice

1. $1000

2. a. $1296   **b.** $2820   **c.** $42
   d. $4.05   **e.** $7617.58

3. a. $7280   **b.** $39 780   **c.** $455

4. B

5. D

6. B

7. 6.5% p.a.

8. 15 months

9. D

10. D

11. B

12. a.

No. of years	1	2	3	4	5
Interest	$450	$900	$1350	$1800	$2250

b.

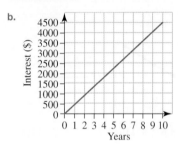

   c. 450
   d. $4500

13. $2778.30

14. $5700.47

15. a. $3932.39   **b.** $12 596.90   **c.** $14 457.96
   d. $23 851   **e.** $5334.67

16. $756.94

17. a. $12 024.02
   b. $12 052.04
   c. Compounding interest

***18. a.**

Principal	Annual interest rate	Time (years)	Simple interest	Compound interest
$1000	10.5%	3	$315.00	$349.23
$1000	10.5%	6	$630.00	$820.43
$1000	10.5%	9	$945.00	$1456.18
$1000	10.5%	12	$1260.00	$2313.96

18. a. * See the table at the bottom of the page.

b.

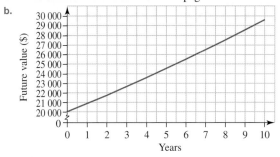

c. $29 500

19. Choose 4.1% p.a. compounding monthly since the amount of interest earned is higher.

20. An investor should choose 3.895% p.a. compounding monthly, as it yields the highest amount of interest.

---

*18. a.

No. of years	1	2	3	4	5
Future value	$20 800	$21 632	$22 497	$23 397	$24 333

---

13 Reducing balance loans

13.1 Overview

LEARNING SEQUENCE

13.1 Overview
13.2 Modelling reducing balance loans with technology
13.3 The effect of interest rate and repayment amount on reducing balance loans
13.4 Review: exam practice

CONTENT

In this chapter, students will learn to:
- understand that reducing balance loans are compound interest loans with periodic repayments
- use technology (online calculator) to model a reducing balance loan
- use technology (spreadsheet) to model a reducing balance loan [complex]
- use technology (online calculator) to investigate the effect of the interest rate and repayment amount on the time taken to repay a loan
- use technology (spreadsheet) to investigate the effect of the interest rate and repayment amount on the time taken to repay a loan [complex].

Fully worked solutions for this chapter are available in the Resources section of your eBookPLUS at www.jacplus.com.au.

13.2 Modelling reducing balance loans with technology

13.2.1 Reducing balance loans

When we invest money with a financial institution the institution pays us interest because it is using our money to lend to others. Conversely, when we borrow money from an institution, we are using the institution's money and so it charges us interest.

In reducing balance loans, interest is usually charged every month by the financial institution and repayments are made by the borrower on a regular basis. The original amount of money borrowed is called the principal and the amount of money owed is called the balance of the loan.

A brief overview of the way a reducing balance loan works is outlined below.

- An amount of money is borrowed from a bank or financial institution.
- The bank will charge you interest for borrowing the money (the amount of interest charged mainly depends on the current interest rate and the amount still owing). The amount of interest is calculated using the compound interest formula.
- The amount of interest is then added to the balance of the loan, effectively making the balance of the loan higher.
- To reduce the balance of the loan to zero over time, a regular payment is made to reduce the amount owed. In order to reduce the balance, these payments must be higher than the amount of interest charged.

Reducing balance loans are compound interest loans with periodic repayments. That is, the balance of the loan will increase exponentially due to the compounding interest being charged, but then slowly reduce over time due to the regular payment being made to the bank.

If we graphed the amount owing against time for a loan it would look like the graph shown at right.

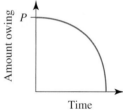

13.2.2 Using technology to model a reducing balance loan

The amount owing in a reducing balance loan can be modelled using a mathematical formula (known as the annuities formula) or recurrence relations. In this course, however, we will only consider solving problems using an online financial calculator. The online calculator used in chapter 12 will also be used in this topic. The calculator can be found at:

http://www.fncalculator.com/financialcalculator?type=tvmCalculator.

Recall from chapter 12 the meaning of the inputs for this calculator. These are defined as follows:

- *Present Value* is the current (present) value of the loan or investment.
- *Payment* is the amount paid at each time period.
- *Future Value* is the value of the investment or loan after a specified number of compounding periods.
- *Annual Rate (%)* is the annual interest rate.
- *Periods* is the total number of compounding periods.
- *Compounding* is the number of times the interest is compounded per year.
- *Mode* is used to indicate whether the interest is compounded at the end or at the beginning of the time period. Leave this set at END.

We also need to be mindful of sign conventions to be used for reducing balance loan calculations.

Sign conventions for reducing balance loans

- Present value should be positive (the bank gives you money).
- Payment should be negative (you repay the loan by making regular payments).
- Future value can be negative, zero or positive:
  - Negative indicates you still owe the bank money.
  - Zero indicates the loan is fully paid out.
  - Positive indicates you have overpaid the loan and the bank needs to repay you some money.

WORKED EXAMPLE 1

A loan of $80 000 is taken out over a 20-year period at a rate of 5% p.a. (interest charged monthly) and is to be paid back with monthly repayments of $527.96. Determine the amount owing after 8 years.

THINK

1. After 8 years a total of 96 payments have been made, since a payment is made each month. Input the following values into the online calculator:
   Present Value: 80 000
   Payment: −527.96
   Rate: 5
   Periods: 96
      (8 years × 12 each year)
   Compounding: Monthly
   To calculate the future value, click on the FV button.

2. Write the answer.

WRITE

Mode	◉ End ○ Beginning	
Present Value	80,000	PV
Payment	−527.96	PMT
Future Value	−57,084.56	FV
Annual Rate (%)	5	Rate
Periods	96	Periods
Compounding	Monthly ▾	

The amount owing after 8 years is $57 084.56.

WORKED EXAMPLE 2

Fredrick wants to borrow $4800 for a new TV at 6% p.a., interest adjusted monthly. What would be Fredrick's monthly repayment if the loan is paid off in 3 years?

THINK	WRITE

THINK

1. After 3 years a total of 36 payments have been made, since a payment is made each month. If a loan is fully paid off this implies that the future value is zero.
   Input the following values into the online calculator:
   PV: 4800
   Future Value: 0
   Rate: 6
   Periods: 36
      (3 years × 12 each year)
   Compounding: Monthly
   To determine the value of the monthly payment, click on the PMT button.

2. Write the answer.

WRITE

Mode ◉ End ○ Beginning	
Present Value 4,800	PV
Payment −146.03	PMT
Future Value 0	FV
Annual Rate (%) 6	Rate
Periods 36	Periods
Compounding Monthly ▼	

The monthly repayment is $146.03 over 3 years.

When borrowing money, it is often wise to have an idea of the total amount of interest you will have to pay and the total cost to you to repay the loan. With this information you can make clearer financial decisions.

The following formulas display some important properties of a reducing balance loan.

Properties of reducing balance loans

$$\text{cost of repaying the loan} = \text{the sum of all payments}$$
$$\text{total interest paid} = \text{total payments} - \text{principal}$$

WORKED EXAMPLE 3

A reducing balance loan of \$50 000 is to be repaid with monthly repayments of \$421.93 at an interest rate of 6% p.a. (debited monthly). Determine:

a. the number of monthly repayments required to fully repay the loan

b. the number of years to fully repay the loan

c. the total interest charged.

THINK

a. 1. If a loan is fully paid off this implies that the future value is zero.
   Input the following values into the online calculator:
   Present Value: 50 000
   Payment: −421.93
   Future Value: 0
   Rate: 6
   Compounding: Monthly
   To determine the number of periods needed to repay the loan, click on the Periods button.

WRITE

a.

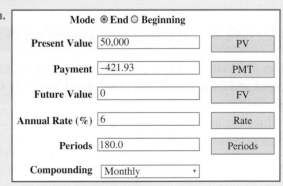

Mode ◉ End ○ Beginning	
Present Value 50,000	PV
Payment −421.93	PMT
Future Value 0	FV
Annual Rate (%) 6	Rate
Periods 180.0	Periods
Compounding Monthly ▼	

2. Write the answer.	180 monthly payments are required to fully repay the loan.
b. 1. Convert the number of time periods to years by dividing by 12.	**b.** $\dfrac{180}{12} = 15$ The loan will be paid off in 15 years.
c. 1. Total interest = total payments − principal	**c.** Total interest = $421.93 \times 180 - 50\,000$ $= 75\,947.40 - 50\,000$ $= \$25\,947.40$
2. Write the answer.	The total interest paid is $\$25\,947.40$.

WORKED EXAMPLE 4

A while ago Susie borrowed $16 000. Interest was charged at 8.8% p.a. on the reducing balance loan (adjusted monthly) and she has been paying $200.95 each month to pay off the loan. Currently she still owes $8106.50. How long ago did Susie start the loan?

THINK

1. We can determine the number of periods (monthly repayments) Susie has already paid by inputting the following values into the online calculator:
   Present Value: 16 000
   Payment: −200.95
   Future Value: −8106.50
      (negative FV because Susie still owes the bank this amount)
   Rate: 8.8
   Compounding: Monthly
   To determine the number of periods, click on the Periods button.

2. Convert the number of time periods to years by dividing by 12.

3. Write the answer.

WRITE

Mode	⦿ End ○ Beginning	
Present Value	16,000	PV
Payment	−200.95	PMT
Future Value	−8,106.50	FV
Annual Rate (%)	8.8	Rate
Periods	72.0	Periods
Compounding	Monthly ▾	

Susie has already made 72 monthly payments.

$\dfrac{72}{12} = 6$

Susie has had the loan for the past 6 years.

WORKED EXAMPLE 5

A loan of $14 000 is being paid off with equal monthly repayments of $200.86 and interest being charged at 12% p.a. (debited monthly). Currently the amount owing is $10 273.99. How much longer will it take to:

a. reduce the amount owing to $7627.35

b. repay the loan in full?

THINK	WRITE

a. 1. The present value of the loan is $10 273.99 and the future value is required to be $7627.35.
Input the following values into the online calculator:
Present Value: 10 273.99
Payment: −200.86
Future Value: −7627.35
    (negative FV because this amount still needs to be repaid)
Rate: 12
Compounding: Monthly
To determine the number of periods, click on the Periods button.

a.
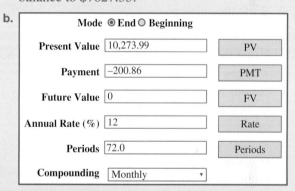

2. Convert the number of time periods to years by dividing by 12.

3. Write the answer.

A further 24 monthly payments are needed to reduce the balance to $7627.35.

$$\frac{24}{12} = 2$$

It will take another 2 years to reduce the loan balance to $7627.35.

b. 1. To repay the loan in full implies that the future value is zero.
Input the following values into the online calculator:
Present Value: 10 273.99
Payment: −200.86
Future Value: 0
Rate: 12
Compounding: Monthly
To determine the number of periods, click on the Periods button.

b.

Mode	◉ End ○ Beginning	
Present Value	10,273.99	PV
Payment	−200.86	PMT
Future Value	0	FV
Annual Rate (%)	12	Rate
Periods	72.0	Periods
Compounding	Monthly ▾	

2. Convert the number of time periods to years by dividing by 12.

3. Write the answer.

A further 72 monthly payments are needed to fully repay the loan.

$$\frac{72}{12} = 6$$

It will take another 6 years to fully pay off the loan.

13.2.3 Changing the repayment

Some loans, such as home loans, are taken over a long period of time. During this time the borrower's financial situation may change. A situation could arise whereby the borrower is able to make a higher repayment (due to an increase in their income) or have to reduce the repayment amount.

WORKED EXAMPLE 6

A reducing balance loan of **$22 000** is taken over 6 years. It is to be repaid with monthly repayments at a rate of 6.4% p.a. (debited monthly).
a. Determine the repayment value.
b. What will be the term of the loan if the repayment is increased to $429.43?
c. Calculate the total interest paid on repayments of $429.43.

THINK

a. 1. Input the following values into the online calculator:
Present Value: 22 000
Future Value: 0
Rate: 6.4
Periods: 72
    (6 × 12)
Compounding: Monthly
To determine the monthly payment, click on the PMT button.

2. Write the answer.

b. 1. Using the new payment value of $429.43 we need to determine the number of periods it will take to fully repay the loan. Input the following values into the online calculator:
Present Value: 22 000
Payment: −429.43
Future Value: 0
Rate: 6.4
Compounding: Monthly
To determine the number of periods, click on the Periods button.

2. Convert the number of time periods to years by dividing by 12.

3. Write the answer.

c. 1. Total interest = total payments − principal

2. Write the answer.

WRITE

a.

Mode ⦿ End ◯ Beginning	
Present Value 22,000	PV
Payment −368.77	PMT
Future Value 0	FV
Annual Rate (%) 6.4	Rate
Periods 72	Periods
Compounding Monthly ▾	

The monthly repayment is $368.77 over 6 years.

b.

Mode ⦿ End ◯ Beginning	
Present Value 22,000	PV
Payment −429.43	PMT
Future Value 0	FV
Annual Rate (%) 6.4	Rate
Periods 60.0	Periods
Compounding Monthly ▾	

Periods = 60

$$\frac{60}{12} = 5$$

When making a repayment of $429.43 per month, the loan will be fully paid off after 5 years.

c. Total interest = 429.43 × 60 − 22 000
    = 25 765.80 − 22 000
    = $3765.80

The total interest paid is $3765.80.

Borrowers wanting to pay off their loan as soon as possible can do so by increasing the repayments. Increasing your payments can reduce the period of your loan, while at the same time reducing the total interest you need to pay back on your loan.

Ryan borrowed \$25 000 to renovate his kitchen and agreed to repay the loan over 10 years with quarterly repayments of \$866.50 and interest debited at 6.8% p.a. However, 5 years into the loan Ryan made the decision to increase his repayments to \$1179.65. Calculate:

a. the actual term of the loan

b. the total interest paid

c. the interest saved by increasing the size of the repayments.

THINK	WRITE
a. 1. Firstly, we need to find the balance of the loan after 5 years (before the new payments commence). Also note that interest on this loan is compounding quarterly. Input the following values into the online calculator: Present Value: 25 000 Payment: −866.50 Rate: 6.8 Periods: $20\,(5 \times 4)$ Compounding: Quarterly To determine the value of the loan after 5 years, click on the FV button.	**a.**

Mode ⦿ End ○ Beginning

Present Value	25,000	PV
Payment	−866.50	PMT
Future Value	−14,587.39	FV
Annual Rate (%)	6.8	Rate
Periods	20	Periods
Compounding	Quarterly ▾	

Future value = \$14 587.39

2. Write the answer.

The balance of the loan after the first 5 years is \$14 587.39.

3. We now need to determine the number of periods (quarters) it will take for a balance of \$14 587.39 to reduce to zero. We must use the new (higher) payment amount. Input the following values into the online calculator:
Present Value: 14 587.39
Payment: −1179.65
Future Value: 0
Rate: 6.8
Compounding: Quarterly
To determine the number of periods, click on the Periods button.

Mode ⦿ End ○ Beginning

Present Value	14,587.39	PV
Payment	−1,179.65	PMT
Future Value	0	FV
Annual Rate (%)	6.8	Rate
Periods	14.0	Periods
Compounding	Quarterly ▾	

A further 14 quarters are required to repay the loan in full.

4. Answer the question by combining the two time periods. Convert into years by dividing by 4.

$$\begin{aligned}
\text{Total term of loan} &= 5 \text{ years} + 14 \text{ quarters}\\
&= 20 \text{ quarters} + 14 \text{ quarters}\\
&= 34 \text{ quarters}\\
&= \frac{34}{4} \text{ years}\\
&= 8.5 \text{ years}
\end{aligned}$$

▸

b. 1. For the two-payment case:
Total interest = total payments −
principal

b. Total interest = $(866.50 \times 20 + 1179.65 \times 14) - 25\,000$
$= (17\,330 + 16\,515.10) - 25\,000$
$= \$8845.10$

2. Write the answer.

The total interest paid is $8845.10.

c. 1. Using the financial calculator,
determine the number of periods it
would take to fully repay the loan
making payments of $866.50 only.

c. Periods = 40

2. Calculate the difference between
the two cases.

Interest difference = $(866.50 \times 40 - 25\,000) - (8845.10)$
$= 9660 - 8845.10$
$= \$814.90$

3. Write the answer.

Ryan will save $814.90 in interest if he increases his
repayments.

Calculations involving reducing balance loans can also be computed through the use of a spreadsheet
rather than a financial calculator. In the Resources section of your eBookPLUS you will find a
downloadable spreadsheet that can be used to assist in the calculations of a reducing balance loan.

 Resources

 Digital document SpreadSHEET Loan repayments (doc-30968)

Exercise 13.2 Modelling reducing balance loans with technology

1. **WE1** A loan of $75 000 is taken out over a 15-year period at a rate of 4.5% p.a. (interest charged
monthly) and is to be paid back with monthly repayments of $573.74. Determine the amount owing
after 10 years.

2. A loan of $60 000 is taken out over a 20-year period at a rate of 6% p.a. (interest charged monthly) and
is to be paid back with monthly repayments of $429.86. Determine the amount owing after 6 years.

3. **WE2** Danni wants to borrow $3700 for a new sound system at 6.5% p.a., interest adjusted monthly.
What would be Danni's monthly repayment if the loan is paid off in 2 years?

4. Adam has borrowed $15 000 to buy a car. He agrees
to pay the loan off over 4 years with monthly
repayments at 7.6% p.a. (adjusted monthly).
Calculate the monthly repayment value.

5. Sonya has borrowed $6000 to buy a ride-on mower.
She agrees to pay the loan off over 3 years
with equal monthly repayments of $183.62.
Determine the annual compounding rate of
interest correct to one decimal place.

6. **WE3** A reducing balance loan of $24 000 is to be
repaid with monthly repayments of $339.28 at an
interest rate of 8% p.a. (debited monthly). Determine:
   a. the number of monthly repayments required to fully repay the loan
   b. the number of years to fully repay the loan
   c. the total interest charged.

7. A loan of $3600 is taken out with a reducing balance interest rate of 5.6% p.a. with interest debited monthly. The borrower wishes to pay instalments of $125.01 per month. Determine how long until the loan is paid off in full, to the nearest month.

8. **WE4** A while ago James borrowed $14 000. Interest was charged at 7.4% p.a. on the reducing balance loan (adjusted monthly) and he has been paying $214.05 each month to pay off the loan. Currently he still owes $4761.42. How long ago did James start the loan?

9. Stephanie borrowed $13 500. Interest on this reducing balance loan has been charged at 7.5% p.a. (adjusted monthly) and she has been paying $207.07 each month to service the loan. Currently she has $8563.94 remaining on the loan, how long ago did Stephanie take out the loan?

10. **WE5** A loan of $16 000 is being paid off with equal monthly repayments of $195.86 and interest being charged at 10.5% p.a. (debited monthly). Currently the amount owing is $10 430.14. How much longer will it take to:
    a. reduce the amount owing to $6859.60
    b. repay the loan in full?

11. A loan for $8500 is being repaid with equal monthly repayments of $146.55 and interest being charged at 7.4% p.a. (debited monthly). Currently the amount owing is $4002.91. How much longer will it take to:
    a. reduce the amount owing to $2749.88
    b. repay the loan in full?

12. **WE6** A reducing balance loan of $16 000 is taken over 4 years. It is to be repaid with monthly repayments at a rate of 7.6% p.a. (debited monthly).
    a. Determine the repayment value.
    b. What will be the term of the loan if the repayment is changed to $350.77?
    c. Calculate the total interest paid with repayments of $350.77.

13. A reducing balance loan of $28 000 is taken over 6 years. It is to be repaid with monthly repayments at a rate of 6.2% p.a. (debited monthly).
    a. Determine the repayment value.
    b. What will be the term of the loan if the repayment is increased to $543.93?
    c. Calculate the total interest paid with repayments of $543.93.

14. **WE7** Tom borrowed $22 500 for a new car and agreed to repay the loan over 12 years with quarterly repayments of $653.90 and interest debited at 5.8% p.a. However, after 7 years into the loan Tom made the decision to increase his repayments to $1220.15. Calculate:
    a. the actual term of the loan
    b. the total interest paid
    c. the interest saved by increasing the size of the repayments.

15. Leah borrowed $19 500 to renovate her bathroom and agreed to repay the loan over 8 years with monthly repayments of $258.16 and interest debited at 6.2% p.a. However, after 5 years into the loan Leah made the decision to increase her repayments to $375.75. Calculate:
    a. the actual term of the loan
    b. the total interest paid
    c. the interest saved by increasing the size of the repayments.

16. **MC** A reducing balance loan of $62 000 is taken out at 6.3% p.a. (adjusted monthly). It is to be repaid with monthly instalments of $533.29. The loan will be paid in full in
    A. 5 years.　　　　**B.** 10 years.　　　　**C.** 15 years.　　　　**D.** 20 years.

17. Paul's reducing balance loan of $8250 is to be repaid with monthly repayments of $168.70 with interest charged at 8.25% p.a. (debited monthly).
    a. If the amount owing is $5359.03. How much longer will it take to:
       i. reduce the amount owing to $1931.76
       ii. repay the loan in full?
    b. At a later time, the amount owing has dropped to $4547.42. How much longer will it take to:
       i. reduce the loan to $3134.18
       ii. repay the loan in full?
18. Jeanie wanted to borrow $60 000 and she was offered a reducing balance loan over 15 years at 6.25% p.a. (adjusted monthly) with monthly instalments.
    a. What will be her monthly instalments?
    b. What would be the term of the loan if instead the repayment was:
       i. increased to $593.30
       ii. decreased to $450.24?

13.3 The effect of interest rate and repayment amount on reducing balance loans

13.3.1 Frequency of repayments

We will look at investigating the effect on the term of the loan, and on the total amount of interest charged if the frequency of the repayments is changed. The value of the repayments will change but the total outlay will stay the same. For example, a $3000 quarterly (every three months) repayment will be compared to a $1000 monthly repayment. That is, the same amount is repaid during the same period of time in each case.

WORKED EXAMPLE 8

Janette borrows $33 000 to renovate her jewellery shop. The loan is charged at 7.5% p.a. and she can afford quarterly repayments of $1180.05. This will pay the loan off in exactly 10 years.

One-third of the quarterly repayment gives the equivalent monthly repayment of $393.35. The equivalent fortnightly repayment is $181.55.

Determine:

i. the term of the loan

ii. the amount still owing prior to the last payment

if Janette made repayments:

a. monthly

b. fortnightly.

THINK	WRITE

a. i. 1. Determine the number of periods using monthly compounding.
Input the following values into the online calculator:
Present Value: 33 000
Payment: −393.35
Future Value: 0
Rate: 7.5
Compounding: Monthly
To determine the number of periods, click on the Periods button.

a. i.

Mode	⦿ End ○ Beginning	
Present Value	33,000	PV
Payment	−393.35	PMT
Future Value	0	FV
Annual Rate (%)	7.5	Rate
Periods	119.26	Periods
Compounding	Monthly ▾	

119.26 payments are required.

2. Interpret result and convert into years by dividing by 12.

119.26 monthly payments are required meaning that a 120th repayment is required.

$$\frac{120}{12} = 10$$

3. Write the answer.

The term of the loan is 10 years.

ii. 1. To find the amount still owing prior to the last payment, set the number of periods to 119.
Input the following values into the online calculator:
Present Value: 33 000
Payment: −393.35
Rate: 7.5
Periods: 119
Compounding: Monthly
To determine the future value, click on the FV button.

ii.

Mode	⦿ End ○ Beginning	
Present Value	33,000	PV
Payment	−393.35	PMT
Future Value	−101.95	FV
Annual Rate (%)	7.5	Rate
Periods	119	Periods
Compounding	Monthly ▾	

2. Write the answer.

The amount still owing prior to the last payment is $101.95.

b. i. 1. Determine the number of periods using fortnightly compounding.
Input the following values into the online calculator:
Present Value: 33 000
Payment: −181.55
Future Value: 0
Rate: 7.5
Compounding: Bi-Weekly (fortnightly)
To determine the number of periods, click on the Periods button.

b. i.

Mode	⦿ End ○ Beginning	
Present Value	33,000	PV
Payment	−181.55	PMT
Future Value	0	FV
Annual Rate (%)	7.5	Rate
Periods	257.96	Periods
Compounding	Bi-Weekly ▾	

2. Write the answer.

The term of the loan is 257.96 fortnights, which is 258 fortnightly repayments to cover the loan.

The loan will take $\frac{258}{26} = 9.92 \approx 10$ years to be fully paid off.

▶

ii. 1. To find the amount still owing prior to the last payment, set the number of periods to 257.

Input the following values into the online calculator:
Present Value: 33 000
Payment: −181.55
Rate: 7.5
Periods: 257
Compounding: Bi-Weekly
To determine the future value, click on the FV button.

2. Write the answer.

ii.

Mode ⦿ End ○ Beginning	
Present Value 33,000	PV
Payment −181.55	PMT
Future Value −173.11	FV
Annual Rate (%) 7.5	Rate
Periods 257	Periods
Compounding Bi-Weekly ▾	

The amount still owing prior to the last payment is $173.11.

Observing Worked example 8 it can be seen that even paying the same outlay there may be a slight reduction in the term of the loan when repayments are made more regularly. Let us now investigate what saving can be made in these situations. In this situation we should consider the final (partial) payment separately because the amount of interest that it attracts is less than a complete repayment. The calculation of the total interest paid is now calculated as usual: total interest paid = total payments − principal.

WORKED EXAMPLE 9

In Worked example 8, Janette's $33 000 loan at 7.5% p.a. gave the following three situations:
1. Quarterly repayments of $1180.05 for 10 years
2. Monthly repayments of $393.35 for 119 months with $101.95 still outstanding
3. Fortnightly repayments of $181.55 for 257 fortnights with $173.11 still outstanding
Compare the total interest paid by Janette if she repaid her loan:
a. quarterly   **b. monthly**   **c. fortnightly.**

THINK

a. 1. Making quarterly payments over 10 years means that $40 \, (10 \times 4)$ individual payments of $1180.05 are made. Use the formula: total interest = total payments − principal.

2. Write the answer.

b. 1. For monthly repayments, first determine the monthly interest rate.

2. Calculate the interest to be paid on the amount still owing ($101.95) and determine the amount of the final payment.

WRITE

a. For quarterly repayments:
$$\text{Total interest} = 1180.05 \times 40 - 33\,000$$
$$= \$14\,202$$

The total interest paid by Janette when repaying her loan quarterly is $14 202.

b. $\text{Rate} = \dfrac{7.5}{12} = 0.625\%$

$$\text{Interest} = 0.625\% \text{ of } 101.95$$
$$= 0.006\,25 \times 101.95$$
$$= \$\,0.64$$
$$\text{Final payment} = 101.95 + 0.64$$
$$= \$102.59$$

3. To calculate the total interest paid, use the formula:
total interest = total payments − principal.

Total interest $= 393.35 \times 119 + 102.59 - 33\,000$
$$= \$13\,911.24$$

c. 1. For fortnightly repayments, first determine the fortnightly interest rate.

c. Rate $= \dfrac{7.5}{26} = 0.2885\%$

2. Calculate the interest to be paid on the amount still owing ($173.11) and determine the amount of the final payment.

Interest $= 0.2885\%$ of 173.11
$$= 0.002\,885 \times 173.11$$
$$= \$0.50$$
Final payment $= 173.11 + 0.50$
$$= \$173.61$$

3. To calculate the total interest paid, use the formula:
total interest = total payments − principal.

Total interest $= 181.55 \times 257 + 173.61 - 33\,000$
$$= \$13\,831.96$$

4. Compare the interest paid.

Examining all 3 scenarios, Janette would be better off making fortnightly repayments as she would pay less interest overall.

13.3.2 Changing the rate

The Reserve Bank meets on the first Tuesday of every month and makes a decision on interest rates. This informs the decisions of individual lenders on what rate they will pass on to borrowers. Due to each monthly meeting, the interest rate can change many times over the duration of the loan.

We will look at comparing loan situations by taking into account the variation of the rate.

WORKED EXAMPLE 10

A reducing balance loan of $22 000 is taken over 6 years. It is to be repaid with monthly repayments of $368.77 at a rate of 6.4% p.a. (debited monthly).
a. What is the total interest paid?
b. If instead the rate was 7.2% p.a. (adjusted monthly) and the repayments remained the same, what would be:
  i. the term of the loan
  ii. the total amount of interest paid?

THINK

a. 1. For the rate of 6.4% p.a., making monthly payments over 6 years means that $72\,(6 \times 12)$ individual payments of $368.77 are made. Use the formula:
total interest = total payments − principal.

2. Write the answer.

WRITE

a. Total interest $= 368.77 \times 72 - 22\,000$
$$= \$4551.44$$

The total interest paid over the 6 years is $4551.44.

b. i. 1. Using the new interest rate of 7.2%, we need to calculate the number of periods it will take to fully pay off the loan.
Input the following values into the online calculator:
Present Value: 22 000
Payment: −368.77
Rate: 7.2
Compounding: Monthly
To determine the number of periods, click on the Periods button.

b. i.

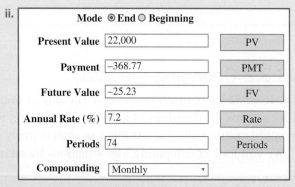

Mode ◉ End ○ Beginning	
Present Value 22,000	PV
Payment −368.77	PMT
Future Value 0	FV
Annual Rate (%) 7.2	Rate
Periods 74.07	Periods
Compounding Monthly ▾	

2. Interpret the result.

74.07 periods means that 74 full payments need to be made plus one smaller payment. Therefore 75 payments are required to repay the loan.

3. Answer the question.

The term of the loan is 75 months or 6 years and 3 months.

ii. 1. To determine the total amount of interest paid we need to calculate the interest accrued on the final (smaller) payment. To do this we need to determine the balance of the loan after 74 payments; that is the balance of the loan before the final payment.
Input the following values into the online calculator:
Present Value: 22 000
Payment: −368.77
Rate: 7.2
Periods: 74
Compounding: Monthly
To determine the future value, click on the FV button.

ii.

Mode ◉ End ○ Beginning	
Present Value 22,000	PV
Payment −368.77	PMT
Future Value −25.23	FV
Annual Rate (%) 7.2	Rate
Periods 74	Periods
Compounding Monthly ▾	

The balance of the loan after 74 payments is $25.23.

2. Calculate the interest rate per month.

$$\text{Interest rate per month} = \frac{7.2}{12} = 0.6\%$$

3. Calculate the interest on the final payment and the value of the final payment.

$$\begin{aligned}\text{Interest on final payment} &= 0.6\% \text{ of } \$25.23 \\ &= 0.006 \times 25.23 \\ &= \$0.15 \\ \text{Final payment} &= 25.23 + 0.15 \\ &= \$25.38\end{aligned}$$

4. Determine the total amount of interest paid using the formula:
total interest = total payments − principal.

$$\begin{aligned}\text{Total interest} &= 368.77 \times 74 + 25.38 - 22\,000 \\ &= \$5314.36\end{aligned}$$

Some important concepts to take away from these investigations are as follows.
1. Increasing the size of the repayment decreases the amount of interest paid and decreases the term of the loan.
2. Increasing the frequency of the repayment decreases the total interest paid and may decrease the term of the loan.
3. Increasing the interest rate increases the total interest paid and generally increases the term of the loan.
4. To maintain the term of the loan when the interest rate increases, increase the regular payment.

Exercise 13.3 The effect of interest rate and repayment amount on reducing balance loans

1. **WE8** Karen borrows $21 000. The loan is charged at 6.2% p.a. and she can afford quarterly repayments of $708.39. This will pay the loan off in exactly 10 years.
   One-third of the quarterly repayment gives the equivalent monthly repayment of $236.13. The equivalent fortnightly repayment is $108.98.
   Determine:
   i. the term of the loan
   ii. the amount still owing prior to the last payment
   if Karen made repayments:
   a. monthly   b. fortnightly.

2. Peter borrows $17 000. The loan is charged at 6.9% p.a. and he can afford quarterly repayments of $871.05. This will pay the loan off in exactly 6 years.
   One-third of the quarterly repayment gives the equivalent monthly repayment of $290.35. The equivalent fortnightly repayment is $134.01.
   Determine:
   i. the term of the loan
   ii. the amount still owing prior to the last payment
   if Peter made repayments:
   a. monthly   b. fortnightly.

3. **WE9** In question 1, Karen's $21 000 loan at 6.2% p.a. gave the following three situations.
   1. Quarterly repayments of $708.39 for 10 years
   2. Monthly repayments of $236.13 for 119 months with $91.16 still outstanding
   3. Fortnightly repayments of $108.98 for 258 fortnights with $35.30 still outstanding
   Compare the total interest paid by Karen if she repaid her loan:
   a. quarterly   b. monthly   c. fortnightly.

4. In question 2, Peter's $17 000 loan at 6.9% p.a. gave the following three situations.
   1. Quarterly repayments of $871.05 for 6 years
   2. Monthly repayments of $290.35 for 71 months with $170.93 still outstanding
   3. Fortnightly repayments of $134.01 for 154 fortnights with $116.61 still outstanding
   Compare the total interest paid by Peter if he repaid his loan:
   a. quarterly   b. monthly   c. fortnightly.

5. **WE10** A reducing balance loan of $13 000 is taken over 6 years. It is to be repaid with monthly repayments of $218.53 at a rate of 6.5% p.a. (debited monthly).
   a. What is the total interest paid?
   b. If instead the rate was 7.0% p.a. (adjusted monthly) and the repayments remained the same, what would be:
   i. the term of the loan   ii. the total amount of interest paid?

6. A reducing balance loan of $19 000 is taken over 7 years. It is to be repaid with monthly repayments of $288.62 at a rate of 7.2% p.a. (debited monthly).
   a. What is the total interest paid?
   b. If instead the rate was 8.0% p.a. (adjusted monthly) and the repayments remained the same, what would be:
      i. the term of the loan
      ii. the total amount of interest paid?

7. Abby borrows $1800 to purchase new ski gear. The loan is to be paid in full over 1 year with quarterly repayments at an interest rate of 7.2% p.a.
   a. Calculate the quarterly payment required.
   b. How much is owing on the loan after the third payment?
   c. Calculate the total cost of repaying the loan.
   d. Calculate the interest paid on the loan.

8. Craig borrows $2500 to purchase new golf clubs. The loan is to be paid in full over 1 year with quarterly repayments at an interest rate of 9.5% p.a.
   a. Calculate the quarterly payment required. If Craig shopped around he could get a loan at 8.8% p.a. (debited quarterly).
   b. Calculate the new quarterly repayments.
   c. Calculate the savings when taking the reduced rate.

9. Brigid borrows $6500 and the loan is to be paid in full over 2 years with monthly repayments at an interest rate of 10.5% p.a.
   a. Calculate the monthly payment required. If Brigid shopped around, she could get a loan at 9.3% p.a. (debited monthly).
   b. Calculate the new monthly repayments.
   c. Calculate the savings when taking the reduced rate.

10. A $7200 loan is taken out and is to be fully paid off in 2 years with monthly repayments at a rate of 11.25% p.a.
    a. Calculate the monthly payment required.
    b. If the monthly repayments were increased by $40 per month, what effect does it have on the term of the loan?

11. A loan of $18 000 attracts interest at 7.5% p.a. on the outstanding balance and the following three options are available.
    a. Half-yearly repayments of $1890.22 for 6 years
    b. Quarterly repayments of $945.11 for 5.75 years with $726.17 still owing
    c. Monthly repayments of $315.04 for 70 months with $282.68 still owing
    Calculate the interest paid on each of the three options.

12. Harrison takes out a $12 000 loan for 4 years at 7.8% p.a. (debited monthly).
    a. Calculate the monthly repayments.
    b. If Harrison made fortnightly repayments, what would they be?
    c. Calculate the interest saved by paying the loan with fortnightly repayments.

The following information refers to questions **13** and **14**.

Grace has borrowed $50 000 to finance developments to her business. Grace chooses to repay the loan, which attracts interest at 8.6% p.a. on the outstanding balance, by fortnightly repayments of $224.79 rather than the equivalent monthly repayment of $487.05.

13. **MC** The term of the loan will be
   A. 14 years 6 months.   **B.** 15 years.   **C.** 15 years 3 months.   **D.** 15 years 6 months.
14. **MC** The amount Grace will save is closest to
   A. $210.   **B.** $240.   **C.** $170.   **D.** $130.

13.4 Review: exam practice

13.4.1 Reducing balance loans: summary

Modelling reducing balance loans with technology

- Reducing balance loans are compound interest loans with periodic repayments.
- A financial calculator can be used to solve problems involving reducing balance loans.
- The definitions of the terms used in a financial calculator are as follows:
  - *Present Value* is the current (present) value of the loan or investment.
  - *Payment* is the amount paid at each time period.
  - *Future Value* is the value of the investment or loan after a specified number of compounding periods.
  - *Annual Rate* (%) is the annual interest rate.
  - *Periods* is the total number of compounding periods.
  - *Compounding* is the number of times the interest is compounded per year.
  - *Mode* is used to indicate whether the interest is compounded at the end or at the beginning of the time period.
- When entering values into a financial calculator it is extremely important to have the correct sign. For a reducing balance loan, the following sign convention should be used.
  - Present value should be positive (the bank gives you money).
  - Payment should be negative (you repay the loan by making regular payments).
  - Future value can be negative, zero or positive:
    - negative indicates you still owe the bank money
    - zero indicates the loan is fully paid out
    - positive indicates you have overpaid the loan and the bank needs to repay you some money.
  - Cost of repaying the loan = the sum of all payments
  - Total interest paid = total payments – principal

The effect of interest rate and repayment amount on reducing balance loans

- Increasing the size of the repayment decreases the amount of interest paid and decreases the term of the loan.
- Increasing the frequency of the repayment decreases the total interest paid and may decrease the term of the loan.
- Increasing the interest rate increases the total interest paid and generally increases the term of the loan.
- To maintain the term of the loan when the interest rate increases, increase the regular payment.

Exercise 13.4 Review: exam practice

Simple familiar

1. A loan of $75 000 is taken out over a 20-year period at a rate of 6.2% p.a. (interest debited monthly) and with monthly repayments of $546.01. Calculate the amount owing after 8 years.

2. Noel took out a reducing balance loan of $5600. Interest was charged at a rate of 7.6% p.a. compounding monthly. His loan is to be repaid fully in 3 years. Calculate the value of the monthly repayment.

3. Calculate the monthly loan repayments on each of the following loans.
   a. $25 000 over a 10-year period at a rate of 9% p.a. compounding monthly
   b. $45 000 over a 15-year period at a rate of 14% p.a. compounding monthly
   c. $164 750 over a 25-year period at a rate of 15% p.a. compounding monthly
   d. $425 000 over a 15-year period at a rate of 12% p.a. compounding monthly

4. Mr and Mrs Warne borrow $125 000 to purchase a home unit. The interest rate is 12% p.a. and the monthly repayments are $1376.36. Calculate:
   a. the first month's interest on the loan
   b. the balance of the loan after the first month.

5. **MC** Leslie borrowed $95 000 from a bank. Interest is charged at the rate of 8% p.a. on the reducing balance. The loan is to be repaid over a 15-year period with monthly payments of $907.87. The total amount of interest paid on this loan is
   A. $7600.   **B.** $68 416.60.   **C.** $102 600.   **D.** $114 000.

6. Fernando wants to borrow $5200 for a new sound system at 6.4% p.a., interest adjusted monthly.
   a. What would be Fernando's monthly repayment if the loan is paid off in 4 years?
   b. What would be the total interest charged?

7. A reducing balance loan of $30 000 is to be repaid with monthly repayments of $333.06 at an interest rate 6% p.a. (debited monthly). Find:
   a. the number of monthly repayments and hence to the nearest year the number of years to fully repay the loan
   b. the total interest charged.

8. A while ago Monique borrowed $21 000. Interest was charged at 7.8% p.a. on the reducing balance loan (adjusted monthly) and she has been paying $252.57 each month to pay off the loan. Currently she still owes $5596.37. How long ago did Monique start the loan?

9. Julie-Ann and Brent borrow $40 000 to renovate their coffee shop. Interest on the unpaid balance is charged to the loan account monthly. The $40 000 is to be fully repaid in equal monthly repayments of $481.10 for 10 years. Determine the annual compounding rate of interest correct to two decimal places.

10. Kristen and Adrian borrow $150 000 for their home. They have the choice of two loans.
    Loan 1: At 8% p.a. interest over 25 years with fixed monthly repayments of $1157.72.
    Loan 2: At 8.25% p.a. interest over 25 years with minimum monthly repayments of $1182.68 and an $8 per month account management fee.
    Kristen and Adrian believe they can afford to pay $1500 per month. If they do, Loan 2 will be repaid in 14 years and 2 months. Which loan should Kristen and Adrian choose if they can afford to pay the extra each month?

11. Mr and Mrs Stone borrow $225 000 for their home. The interest rate is 9.6% p.a. and the term of loan is 25 years. The monthly repayment is $1989.48.
    a. Calculate the total repayments made on this loan.
    b. If Mr and Mrs Stone increase their monthly payments to $2000, the loan will be repaid in 24 years and 1 month. Calculate the amount they will save in repayments with this increase.

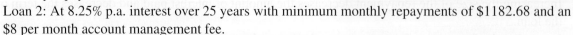

12. Mr and Mrs Rowe take out a $233 000 home loan at 12% p.a. over a 25-year term.
    a. Calculate the amount of each monthly repayment.
    b. After 3 years the balance on the loan has been reduced to $227 657. The interest rate then rises to 13% p.a. Calculate the new monthly repayment required to complete the loan within the existing term.

Complex familiar

13. A $25 000 reducing balance loan is taken out over 8 years charged at 9.8% p.a. (debited fortnightly) and paid with fortnightly repayments.
    a. Find the fortnightly repayments.
    b. What will be the term of the loan if repayments are $200.28?
    c. What would the fortnightly repayments need to be if the loan is to be paid off in 6 years?

14. A loan for $16 000 is being paid off at a rate of $312.80 per month with interest being charged at 12% p.a. (debited monthly). Currently the amount owing is $6645.19. How much longer will it take to:
    a. reduce the amount owing to $3520.87
    b. repay the loan in full?

15. A reducing balance loan of $13 000 is taken over 6 years. It is to be repaid with monthly repayments at a rate of 7.6% p.a. (debited monthly).
    a. Calculate the value of the monthly repayment.
    b. Determine the term of the loan if the repayment is increased to $261.11.
    c. Calculate the total interest paid with repayments of $261.11.

16. Fraser borrows $27 000 to purchase a new car. The loan is charged at 8.8% p.a. and she can afford quarterly repayments of $995.68 and this will pay the loan off in 10 years.
    One-third of the quarterly repayment gives the equivalent monthly repayment of $331.89. The equivalent fortnightly repayment is $153.18.
    Determine:
    i. the term of the loan
    ii. the amount still owing prior to the last payment
    if Fraser made repayments:
    a. monthly   b. fortnightly.

Complex unfamiliar

17. A reducing balance loan of $42 000 is taken over 8 years. It is to be repaid with monthly repayments of $549.90 at a rate of 5.9% p.a. (debited monthly).
    a. Calculate the total interest paid.
    If, instead the rate was 6.6% p.a. (adjusted monthly) and the repayments remained the same, calculate:
    b. the term of the loan
    c. the total amount of interest paid.

18. Shenna takes out a loan for $27 000 to purchase a new car. She wants to pay the loan off in 3 years with monthly repayments. If Shenna chose a loan that charged 7.2% p.a. (debited monthly), determine:
    a. the monthly repayments
    b. the total cost of repaying the loan.

19. Xavier purchased a new kitchen for $24 500. He was able to save a $5000 deposit to reduce the size of the loan he had to take out. Xavier borrowed the remainder of the loan at 10.5% p.a. (debited fortnightly) over 4 years.
    a. What were Xavier's fortnightly repayments?
    b. If his repayments were doubled, what would be the new term of the loan?

20. Shirly has been saving to buy a city apartment. She has been able to save a $35 000 deposit. Shirly negotiated with the real estate agent to reduce the purchase price of the apartment to $380 000. She paid the full deposit and then took out a reducing balance loan at 5.8% p.a. (debited monthly) over 20 years.
a. What are Shirly's monthly repayments?
b. Determine the total cost of repaying the loan.
Shirly was investigating if she could afford to pay the loan off in 15 years.
c. What would be her monthly repayments to pay off the loan in 15 years?
d. Determine the total cost of repaying the loan over the 15 years.
e. How much would Shirly save if she could afford to make it a 15-year loan, compared to 20 years?

Answers

13 Reducing balance loans

Exercise 13.2 Modelling reducing balance loans with technology

1. $30 776.08
2. $48 778.95
3. $164.82
4. $363.38
5. 6.4%
6. a. 96 monthly payments   b. 8 years
   c. $8570.88
7. 31 months
8. 60 months or 5 years
9. 36 months or 3 years
10. a. 30 months or 2.5 years   b. 72 months or 6 years
11. a. 10 months   b. 30 months or 2.5 years
12. a. $387.61   b. 54 months or 4.5 years
    c. $2941.58
13. a. $466.69   b. 60 months or 5 years
    c. $4635.80
14. a. 9.5 years   b. $8010.70
    c. $876.50
15. a. 84 months or 7 years   b. $5007.60
    c. $275.76
16. C
17. a. i. 24 months or 2 years
    ii. 35.96 months, so 36 monthly repayments, or 3 years
    b. i. 9.97 months, so 10 monthly repayments
    ii. 29.91 months, so 30 monthly repayments, or 2.5 years
18. a. $514.45
    b. i. 144 months or 12 years
    ii. 228 months or 19 years

Exercise 13.3 The effect of interest rate and repayment amount on reducing balance loans

1. a. i. 119.39 months, so 120 monthly repayments
   ii. $91.16
   b. i. 258.32 fortnights, so 259 fortnightly payments
   ii. $35.30
2. a. i. 71.59 months, so 72 monthly repayments
   ii. $170.93
   b. i. 154.87 fortnights, so 155 fortnightly repayments
   ii. $116.61
3. Total interest if paid quarterly: $7335.60; total interest if paid monthly: $7191.10; total interest if pay fortnightly: $7152.22. The interest saving by paying monthly compared to quarterly is $144.50. The interest saving by paying fornightly compared to quarterly is $183.38.
4. Total interest if paid quarterly: $3905.20; total interest if paid monthly: $3786.76; total interest if pay fortnightly: $3754.46. The interest saving by paying monthly compared to quarterly is $118.44. The interest saving by paying fornightly compared to quarterly is $150.74.
5. a. $2734.16
   b. i. 73.28 months   ii. $3013.16
6. a. $5244.08
   b. i. 86.96 months   ii. $6098.14
7. a. $470.43   b. $462.11
   c. $1881.72   d. $81.72
8. a. $662.54   b. $659.75   c. $11.16
9. a. $301.44   b. $297.85   c. $86.16
10. a. $336.41
    b. Reduces the term of the loan to 21.18 months
11. a. $4682.64   b. $4477.32   c. $4337.25
12. a. $291.83   b. $134.49   c. $20.88
13. D
14. C

13.4 Review: exam practice

1. $55 363.86
2. $174.45
3. a. $316.69   b. $599.28
   c. $2110.17   d. $5100.71
4. a. $1250   b. $124 873.64
5. B
6. a. $123.08   b. $707.84
7. a. 120 months or 10 years   b. $9967.20
8. 96 months or 8 years
9. 7.8%
10. Loan 2
11. a. $596 844   b. $18 844
12. a. $2454.01   b. $2618.57
13. a. $173.62   b. 169 months or 6.5 years
    c. $212.26
14. a. 12 months or 1 year   b. 24 months or 2 years
15. a. $225.40   b. 60 months or 5 years
    c. $2666.60
16. a. i. 124.24 months   ii. $80.02
    b. i. 268.66 fortnights   ii. $101.25
17. a. $10 790.40   b. 99.34 months
    c. $12 625.99
18. a. $836.15   b. $30 101.40
19. a. $230.00   b. 46.59 fortnights
20. a. $2432.05   b. $583 692.00
    c. $2874.16   d. $517 348.80
    e. $66 343.20

PRACTICE ASSESSMENT 3

Essential Mathematics: Problem solving and modelling task

Unit
Unit 4: Graphs, chance and loans

Topic
Fundamental topic: Calculations
Topic 1: Bivariate graphs

Conditions

Duration	Mode	Individual/group
5 weeks	Written report	Individual

Resources permitted	Length
The use of technology is required, for example: • non-CAS graphics calculator • spreadsheet software • other mathematical software.	• Up to 8 pages (including tables, figures and diagrams) • Maximum of 1000 words • Appendixes can include raw data, repeated calculations, evidence of authentication and student notes (appendixes are not to be marked)

Criterion	Grade
Formulate *Assessment objectives 1, 2, 5	
Solve *Assessment objectives 1, 6	
Evaluate and verify *Assessment objectives 4, 5	
Communicate *Assessment objective 3	
Milestones	
Week 1	
Week 2	
Week 3	
Week 4	
Week 5 (assessment submission)	

* © State of Queensland (Queensland Curriculum & Assessment Authority), *Essential Mathematics Applied Senior Syllabus 2019 v1.1*, Brisbane. For the most up-to-date assessment information, please see www.qcaa.qld.edu.au/senior.

Context

Statistical data can be communicated in the form of written prose, tables and graphs. It is difficult to compare and see trends in written prose. To gain a full understanding of trends and comparisons of data presented in table form, we often need to perform additional calculations. The visual impact of graphs is immediate; however, often the impression we receive is a misleading one if we fail to look carefully at the graph. Graphs can be manipulated to produce the desired effect. Consider the two graphs shown.

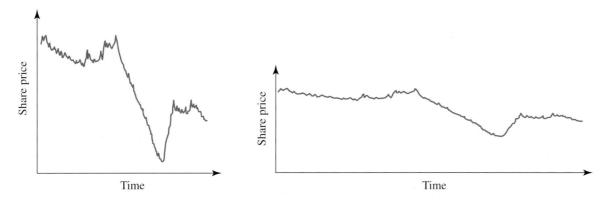

These graphs both display the same data (the value of shares over the same period of time), yet they look quite different. If the objective of the graph is to create the impression that the value of the shares has dropped quite dramatically, the first graph would serve the purpose. On the other hand, if we are to be convinced that the shares have not changed very much in value, the second graph reflects this point of view.

Task

The following table shows the times for the men's and women's 100-m sprints in the Summer Olympic Games. A sports journalist wants to convince readers that, since 1928, the women have been more successful in reducing their time than the men.

Year	100-m men	Men's time (seconds)	100-m women	Women's time (seconds)
1928	Percy Williams, CAN	10.80	Betty Robinson, USA	12.20
1932	Eddie Tolan, USA	10.30	Stella Walsh, POL	11.90
1936	Jesse Owens, USA	10.30	Helen Stephens, USA	11.50
1948	Harrison Dillard, USA	10.30	Fanny Blankers-Koen, HOL	11.90
1952	Lindy Remigino, USA	10.40	Marjorie Jackson, AUS	11.50
1956	Bobby Morrow, USA	10.50	Betty Cuthbert, AUS	11.50
1960	Armin Hary, GER	10.20	Wilma Rudolph, USA	11.00
1964	Bob Haryes, USA	10.00	Wyomia Tyus, USA	11.40
1968	Jim Hines, USA	9.95	Wyomia Tyus, USA	11.08
1972	Valery Borzov, URS	10.14	Renate Stecher, GDR	11.07
1976	Hasely Crawford, TRI	10.06	Annegret Richter, FRG	11.08
1980	Allan Wells, GBR	10.25	Lyudmila Kondratyeva, URS	11.06
1984	Carl Lewis, USA	9.99	Evelyn Ashford, USA	10.97
1988	Carl Lewis, USA	9.92	Florence Griffith Joyner, USA	10.54
1992	Linford Christie, GBR	9.96	Gail Devers, USA	10.82
1966	Donovan Bailey, CAN	9.84	Gail Devers, USA	10.94
2000	Maurice Greene, USA	9.87	Marion Jones, USA	10.75
2004	Justin Gatlin, USA	9.85	Yuliya Nesterenko, BLR	10.93
2008	Usain Bolt, JAM	9.69	Shelley-Ann Fraser, JAM	10.78

Note: The Summer Olympic Games were not held in 1940 and 1944.

- Draw a standard graph of the two sets of data and comment on the differences noticed.
- On a separate sheet of graph paper, draw two graphs to support the sports journalist's view.
- Describe the features on your graphs that helped you present the journalist's view effectively.
- Imagine you are a sports journalist and you are about to publish the graphs in a paper. Write a short article that would accompany the graphs.

Using the standard graphs, complete the following:
- Find the equations of the lines of best fit for the women and the men.
- If you were looking for being more successful in reducing the time over the 80-year period what would you look for in the equations?
- Use your equations to predict the times expected for both men and women at the 2016 Olympic Games and compare to the real values.
- Use your equations to predict the times expected in 2036. What type of predicting is this? Explain if it is a reliable prediction.
- Use your equations to predict the times that might have been run in 1940 if the Olympic Games were run. What type of predicting is this?
- Give a general summary of the limitations to the linear model and its ability to predict into the future.

To complete this task, you must:
- Use the problem-solving and mathematical modelling approach to develop your response.
- Respond with a range of understanding and skills, such as using mathematical language, appropriate calculations, tables of data, graphs and diagrams.
- Provide a response that highlights the real-life application of mathematics.
- Respond using a written report format that can be read and interpreted independently of the instrument task sheet.
- Develop a unique response.
- Use both analytic procedures and technology.

Approach to problem solving and modelling

Formulate

In this task you will investigate Olympic Games 100-m sprint times.

Graph the data using a standard graph and then graph the data so that it favours the view that women have made better inroads in improving their times over this period. Remember to state the necessary assumptions, variables and observations. You must also explain how you will make use of technology.

Solve

Construct the graphs and investigate their differences and why they differ. Review and make any refinements to the graphs. Link the analysis back to the context of the question. You will make further refinements and comparisons as necessary.

You must use technology efficiently and show detailed calculations demonstrating the procedures used.

Is it solved?

Evaluate and verify

Evaluate the reasonableness of your original solution.

Based on your graphs and predictions, consider whether there could be a more accurate method.

Justify and explain all procedures you have used and decisions you have made. Considering the original task, how valid is your solution?

Is the solution verified?

Communicate

Once you have completed all necessary work, you should consider how you have communicated all aspects of your report. Communicate using appropriate language that refers to the calculations, tables and graphs included in previous sections. Your response should be coherently and concisely organised.

Ensure you have:
- used mathematical, statistical and everyday language
- considered the strengths and limitations of your solution
- drawn conclusions by discussing your results
- included recommendations.

PRACTICE ASSESSMENT 4

Essential Mathematics: Unit 4 examination

Unit
Unit 4: Graphs, chance and loans

Topic
Fundamental topic: Calculations
Topic 1: Bivariate graphs
Topic 2: Probability and relative frequencies
Topic 3: Loans and compound interest

Conditions

Technique	Response type	Duration	Reading
Paper 1: Simple (27 marks) Paper 2: Complex (13 marks)	Short response	60 minutes	5 minutes

Resources	Instructions
• QCAA formula sheet • Notes not permitted • Scientific calculator permitted	• Show all working. • Write responses using a black or blue pen. • Unless otherwise instructed, give answers to **two decimal places**.

Criterion	Marks allocated	Result
Foundational knowledge and problem-solving *Assessment objectives 1, 2, 3, 4, 5 and 6	40	

* © State of Queensland (Queensland Curriculum & Assessment Authority), *Essential Mathematics Applied Senior Syllabus 2019 v1.1*, Brisbane. For the most up-to-date assessment information, please see www.qcaa.qld.edu.au/senior.

A detailed breakdown of the examination marks summary can be found in the PDF version of this assessment instrument in your eBookPLUS.

Question 1 (4 marks)

Write down the gradient and y-intercept of the following graphs.

a. $y = \dfrac{4x - 3}{2}$

_____

_____

_____

b. $6x - 4y + 48 = 0$

_____

_____

_____

Question 2 (3 marks)

Pearson's correlation coefficient for a scatterplot was found to be -0.7845.
Describe what this value tells you about the association between the two variables in terms of strength, direction and form.

_____

_____

_____

_____

_____

_____

_____

_____

Question 3 (5 marks)

Shane owes his parents $6000 and agrees to pay them back $120 per week.

a. State the linear rule that demonstrates this reducing-debt schedule.

_____

_____

_____

_____

b. How many weeks does it take Shane to repay the debt?

_____

_____

_____

c. How much does he still owe after 20 weeks?

_____

_____

_____

d. After how many payments does he still owe $1400?

_____

_____

_____

_____

Question 4 (6 marks)

Amy invests $27 500 at 4.5% simple interest for 3 years.

a. Calculate the interest earned by Amy.

_____

_____

_____

_____

b. Calculate the total value of Amy's investment after 3 years.

_____

_____

_____

_____

c. If the investment paid interest in quarterly instalments, calculate the value of each interest payment.

_____

_____

_____

_____

_____

_____

_____

_____

_____

Question 5 (4 marks)

a. The general form of a linear equation is $y = mx + c$ where m is the gradient and c is the y-intercept. Calculate the value of the gradient and y-intercept of the straight line shown, and hence write down the equation of the line.

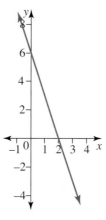

b. Use the equation to complete the following table.

x	y
-3	
0	
5	

Question 6 (3 marks)

a. What is the probability of rolling a 3 on a 6-sided die?

b. What is the complement of the event in part **a**?

c. What is the probability of the complement occurring?

Question 7 (2 marks)

a. Calculate the x-intercept of the equation $3x + 4y + 12 = 0$.

Question 8 (4 marks)

Using the data in the table below to complete the following tasks.

x	1	2	3	4	5	6	7	8	9	10
y	−2	0	1	5	7	8	9	11	15	16

a. Plot the data on a scatterplot.

b. Comment on the direction and strength of the data.

_____

_____

_____

c. Draw a line of best fit and determine its equation.

_____

_____

_____

_____

_____

_____

Question 9 (6 marks)

Jane and Brad each invest $12 500 for a period of 5 years.

a. Jane invests her money at 9.9% p.a. with interest compounded annually. Calculate the compounded value of her investment.

_____

_____

_____

_____

_____

b. Brad invests his money at 9.6% p.a. with interest compounded quarterly. Calculate the compounded value of his investment.

_____

_____

_____

_____

_____

c. Who has greater compounded value? Explain why.

_____

_____

_____

_____

_____

_____

Question 10 (3 marks)

Sonya has forgotten her last two digits of her bank PIN number.

a. What is the probability that she correctly guesses either one of the two final digits (but not both)?

_____

_____

_____

_____

_____

_____

_____

b. What is the probability that she guesses both of the two final digits?

_____

_____

_____

_____

_____

c. What is the probability that she incorrectly guesses both of the two final digits?

_____

_____

_____

_____

_____

APPENDIX
Maths skills workbook

Introduction

We use mathematical skills every day, quite often without even realising it. This book contains activities that help to develop and refine some more commonly used mathematical processes.

There are many different ways to approach questions, and we have all learned different ways of getting to the same answer. If you have your own methods that are mathematically correct, you may wish to stay with these more familiar steps. Check with your teacher that your processes are acceptable.

CONTENTS

KEY SKILL 1 Fractions

Every day of our lives we make decisions that involve fractions. We might be cutting a cake into portions or dividing money among a group of friends. Changing a recipe to cater for more or fewer people requires an understanding of fractions. Many occupations require a knowledge of how to calculate fractions of quantities.

Language of fractions

Numerator: the top number

Denominator: the bottom number

Simplest form: when there are no numbers other than 1 that can divide exactly into the numerator and denominator

Proper fraction: a fraction with a numerator smaller than the denominator, e.g. $\dfrac{2}{5}$

Improper fraction: a fraction with a numerator larger than the denominator, e.g. $\dfrac{10}{7}$

Mixed number: a whole number and a proper fraction, e.g. $3\dfrac{1}{4}$

Questions

1. Write the fraction for the amount shown.

   a.

   b.

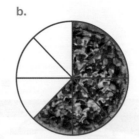

   c.

2. Write out the fraction that is described, then cancel down this fraction into its simplest form.

   a. White squares

   b. Blue buttons

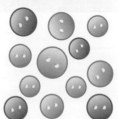

   c. Orange fish

3. Cancel each of these fractions into their simplest form by finding the largest number that divides into both the numerator and denominator.

   a. $\dfrac{10}{15} =$ **b.** $\dfrac{20}{40} =$ **c.** $\dfrac{21}{35} =$

   d. $\dfrac{40}{25} =$ **e.** $\dfrac{25}{75} =$ **f.** $\dfrac{160}{400} =$

KEY SKILL 2 Equivalent fractions

Equivalent fractions are equal to each other: they are worth the **same** amount. Equivalent fractions can be found by either multiplying or dividing both numerator and denominator by the same amount.

 $\dfrac{1}{2}$

 $\dfrac{2}{4}$

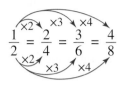

$$\frac{1}{2} = \frac{2}{4} = \frac{3}{6} = \frac{4}{8}$$

WORKED EXAMPLE

Fill in the missing values: $\dfrac{2}{5} = \dfrac{4}{-} = \dfrac{6}{15} = \dfrac{}{30}$.

THINK

1. To make a 2 into a 4 you need to $\times 2$.

2. To make a 5 into a 30 you need to $\times 6$.

WRITE

$$\overset{\times 2}{\underset{\times 2}{\frac{2}{5} = \frac{4}{10}}}$$

$$\overset{\times 6}{\underset{\times 6}{\frac{2}{5} = \frac{12}{30}}}$$

Questions

Draw in your own links and find the missing numbers.

1. $\dfrac{4}{5} = \dfrac{}{10} = \dfrac{}{20} = \dfrac{24}{}$

2. $\dfrac{2}{3} = \dfrac{4}{} = \dfrac{}{12} = \dfrac{20}{}$

3. $\dfrac{2}{5} = \dfrac{}{10} = \dfrac{}{25} = \dfrac{12}{}$

4. $\dfrac{2}{7} = \dfrac{4}{} = \dfrac{16}{} = \dfrac{20}{}$

5. $\dfrac{5}{9} = \dfrac{}{18} = \dfrac{}{45} = \dfrac{50}{}$

6. $\dfrac{7}{11} = \dfrac{14}{} = \dfrac{35}{} = \dfrac{}{121}$

7. Ben ate 15 chocolates from a box of 25 chocolates.
   a. Write this as a fraction.

   _____

   _____

   b. Write the fraction in its simplest form.

   _____

   _____

   c. What fraction is left?

   _____

   _____

KEY SKILL 3 Multiplying fractions

Multiplying fractions is an important skill to have as it will not just be used for fractions but also when calculating **percentages.**

Being able to cancel down is important because it minimises the size of the numbers that are being used.

Remember that you can cancel the **numerator** and the **denominator** from one fraction or cancel **diagonally** across two fractions.

Remember the saying: **cancel one from the top with one from the bottom**.

WORKED EXAMPLE

a. **Multiply** $\dfrac{3}{4} \times \dfrac{5}{7}$.

b. **Multiply** $\dfrac{6}{7} \times \dfrac{5}{12}$.

c. **Multiply** $\dfrac{4}{5} \times 100$.

THINK

a. 1. Can any numbers on the top lines cancel down with numbers on the bottom? No.

2. Write out the two numbers on the top lines together on one single top line, and write out the two bottom numbers together on a single bottom line.

3. Multiply the top numbers, and multiply the bottom numbers.

b. 1. Can any numbers on the top lines cancel down with numbers on the bottom? Yes.

2. Cancel down by finding what number will divide exactly into the top and bottom numbers.

3. Multiply what remains after cancelling down.

4. Multiply the top numbers, and multiply the bottom numbers.

c. 1. Make 100 into a fraction by putting it over 1.

2. Check to see if there is any cancelling to be done.

3. Multiply the top numbers, and multiply the bottom numbers.

WRITE

a. $\dfrac{3}{4} \times \dfrac{5}{7}$

$= \dfrac{3 \times 5}{4 \times 7}$

$= \dfrac{15}{28}$

b. $\dfrac{6}{7} \times \dfrac{5}{12}$

$= \dfrac{\overset{1}{\cancel{6}}}{7} \times \dfrac{5}{\underset{2}{\cancel{12}}}$

$= \dfrac{1}{7} \times \dfrac{5}{2}$

$= \dfrac{5}{14}$

c. $\dfrac{4}{5} \times \dfrac{100}{1}$

$= \dfrac{4}{\underset{1}{\cancel{5}}} \times \dfrac{\cancel{100}^{20}}{1}$

$= \dfrac{4}{1} \times \dfrac{20}{1}$

$= \dfrac{80}{1}$

$= 80$

Questions

1. Check to see what can be cancelled down, then multiply these fractions.

   a. $\dfrac{5}{6} \times \dfrac{3}{10}$

   b. $\dfrac{7}{10} \times \dfrac{5}{21}$

   c. $\dfrac{2}{15} \times \dfrac{10}{11}$

   d. $\dfrac{3}{8} \times \dfrac{12}{13}$

   _____

   _____

   _____

   _____

   e. $\dfrac{25}{63} \times \dfrac{9}{35}$

   f. $\dfrac{15}{42} \times \dfrac{14}{25}$

   g. $\dfrac{20}{50} \times \dfrac{100}{1}$

   h. $\dfrac{25}{75} \times \dfrac{6}{7}$

   _____

   _____

   _____

   _____

2. Multiply each of the fractions with the whole number.

   a. $\dfrac{3}{5} \times 100$

   b. $\dfrac{25}{75} \times 100$

   c. $\dfrac{1}{5} \times 100$

   d. $\dfrac{49}{50} \times 100$

   _____

   _____

   _____

   e. $\dfrac{6}{9} \times 45$

   f. $\dfrac{2}{3} \times 90$

   g. $\dfrac{1}{6} \times 54$

   h. $\dfrac{20}{55} \times 110$

   _____

   _____

   _____

3. a. In a box of 50 matches, only $\dfrac{4}{5}$ of them would strike. How many matches is this?

   b. In a school, $\dfrac{6}{7}$ of 84 Year 11 students have a part-time job. How many is this?

   _____

   _____

   _____

   c. In a bag of 720 lollies, $\dfrac{2}{9}$ of them are jelly beans. How many jelly beans are there?

   d. In a box of 300 nails, $\dfrac{1}{20}$ are faulty and cannot be used. How many nails can be used?

   _____

   _____

   _____

KEY SKILL 4 Percentage skills

The number skills concepts of fractions and percentages are very closely related and the skills used in one topic may be useful for others.

It is useful to remember that when you think of **percentages**, think of the number **100**. It's always the best place to start. For example:

$$50\% \text{ means:}$$
$$50 \text{ out of } 100$$
or
$$\frac{50}{100}$$
or
$$50 \div 100$$
or
$$0.50$$

Questions

Complete the following table. The first two have been done for you.

	Percentage	Fraction	Simplified fraction	Calculator steps	Decimal number
1.	10%	$\frac{10}{100}$	$\frac{1}{10}$	$10 \div 100$	0.1
2.	20%	$\frac{20}{100}$	$\frac{1}{5}$	$20 \div 100$	0.2
3.	25%				
4.	30%				
5.	50%				
6.	75%				
7.	80%				
8.	90%				
9.	12.5%				
10.	150%				
11.	33.33%				
12.	66.67%				

13. Mara collected shells at the beach. She collected 100 shells, of which 70 were in perfect condition.
    a. What percentage was in perfect condition?
    b. Write as a fraction the number of perfect shells out of the total number of shells.
    c. Write as a decimal the number of perfect shells out of the total number of shells.

KEY SKILL 5 Percentages

Quantities are often expressed as a percentage of an amount; for example, 2% of pet owners have a rabbit. This statement gives a proportion; for example, 2 out of 100 pet owners have a rabbit. In many applications of mathematics the word *of* refers to multiplication.

WORKED EXAMPLE

a. Find 40% of 620.   **b. Find 12.5% of 180.**

THINK	WRITE
a. 1. Write 40% as a fraction.	a. $\dfrac{40}{100}$
2. Multiply this by 620.	$\dfrac{40}{100} \times 620$
3. Use a calculator for this calculation.	$40 \div 100 \times 620 = 248$
b. 1. Write 12.5% as a fraction.	b. $\dfrac{12.5}{100}$
2. Multiply this by 180.	$\dfrac{12.5}{100} \times 180$
3. Use a calculator for this calculation.	$12.5 \div 100 \times 180 = 22.5$

Questions

1. Fill in the missing numbers and solve these problems.
   a. Find 20% of 380.

   $= \qquad \times$

   $=$

   b. Find 15% of 175.

   $= \qquad \times$

   $=$

2. Now create your own working out for the following.
   a. 60% of 80

   _____

   b. 12.5% of 204

   _____

   c. 250% of 84

   _____

   d. 18% of 44

   _____

   e. 15.5% of 360

   _____

   f. 33.33% of 180

   _____

   g. $55\frac{3}{4}\%$ of 96

   _____

   h. $18\frac{1}{4}\%$ of 688

   _____

KEY SKILL 6 Fractions into percentages

A percentage is another way of writing a fraction, but in this case the denominator is always 100. It is a useful way to make comparisons.

As an example, we can say that 30% is the same as $\frac{30}{100}$ or 90% is the same as $\frac{90}{100}$. To change a fraction to a percentage, multiply the fraction by 100, then simplify.

WORKED EXAMPLE

What percentage is 15 out of 50?

THINK

1. 15 out of 50 means 15 divided by 50.

2. Now multiply by 100 to make a percentage.

3. Use a calculator for this calculation.

WRITE

$15 \div 50$

$\frac{15}{50} \times 100$

$5 \div 50 \times 100 = 30\%$

Questions

1. What percentage is:
   a. 45 out of 90

   _____

   b. 5 out of 8

   _____

   c. 12 out of 60?

   _____

2. The table below shows the results for two Maths tests. Turn each score into a percentage, then shade in or circle the *best* test score for each student. Round off to two decimal places if required.

Student	Test 1 (out of 80)	Working out	Score (%)	Test 2 (out of 60)	Working out	Score (%)
Michelle	75	$75 \div 80 \times 100 =$	93.75%	54		
Yang	15			15		
Simon	40			32		
Alba	49			45		
Christina	65			55		

3. Convert the following amounts into percentages. (Round off to two decimal places if required.)
   a. 12 seconds out of 60 seconds

   _____

   b. 15 kg out of 75 kg

   _____

   c. 35 cents out of $2.00

   _____

KEY SKILL 7 Ratios

A ratio is a comparison between two or more amounts (or values). The symbol : is used to separate the values.

A ratio could be written as 5 : 4. This ratio means 'five **compared** to four', or we might say 'five to four'. The ratio 7 : 10 : 9 could be described as 'seven to ten to nine'.

A ratio written should contain only whole numbers and the order of the numbers is very important.

A ratio can also be written as a fraction. For example, the ratio 5 : 4 is the same as the fraction $\frac{5}{4}$.

WORKED EXAMPLE

The Essential Mathematics class has 9 boys and 13 girls in it. Write this as a ratio.

THINK

1. The values, in the order of the question, are 9 and 13.

2. Use the ratio symbol to separate the values.

WRITE

9 boys and 13 girls

9 : 13

Questions

1. Write the following as ratios.
    a. 10 compared to 13 = _____ : _____
    b. Five compared to seven = _____ : _____
    c. 100 compared to 1 = _____ : _____
    d. 11 to 3 = _____ : _____
    e. 6 to 30 = _____ : _____
    f. 12 to 7 = _____ : _____

2. Write the following as ratios.
    a. A farmer had 350 cows and 7 bulls. _____ : _____
    b. A car travelled 150 km in 2 hours. _____ : _____
    c. Peter spent $3 on 2 ice-creams. _____ : _____
    d. A cake requires 3 cups of flour and 2 eggs. _____ : _____
    e. Cordial needs 4 parts of water and 1 part of cordial syrup. _____ : _____
    f. A jet travelled 640 km in one hour. _____ : _____
    g. A bag of lollies had 10 jelly beans, 15 snakes and 9 mint leaves in it.

    _____ : _____ : _____

    h. A bowl of fruit contained 4 bananas, 2 apples, 35 grapes and 2 oranges.

    _____ : _____ : _____

KEY SKILL 8 Using ratios

A common example of a ratio is the **cement to water** ratio of 4 : 1, which is used in making concrete. This means that the amount of cement used is four times greater than the amount of water used. The actual amounts of cement and water used or the amount of cement being made is not stated, because the ratio is a comparison of the two quantities.

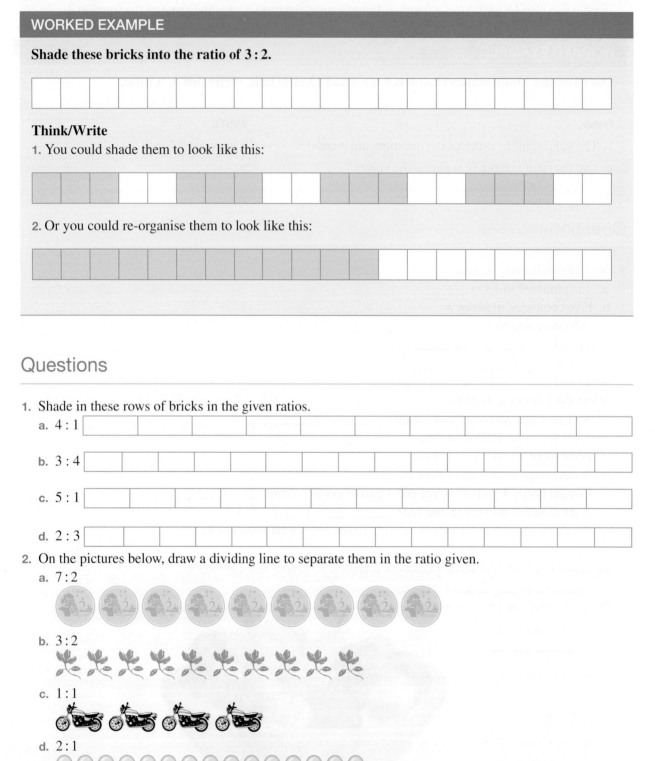

WORKED EXAMPLE

Shade these bricks into the ratio of 3 : 2.

Think/Write

1. You could shade them to look like this:

2. Or you could re-organise them to look like this:

Questions

1. Shade in these rows of bricks in the given ratios.

    a. 4 : 1

    b. 3 : 4

    c. 5 : 1

    d. 2 : 3

2. On the pictures below, draw a dividing line to separate them in the ratio given.

    a. 7 : 2

    b. 3 : 2

    c. 1 : 1

    d. 2 : 1

KEY SKILL 9 Simplifying ratios

Ratios are like fractions — sometimes they can be cancelled down into smaller numbers (or simplest form).

WORKED EXAMPLE

Cancel down 12 : 2 into its simplest form.

THINK

1. Find the highest common factor (the biggest number that divides into all parts of the ratio).

2. Divide all parts of the ratio by this number.

WRITE

12 and 2 are both divisible by 2.

$$\frac{12}{2} \div \frac{2}{2}$$

$$= 6 : 1$$

Questions

1. Cancel these down into their simplest form.

   a. $6 : 2$

   b. $15 : 10$

   c. $9 : 3$

   d. $30 : 12$

   e. $9 : 6$

   f. $20 : 16 : 12$

2. On a necklace there are 15 black beads and 10 white beads. What is the ratio of black to white beads in simplest form?

3. In a fruit bowl there are 6 mandarins and 2 bananas. What is the ratio of mandarins to bananas in simplest form?

4. A school has 320 boys and 360 girls. What is the ratio of boys to girls in simplest form?

INVESTIGATION Crash statistics

Reducing drink driving is a prominent public safety campaign in all Australian states and territories. The police conduct regular random breath tests (RBTs) to screen for drivers abusing alcohol and drugs. Standard police cars can also act as Booze Buses and conduct random breath tests.

In 2016, Queensland Police breath tested 1.37 million drivers and riders from Booze Bus operations. Over 5500 drivers and riders were caught with an illegal blood alcohol concentration (BAC) over this period. Of the 34 drivers and motorcyclists killed in 2016 in Queensland with a BAC of 0.05 or over:

- 91% were males
- 47% were between 21 and 29 years of age, 29% were aged between 30 and 39 years, 18% were aged over 40, and the remaining 6% were 20 years of age or younger
- 79% were involved in single-vehicle crashes
- 62% of fatalities occurred on country roads
- 82% died in crashes that occurred between the hours of 6 pm and 6 am.

Questions

1. What percentage of fatalities occurred on city roads in 2016?
2. If 82% of deaths occurred in accidents between the hours of 6 pm and 6 am:
   a. what is the decimal equivalent
   b. what percentage of accidents occurred at other times
   c. what times did the other accidents in b occur between?
3. What percentage of drivers were caught with an illegal BAC in 2016?
   In 2017, more than one in four drivers and motorcyclists killed in Queensland tested at or over 0.05. Approximately 80% of those killed were male, and the majority killed were aged 21 to 39 years.
4. Express the ratio of drivers and motorcyclists killed in 2017 as a fraction, a percentage and a decimal.
5. In 2017, 80% of the drivers killed were male. There were 40 people killed on the roads that year — how many males does this percentage represent?

6. Below are statistics on what daytime road accidents have occurred. Find the percentage of crashes on each day. (Hint: First find the total amount of accidents and round to one decimal place.)

Day of the week	Number of accidents	Percentage
Mon.	117	
Tues.	138	
Wed.	138	
Thurs.	117	
Fri.	166	
Sat.	186	
Sun.	154	
Total		100.0%

Answers

Appendix Maths skills workbook

KEY SKILL 1 Fractions

1. a. $\dfrac{7}{10}$ b. $\dfrac{5}{8}$ c. $\dfrac{1}{6}$

2. a. $\dfrac{1}{2}$ b. $\dfrac{1}{3}$ c. $\dfrac{3}{4}$

3. a. $\dfrac{2}{3}$ b. $\dfrac{1}{2}$ c. $\dfrac{3}{5}$

   d. $\dfrac{8}{5}$ e. $\dfrac{1}{3}$ f. $\dfrac{2}{5}$

KEY SKILL 2 Equivalent fractions

1. $\dfrac{8}{10}, \dfrac{16}{20}, \dfrac{24}{30}$

2. $\dfrac{4}{6}, \dfrac{8}{12}, \dfrac{20}{30}$

3. $\dfrac{4}{10}, \dfrac{10}{25}, \dfrac{12}{30}$

4. $\dfrac{4}{14}, \dfrac{16}{56}, \dfrac{20}{70}$

5. $\dfrac{10}{18}, \dfrac{25}{45}, \dfrac{50}{90}$

6. $\dfrac{14}{22}, \dfrac{35}{55}, \dfrac{77}{121}$

7. a. $\dfrac{15}{25}$ b. $\dfrac{3}{5}$ c. $\dfrac{2}{5}$

KEY SKILL 3 Multiplying fractions

1. a. $\dfrac{1}{4}$ b. $\dfrac{1}{6}$ c. $\dfrac{4}{33}$ d. $\dfrac{9}{26}$

   e. $\dfrac{5}{49}$ f. $\dfrac{1}{5}$ g. 40 h. $\dfrac{2}{7}$

2. a. 60 b. $\dfrac{100}{3}$ c. 20 d. 98

   e. 30 f. 60 g. 9 h. 40

3. a. 40 b. 72 c. 160 d. 285

KEY SKILL 4 Percentage skills

3. $\dfrac{25}{100}, \dfrac{1}{4}, 1 \div 4, 0.25$

4. $\dfrac{30}{100}, \dfrac{3}{10}, 3 \div 10, 0.3$

5. $\dfrac{50}{100}, \dfrac{1}{2}, 1 \div 2, 0.5$

6. $\dfrac{75}{100}, \dfrac{3}{4}, 3 \div 4, 0.75$

7. $\dfrac{80}{100}, \dfrac{4}{5}, 4 \div 5, 0.8$

8. $\dfrac{90}{100}, \dfrac{9}{10}, 9 \div 10, 0.9$

9. $\dfrac{12.5}{100}, \dfrac{1}{8}, 1 \div 8, 0.125$

10. $\dfrac{150}{100}, \dfrac{3}{2}, 3 \div 2, 1.5$

11. $\dfrac{33.33}{100}, \dfrac{1}{3}, 1 \div 3, 0.33$

12. $\dfrac{66.67}{100}, \dfrac{2}{3}, 2 \div 3, 0.67,$

13. a. 70% b. $\dfrac{7}{10}$ c. 0.7

KEY SKILL 5 Percentages

1. a. 76 b. 26.25
2. a. 48 b. 25.5 c. 210 d. 7.92
   e. 55.8 f. 60 g. 53.52 h. 125.56

KEY SKILL 6 Fractions into percentages

1. a. 50% b. 62.5% c. 20%

2.

	Test 1	Test 2
Michelle	93.75%	90%
Yang	18.75%	25%
Simon	50%	53.33%
Alba	61.25%	75%
Christina	81.25%	91.67%

3. a. 20% b. 20% c. 17.5%

KEY SKILL 7 Ratios

1. a. $10:13$ b. $5:7$ c. $100:1$
   d. $11:3$ e. $6:30$ f. $12:7$
2. a. $350:7$ b. $150:2$ c. $3:2$
   d. $3:2$ e. $4:1$ f. $640:1$
   g. $10:15:9$ h. $4:2:35:2$

KEY SKILL 8 Using ratios

1.

2. Sample responses can be found in the Worked solutions in the online resources.

KEY SKILL 9 Simplifying ratios

1. a. $3:1$ b. $3:2$ c. $3:1$
   d. $5:2$ e. $3:2$ f. $5:4:3$
2. $3:2$
3. $3:1$
4. $8:9$

INVESTIGATION Crash statistics

1. 38%
2. a. 0.82 b. 18% c. 6 am and 6 pm
3. 0.40%
4. $\frac{1}{4}, 25\%, 0.25$

5. 32

6.

Day	Number of accidents	Percentage
Mon.	117	11.5%
Tues.	138	13.6%
Wed.	138	13.6%
Thurs.	117	11.5%
Fri.	166	16.3%
Sat.	186	18.3%
Sun.	154	15.2%
Total	1016	100.0%

GLOSSARY

acute angle an angle that is larger than 0° but smaller than 90°

adjacent in trigonometry, the side of a triangle adjacent to an angle is the line that together with the hypotenuse, forms the angle

adjacent angles two angles that share both a common vertex and a common side

angle of depression the angle measured down from the horizontal line (through the observation point) to the line of vision

angle of elevation the angle measured up from the horizontal line (through the observation point) to the line of vision

angles the space between two intersecting lines, measured in degrees at the point where they meet

approximation an estimate based on the information available, when a precise answer is not necessary

arc part of the circumference of a circle

arc length calculated by finding the circumference of a circle and multiplying by the fraction of the angle that it forms at the centre of the circle: arc length $= \dfrac{\theta}{360} \times 2\pi r$

area the amount of flat surface enclosed by a two-dimensional shape. It is measured in square units, such as square metres, m^2, or square kilometres, km^2.

array a collection of objects or values that have been ordered

BIDMAS the order in which calculations are performed. The order is: Brackets; Index (or power); Division or Multiplication from left to right; Addition or Subtraction from left to right.

bivariate data that contain two variables, where one variable affects the other variable

box plot (or box-and-whisker plot) a graphical representation of the five-number summary

capacity the maximum amount of fluid that can be contained in an object. It is usually applied to the measurement of liquids and is measured in units such as millilitres (mL), litres (L) and kilolitres (kL).

Cartesian plane a region in which any point can be defined by specifying its x- and y-values as an ordered pair: (x, y)

categorical data data that involve grouping or classifying things, not numbers; for example, grouping hair colour

causation the measure of how much the change in one variable is caused by the other.

certain the probability of an event happening is 1

circle a closed two-dimensional shape containing all points that are a given distance away from a certain point, which is the centre of the circle, O

circumference the curve that forms a circle

class intervals a subdivision of a set of data; for example, students' heights may be grouped into class intervals of 150 cm − 154 cm, 155 cm − 159 cm

coefficient a number multiplying a pronumeral.

complement in probability, the likelihood that the event discussed will not happen

complementary angles two angles whose sum is exactly 90°

composite figure a closed figure that comprises two or more different common figures

compound interest interest calculated on the changing value throughout the time period of a loan or investment. Interest is added to the balance before the next interest calculation is made. The amount of the loan or investment can be calculated using $A = P(1 + i)^n$.

compounding factor the factor, R, by which the value of a loan or investment changes over each time period

concave having a surface or side that curves inwards

concave polygon a polygon that has at least one angle greater than 180°

consecutive in geometry, consecutive angles are formed at either ends of the same line segment. Consecutive sides are line segments that meet at the same vertex.

constant term a term that does not have a pronumeral

continuous data numerical data that can take any value that lies within an interval. Continuous data values are subject to the accuracy of the measuring device being used.

convex having a surface or side that curves outwards

convex polygon a polygon where all angles are less than $180°$

coordinates a pair of values (typically x and y) that represent a point

correlation a measure of the strength of the linear relationship between two variables

cosine ratio the ratio of the side length adjacent to an internal angle of a right-angled triangle with the hypotenuse: $\cos \theta = \dfrac{\text{adjacent}}{\text{hypotenuse}}$

cumulative frequency the number of observations in a data set that are above or below the particular value

cumulative frequency curve the curve created when plotting data from a cumulative frequency table

cumulative frequency table a method of recording cumulative frequency

deciles the nine values that split a ranked set of data into ten equal sections

decimal places the number of digits that are to the right of the decimal point

decimal point a point in a number that separates the whole number part from the fractional part

dependent variable a variable whose value changes because of a change in the independent variable

diagonal a line that connects one angle of a quadrilateral to the angle opposite it

diameter the straight line from one point on the circumference of a circle to another on the circumference, passing through the centre

dimension lines thin continuous lines with arrowheads at both ends that show the dimension of a line

discrete data numerical data that are counted in exact values, with the values often being whole numbers

edge straight lines where pairs of faces of a polyhedron meet

equally likely outcomes outcomes that have the same chance of occurring

equation a mathematical statement containing a left- and right-hand side separated by an equals sign. e.g. $y = 5x + 6$ is an equation

estimating approximating an answer when a precise solution is not required

expanded form the form of a number where it is separated into each of its individual parts

exterior angle the angle created on the outside of a shape when the side of a polygon continues past a vertex

extrapolation making a prediction from a line of best fit that appears outside the parameters of the original data set

face a two-dimensional, closed shape that forms the surface of a polyhedron

favourable outcome the desired result in a probability experiment

finite decimal a decimal that reaches a point where it has no remainder

five-number summary a way of summarising data so that it can be put into a box plot using the minimum, lower quartile, median, upper quartile and maximum values

formula an equation that describes the relationship between certain input and output values

frequency the number of times a score occurs in a set of data

front elevation the diagram of an object with the view being from straight in front

front view the diagram of an object with the view being from straight in front

geometric solids another word to describe polyhedrons

gradient also known as the slope; determines the change in the y-value for each change in x-value. This measures the steepness of a line as the ratio $m = \dfrac{\text{rise}}{\text{run}}$. If (x_1, y_1) and (x_2, y_2) are two points on the line, then the gradient, m, can be calculated using $m = \dfrac{y_2 - y_1}{x_2 - x_1}$.

hectare a unit of area equal to the area of a square with side lengths of $100\,\text{m}\,(1\,\text{ha} = 10\,000\,\text{m}^2)$

horizontal in trigonometry, the base line, parallel to the plane of the horizon, from which angles of elevation and depression are measured

hypotenuse the longest side of a right-angled triangle. The hypotenuse will be opposite the right angle.

impossible the probability of an event happening is 0

independent variable a variable that is changed to produce a change in a dependent variable

infinite not finite; never ending, unlimited

infinite recurring decimal a decimal value with a repeating pattern

interest (I) when investing money, interest is a payment earned for having money invested in a bank or financial institution. When borrowing money from a bank or financial institution, interest is charged as a fee for using their money.

interpolation making a prediction from a line of best fit that appears within the parameters of the original data set

interquartile range (IQR) the difference between the upper and lower quartiles of a data set

intersection the point where two lines meet

irregular polygons polygons that do not contain both equal angles and equal lengths

isometric drawing a two-dimensional drawing of a three-dimensional object

isometric paper graph paper that an isometric drawing is done on, containing either dots or lines that form equilateral triangles

linear function a function that is a straight line when drawn

lower fence the lower boundary beyond which a data value is considered to be an outlier: $Q_1 - 1.5 \times IQR$

mass how much matter makes up an object, standard unit is the kilogram (kg)

mean commonly referred to as the average; a measure of the centre of a set of data. The mean is calculated by dividing the sum of the data values by the number of data values.

median the middle value of a data set when the values are placed in numerical order

midpoint the average of the maximum and minimum values for the class interval

modal class the class interval with the highest frequency

mode the category or data value(s) with the highest frequency. It is the most frequently occurring value in a data set.

multiples of 10 numbers that can be divided by 10, such as 10, 20, 30, 100 and so on

net a two-dimensional plan of the surfaces that make up a three-dimensional object

numerical data data that can be counted or measured

obtuse angle an angle that is larger than $90°$ but smaller than $180°$

ogive the curve created when data from a cumulative frequency table is plotted, also referred to as a cumulative frequency curve

opposite in trigonometry, the side of a triangle opposite an angle is the only line that does not form a part of the angle

order of operations the correct sequence for performing the mathematical operations in an expression; often characterised by a mnemonic such as BIDMAS (Brackets; Index (or power); Division or Multiplication from left to right; Addition or Subtraction from left to right)

origin the point on the Cartesian plane at which the x- and y-axes intersect, $(0, 0)$

outcome the result obtained when a probability experiment is conducted

outlier an extreme value or unusual reading in the data set, generally considered to be any value beyond the lower or upper fences

outside surfaces the surfaces of a solid object that make up the total surface area

parallel lines that never intersect and maintain a constant distance between them along their length

parameters the boundary or limiting values associated with populations

Pearson's correlation coefficient, r a measure of the strength of a linear trend that is denoted by the letter r and associated with a numerical value between -1 and $+1$. Values close to $+1$ or -1 indicate strong linear trends; values close to zero indicate weak or no linear trends.

perigon an angle that is a full circle; an angle of exactly $360°$, also referred to as a revolution

perimeter the distance around a closed figure

plan view the diagram of an object with the view being from directly above

Platonic solids another term to describe polyhedrons

polygon a two-dimensional shape consisting of at least three straight sides

polyhedra a plural form of polyhedron

polyhedron a two-dimensional object enclosing a space containing flat surfaces and straight edges

polyhedrons a plural form of polyhedron

powers of 10 numbers that include 10 (10^1), 100 (10^2), 1000 (10^3), 10 000 (10^4) and so on

principal the amount that is borrowed or invested

prism a solid object that has identical opposite polygonal ends that are joined by flat surfaces and a cross-section that is the same along its length

projection lines thin continuous lines drawn perpendicular to the measurement shown, that do not touch the object

pronumerals letters used in place of a number; another name for variables

Pythagoras' theorem the square of the hypotenuse of a right-angled triangle is equal to the sum of the squares of the other two sides, written as $c^2 = a^2 + b^2$

quadrilateral a polygon containing four sides and four vertices

quantiles any number of equal parts that a set of data can be split up into

quartiles these divide a set of data into quarters. The lower quartile (Q_1) is the median of the lower half of an ordered data set. The upper quartile (Q_3) is the median of the upper half of an ordered data set. The middle quartile (Q_2) is the median of the whole data set.

radii the plural form of radius

radius a straight line from a circle's centre to any point on its circumference

random number generators devices or programs that generate random numbers between two given values

range a measure of spread determined by calculating the difference between the lowest and highest values

rate of interest the percentage of the principal that is paid out in a given time period as interest

recurrence relation an equation that recursively defines a sequence; that is, once one or more initial terms are given, each further term of the sequence is defined as a function of the preceding terms

reflex angle an angle that is larger than 180° but smaller than 360°

regression line the equation of the line of best fit for a set of data of the form $y = mx + c$, where m is the gradient and c is the y-intercept

regular polygons polygons that contain equal sides and equal angles

relative frequency the frequency of a particular score divided by the total sum of the frequencies

repeating pattern the same number or numbers being repeated in the same order; for example, 333333 ... or 454545 ...

revolution an angle that is a full circle; an angle of exactly 360°

right angle an angle of exactly 90°

rule an equation that describes the relationship between certain input and output values

sample space in probability, the complete set of outcomes or results obtained from an experiment. It is shown as a list enclosed in a pair of braces, {}, and is denoted by the symbols ξ or S.

scale a series of marks indicating measurement increasing in equal quantities

scale drawings drawings of real-life objects where all dimensions of the drawing are kept in the same ratio as the actual object

scale factor a measure of the relative size of two similar figures. It is the amount of enlargement or reduction and is expressed as integers, fractions or scale ratios.

scatterplot a visual display of bivariate data

side elevation the diagram of an object with the view being from one side; either left or right

side view the diagram of an object with the view being from one side; either left or right

sides the straight lines that form a polygon

simple interest interest calculation based on the original amount borrowed or invested; also known as, 'flat rate' as it is a constant amount: $I = Pin$

simulations outcomes of events that are modelled to represent what would happen in real life

sine ratio the ratio of the side length opposite an internal angle of a right-angled triangle with the hypotenuse: $\sin \theta = \dfrac{\text{opposite}}{\text{hypotenuse}}$

spread how far a data set is spread from the centre or from each other

squared units the units that area and total surface area are measured in (mm^2, cm^2, m^2 or km^2)

standard deviation a measure of the spread of continuous numerical data around the mean

straight angle an angle of exactly 180°

successful trial a trial where the result is the outcome that was desired

summary statistics the boundary or limiting values associated with samples

supplementary angles two angles whose sum is exactly 180°

tangent ratio the ratio of the side length opposite an internal angle of a right-angled triangle with the side length adjacent to it: $\tan \theta = \dfrac{\text{opposite}}{\text{adjacent}}$

term a group of letters and/or numbers

term deposit lending money to a bank for a set amount of time

total surface area the combined total of the external areas of each individual surface that forms a solid object

transversal a line that meets two or more other lines in a plane

tree diagrams branching diagrams that list all the possible outcomes of a probability experiment

trial an experiment performed in the same way every time

triangle a polygon with three sides and three vertices

trigonometry a branch of mathematics that analyses the relationships between angles and sides of triangles

undefined a numerical value that cannot be calculated

upper fence the upper boundary beyond which a data value is considered to be an outlier: $Q_3 + 1.5 \times IQR$

variables letters or symbols in an equation or expression that may take many different values

vertex the point of a polygon where two lines meet, and where the angle is measured from

vertically opposite angles when two lines intersect, four angles are formed at the point of intersection, and two pairs of vertically opposite angles result. Vertically opposite angles are equal.

volume the amount of space a three-dimensional object occupies. The units used are cubic units, such as cubic centimetres (cm^3) and cubic metres (m^3).

INDEX

Note: Figures and tables are indicated by italic *f* and *t*, respectively, following the page reference.